Materials, des

Related titles:

Friction stir welding: from basics to applications
(ISBN 978-1-84569-450-0)
Friction stir welding (FSW) is a solid-state welding process that is gaining wide acceptance in industry, especially in the shipbuilding, aerospace, mass transportation and automotive industries. FSW is particularly suited to those industries that use aluminium and its alloys. This authoritative book provides a comprehensive review of the subject of friction stir welding and covers topics such as process basics, equipment, modelling, inspection and quality control and applications.

Hydroforming for advanced manufacturing
(ISBN 978-1-84569-328-2)
Hydroforming is a process where fluid pressure is applied to ductile metallic blanks (either tubes or sheet metal) to form component shapes. The process is gradually replacing stamped and welded parts in many automotive and other applications. This comprehensive book covers the fundamentals of the process, hydroforming systems, material selection and industrial applications. Process variations such as low-pressure hydroforming and heatform/warm hydroforming are covered, and the future of hydroforming hybrid processes is analysed. It presents current research from leading innovators in the important hydroforming field around the world.

The science and technology of materials in automotive engines
(ISBN 978-1-85573-742-6)
This authoritative book provides an introductory text on the science and technology of materials used in automotive engines. It focuses on reciprocating engines, both four- and two-stroke, with particular emphasis on their characteristics and the types of materials used in their construction. The book considers the engine in terms of each specific part: the cylinder, piston, camshaft, valves, crankshaft, connecting rod and catalytic converter. The materials used in automotive engines are required to fulfil a multitude of functions and the intention here is to describe the metallurgy, surface modification, wear resistance, and chemical composition of these materials. Supplementary notes support the core text. The book is essential reading for engineers and designers of engines, as well as lecturers and graduate students in the fields of combustion engineering, machine design and materials science and anyone looking for a concise, expert analysis of automotive materials.

Details of these and other Woodhead Publishing materials books can be obtained by:

- visiting our web site at www.woodheadpublishing.com
- contacting Customer Services (e-mail: sales@woodheadpublishing.com; fax: +44 (0) 1223 832819; tel.: +44 (0) 1223 499140 ext. 130; address: Woodhead Publishing Limited, 80 High Street, Sawston, Cambridge, CB22 3HJ, UK)

If you would like to receive information on forthcoming titles, please send your address details to: Francis Dodds (address, tel. and fax as above; e-mail: francis.dodds@woodheadpublishing.com). Please confirm which subject areas you are interested in.

Materials, design and manufacturing for lightweight vehicles

Edited by
P. K. Mallick

CRC Press
Boca Raton Boston New York Washington, DC

WOODHEAD PUBLISHING LIMITED
Oxford Cambridge New Delhi

Published by Woodhead Publishing Limited,
80 High Street, Sawston, Cambridge CB22 3HJ, UK
www.woodheadpublishing.com

Woodhead Publishing India Private Limited, G-2, Vardaan House, 7/28 Ansari Road,
Daryaganj, New Delhi – 110002, India
www.woodheadpublishingindia.com

Published in North America by CRC Press LLC, 6000 Broken Sound Parkway, NW,
Suite 300, Boca Raton, FL 33487, USA

First published 2010, Woodhead Publishing Limited and CRC Press LLC;
Reprinted 2011

British Library Cataloguing in Publication Data
A catalogue record for this book is available from the British Library.

Library of Congress Cataloging in Publication Data
A catalog record for this book is available from the Library of Congress.

Woodhead Publishing ISBN 978-0-08-101436-3 (print)
Woodhead Publishing ISBN 978-1-84569-782-2 (online)
CRC Press ISBN 978-1-4398-2972-1
CRC Press order number: N10168

The publishers' policy is to use permanent paper from mills that operate a sustainable forestry policy, and which has been manufactured from pulp which is processed using acid-free and elemental chlorine-free practices. Furthermore, the publishers ensure that the text paper and cover board used have met acceptable environmental accreditation standards.

Typeset by Replika Press Pvt Ltd, India

Contents

Part II Manufacturing and design of lightweight automotive structures

Contributor contact details

(* = main contact)

Editor and Chapters 1, 5, 6 and 8

Dr P. K. Mallick
William E. Stirton Professor of Mechanical Engineering
Center for Lightweighting Automotive Materials and Processing
University of Michigan-Dearborn
4901 Evergreen Road
Dearborn
Michigan
MI 48128 2406
USA

Email: pkm@umich.edu

Chapter 2

Mr C. D. Horvath
General Motors
Mail Code 480-210-3B1
30001 Van Dyke Avenue
Warren
Michigan
MI 48093-2350
USA

Email: curt.d.horvath@gm.com

Chapter 3

Professor J. C. Benedyk
Thermal Processing Technology Center (TPTC)
Illinois Institute of Technology
Chicago
Illinois
IL 60616-3793
USA

Email: benedyk@iit.edu

Chapter 4

Dr B. R. Powell* and Dr A. A. Luo
Light Metals for Powertrain and Structural Subsystems Group
GM Research and Development Center
Materials and Processes Laboratory
Mail Code 480-106-212
30500 Mound Road
Warren
Michigan
MI 48090-9055
USA

Email: bob.r.powell@gm.com; alan.luo@gm.com

Dr P. E. Krajewski
Manager, Novel Alloys and Functional Materials Group
GM Research and Development Center
Materials and Processes Laboratory
Mail Code 480-106-212
30500 Mound Road
Warren
Michigan
MI 48090-9055
USA

Email: paul.e.krajewski@gm.com

Chapter 7

Dr Ghassan T. Kridli*
Industrial and Manufacturing Systems Engineering
University of Michigan–Dearborn
4901 Evergreen Road
Dearborn
Michigan
MI 48128–2406
USA

Email: gkridli@umd.umich.edu

Dr Peter A. Friedman
Technical Expert and Group Leader, Metal Forming Methods
Ford Research and Innovation Center
2101 Village Road
MD 3135/RIC
Dearborn
Michigan
MI 48121–2053
USA

Email: pfriedma@ford.com

Dr James M. Boileau
Technical Expert and Group Leader, Materials Characterization
Ford Research and Innovation Center
2101 Village Road
MD 3182/RIC
Dearborn
Michigan
MI 48121-2053
USA

Email: jboileau@ford.com

Chapter 9

Mr S. Das
Energy and Transportation Science Division
Oak Ridge National Laboratory
2360 Cherahala Blvd
Knoxville
Tennessee 37932-6472
USA
Email: dass@ornl.gov

Chapter 10

Professor A. Deb
Centre for Product Design and Manufacturing
Indian Institute of Science
Bangalore-560012
India

Email: adeb@cpdm.iisc.ernet.in

Preface

The materials for the construction of automobiles are changing from mostly low carbon steels to a combination of steels, light alloys, such as aluminum and magnesium alloys, and polymer matrix composites. Many of these materials are already used in today's vehicles, albeit in smaller volumes. Future vehicles, which will have to be much lighter in weight for improved fuel economy and reduced environmental pollution, will contain much larger volumes of these materials. The selection of materials will not only be influenced by their weight reduction potential, but also by factors such as safety, durability, processing, joining, recycling and cost. These are the issues addressed in this book.

This book contains chapters on materials, their properties, processing technology and design and materials selection issues pertinent to designing lightweight vehicles. Each chapter is written by experts from either industry or academia who have first-hand knowledge and experience of working with these materials. It starts with a broad review of the materials scenario and design considerations for lightweight automotive structures. It is then divided into two major parts: materials, and design and manufacturing. The materials part contains chapters on advanced steels, aluminum alloys, magnesium alloys and polymer matrix composites. Each of these chapters contains information on material properties, processing characteristics and application examples. The design and manufacturing part contains chapters on manufacturing processes for light alloys, joining, crashworthiness considerations, recycling and life-cycle issues. It is my sincere hope that this book will be useful to both practicing engineers in the automotive industry as well as researchers in the materials field.

I would like to thank all the authors who have contributed to this book, and without whom, a book of such a wide scope could not have been written. The concept and the outline of this book were with me ever since I developed and taught a graduate course on designing and manufacturing of lightweight vehicles in the early 2000. However, the book would not have been published if Rob Sitton of Woodhead Publishing Limited had not e-mailed me about two years ago and asked me if I wanted to put together a book on materials for lightweight vehicles. I would like to thank him, and all editorial staff at

Woodhead Publishing Limited who have worked with me to bring this book to completion. Finally, I would also like to thank my wife, Sunanda, for her support and the help she has given me during the writing of my chapters.

P. K. Mallick
Center for Lightweighting Automotive Materials and Processing
University of Michigan-Dearborn

1
Overview

P. K. MALLICK, University of Michigan-Dearborn, USA

Abstract: With increasing demand for fuel economy improvement and emission control, there is a great deal of interest in finding alternative solutions to producing lighter vehicles through material substitution, design modification and efficient manufacturing. The materials available today include high-strength steels, light alloys, such as aluminum and magnesium alloys, and a variety of composites. This chapter gives an overview of these materials and their current and future applications in automobiles. It also provides general guidelines that need to be considered in selecting these materials for lightweight vehicles.

Key words: lightweight vehicles, steel, light alloys, composites, design, manufacturing.

1.1 Introduction

Fuel economy improvement and emission control are the two most important challenges the automotive industry faces today. In the USA, the average fuel economy is regulated by the government-mandated CAFE (Corporate Average Fuel Economy) standard, which is a sales-weighted average fuel economy (expressed in miles per US gallon or mpg, 1 mpg = 0.43 km/liter) of a vehicle manufacturer's fleet of passenger cars or light trucks. For passenger cars, the CAFE standard increased from the initial 18 mpg in 1978 to the current 27.5 mpg. It is destined to increase to 35 mpg in 2020. The fuel economy standards in other countries are either directly or indirectly regulated by their governments or voluntarily maintained by the car companies. These standards are different from the US standard and many of them are much more stringent, but they all have the same dual goals of improving fuel efficiency and reducing environmental pollution and CO_2 emission, the major cause of the greenhouse effect.

Fuel economy of a vehicle is measured using a specific driving cycle (the EPA driving cycle in USA) and depends on many factors, which include vehicle power requirement, vehicle speed, engine and transmission efficiencies, and fuel type. The vehicle power requirement is the sum of the power requirements for vehicle acceleration, driving on a grade, overcoming the rolling resistance at the tire–road interface, overcoming the aerodynamic drag and operating the accessories, such as air conditioner, heaters and entertainment modules. The first three of these power requirements are directly proportional to the vehicle weight. Thus, reducing the vehicle weight can cause a significant reduction in vehicle power requirement, and consequently increase the fuel economy. Aerodynamic drag, tire rolling resistance coefficient and accessory/standby power requirement have much smaller effects on fuel economy. Studies have shown that every 10% reduction in vehicle weight can result in 5 to 8% greater fuel efficiency (Brooke and Evans, 2009). In terms of the greenhouse effect, reducing vehicle weight by 100 kg results in CO_2 reduction up to 12.5 g/km.

In addition to the primary benefit on fuel economy, weight saving has several secondary benefits. For example, when the vehicle weight is reduced, the power needed for acceleration and braking is also reduced, which creates an opportunity to design smaller engine, transmission and braking systems. From the standpoint of vehicle dynamics, vehicle weight reduction does not affect the vehicle stability and control. On the other hand, weight saving in selected components can be utilized to equalize the vehicle mass distribution between the axles and lower the vehicle's center of gravity, both of which improve vehicle handling. Two vehicle attributes that may be negatively affected by vehicle weight reduction are ride comfort and safety. However, they are also influenced by vehicle design and material selection.

Historically, vehicle weight reduction started in the USA in the 1970s, when the vehicle manufacturers started downsizing their vehicles to meet the 1978 CAFE requirement of 18 mpg (Brooke and Evans, 2009). Downsizing was achieved primarily by introducing cars with smaller wheelbases and by shifting from body-on-frame structures to unibody or body-in-white (B-I-W) structures (see Section 1.3.1). Other changes that contributed to vehicle weight reduction include smaller engines (4-cylinder engines instead of 6- and 8-cylinder engines) and front-wheel drive transmission instead of rear-wheel drive transmission. As a result of these design changes, the average vehicle weight of US cars decreased from 1839 kg in 1976 to 1385 kg in 1986. However, as Table 1.1 shows, the average vehicle weight started to creep up in the 1990s due to the addition of several new features, such as safety equipment, emission-control devices and entertainment modules. In the US market, bigger vehicles, such as sports utility vehicles (SUVs) and pick-up trucks, became more popular than the sedans, which also contributed to the increase in vehicle weight. Starting in 2007, as the gasoline price began to increase in the USA, sales of SUVs started to decline and, in response, the vehicle manufacturers started to produce crossover vehicles that were smaller than the full-size SUVs, but bigger than the mid-size sedans. This started a declining trend for the average vehicle weight, which is expected to continue and may even become more critical for future vehicles.

With increasing fuel price, unpredictable long-term availability of petroleum and greater realization that auto emission is hazardous to both the environment and public health, fuel economy improvement has become the top priority

Table 1.1 Average vehicle weights of US cars Source: Brooke and Evans, 2009

Year	Vehicle (curb) weight	
	kg	lb
1976	1839	4059
1986	1385	3057
1996	1600	3532
1999	1683	3716
2002	1742	3846
2004	1820	4018
2007	1775	3918
2009*	1701	3755
2015*	1631	3600

*Forecast

for the vehicle manufacturers around the world. Vehicle weight reduction is considered one of the key elements in the fuel economy improvement strategies. While vehicle downsizing is still an option for achieving significant vehicle weight reduction, at least in the near future, there will be a mix of small and large size vehicles to satisfy the customer base. The other option is to reduce component weights using material substitution, parts consolidation and design optimization. This chapter gives a broad overview of the materials that are being considered for making lightweight components. Many of these materials are covered in much greater details in the remaining chapters of this book.

1.2 Materials scenario

Low carbon steel and cast iron were the workhorse materials in the automotive industry prior to 1970s. As Table 1.2 indicates, even today steel is used in much larger quantities than any other material, although instead of only low carbon steels, there is now a mix of low carbon steels and high strength steels. However, with greater emphasis on vehicle weight reduction, the materials scenario is changing rapidly to include other materials, such as aluminum alloys, magnesium alloys and polymer matrix composites (Powers, 2000). Table 1.3 lists the tensile properties of a few selected materials that are in competition with steels for future vehicle construction. They are all lighter than steels and many of them also provide a good opportunity for parts consolidation, but they are at present not cost-competitive with steels, particularly for large production volumes. Nevertheless, their technical viability and weight saving potential have been demonstrated in many concept vehicles, and now they are appearing in increasing quantities in many production vehicles.

The greatest opportunity for component weight reduction exists in the

Table 1.2 Material distributions in typical automobiles

Material	Percentage of vehicle weight	Major areas of application
Steel	55	Body structure, body panels, engine and transmission components, suspension components, driveline components
Cast iron	9	Engine components, brakes, suspension
Aluminum	8.5	engine block, wheel
Copper	1.5	Wiring, electrical components
Polymers (plastics) and polymer matrix composites	9	interior components, electrical and electronic components, under-the hood components, fuel line components
Elastomers	4	tires, trims, gaskets
Glass	3	Glazing
Other	10	carpes, fluids, lubricants, etc.

Table 1.3 Material property comparisons

Material	Density (g/cm^3) (ρ)	Tensile modulus (E) (GPa) (MPa)	Yield strength (S_y) (MPa)	Tensile strength (S_t) (MPa)	Coeff. of thermal expansion (10^{-6}/°C)
DQ Low carbon steel	7.87	207	186	317	11
DP 400/700 Steel	7.87	207	400	700	11
TRIP 450/800 Steel	7.87	207	450	800	11
5182-H24 Aluminum	2.7	70	235	310	23
6111-T62 Aluminum	2.7	70	320	360	23
AZ91 Magnesium	1.8	45	160	240	26
Ti-6Al-4V Titanium	4.43	114	827	896	9
304 Stainless steel	7.9	200	241	614	17
Nitronic 30 stainless steel	7.86	193	393	862	16
High strength CFRE (unidirectional)	1.55	138	–	1550	–
High modulus CFRE (unidirectional)	1.63	215	–	1240	–0.9 (L) 27 (T)
GFRE (unidirectional)	1.85	39	–	965	6 (L) 19 (T)
CFRE (quasi-isotropic)	1.55	45.5	–	579	0.9
Sheet molding compound (SMC-R50)	1.87	16	–	164	14.8

L is the longitudinal direction and T is the transverse direction.
CFRE is carbon fiber reinforced epoxy, GFRE is glass fiber reinforced epoxy.

body and chassis components, which comprise 60% of a vehicle's weight. Many new materials and manufacturing processes have been developed in the last 20 years to decrease the weight of the body structure, body panels and suspension components. Powertrain weight, which includes both engine and transmission, is between 25 to 30% of the vehicle weight. Several new materials and manufacturing process developments have also occurred to reduce the powertrain weight. This section reviews the new materials and manufacturing processes that either are being used, or have the future potential for use, in achieving the vehicle weight reduction necessary for fuel economy improvement.

1.2.1 Steel

Approximately 55% of a typical US car's weight is due to steel. The greatest advantages of steel over other materials that are either being used or are under consideration for use in automotive structures are as follows:

- Low cost and high modulus (207 GPa, which is higher than all other competitive structural metals)
- Wide range of strength and ductility, which can be achieved by a variety of means, such as alloying, work hardening (for low and medium carbon steels) and heat treatment (for medium carbon, high carbon and alloy steels)
- Excellent formability for low carbon steels and many newly developed high strength steels, such as high strength low alloy (HSLA) steels and dual phase (DP) steels, which make them highly suitable for high production rate forming operations, such as stamping and roll forming

The automotive steel scenario has changed tremendously in the last 25 years, principally because of the challenges posed by lighter weight materials, such as aluminum and plastics. New steel making processes (e.g. vacuum degassing) have made it possible to produce steel more cost effectively with much lower impurity levels (only about 10–20 ppm compared to 200–400 ppm by the traditional processes). Combination of new alloying techniques and improved heat treatment procedures, such as continuous annealing, are now used to produce not only a broad spectrum of strength and ductility, but also better surface qualities and more uniform properties in sheet steels. Better corrosion resistance is achieved by new types of zinc alloy coatings (e.g. Zn–Fe and Zn–Ni) as well as by new methods of applying them onto the steel surface (e.g. by electro-deposition instead of hot dipping). A relatively new process, called galvanneal, is able to produce superior corrosion resistance and formability, as well as weldability, of coated sheet steel. Laminated sheet steel with steel outer skins and a thin viscoelastic constrained layer (typically 0.025 mm thick) is available for noise and vibration control purposes (Yang *et al.*, 2001).

Sheet steel used for body panels and body structures is gradually shifting from traditional drawing quality (DQ) or drawing quality, aluminum-killed

(DQAK) steels to high strength steels. Earlier difficulties with formability of high strength steels have been greatly reduced through better de-oxidation practices, micro-alloying and inclusion control. Currently, there is a wide variety of high strength steels available with yield strength ranging from 210 MPa to 1250 MPa (Table 1.4). One of these high strength steels is Bake Hardening (BH) steel, which gains in yield strength during the paint baking cycle due to strain aging. The increase in yield strength depends on the strain imposed by the stamping process prior to paint baking. Another high strength steel is the High Strength Low Alloy (HSLA) steel, which achieves its high yield strength (300–550 MPa) due to the presence of small amounts (in the range of 0.005%) of carbide and nitride forming alloying elements, such as vanadium, niobium and titanium. The carbon content in HSLA steels is restricted to a maximum of 0.13% for improved formability and weldability. In the next higher level of high strength steels, called the Advanced High Strength Steels (AHSS), are the Dual Phase (DP) steels containing martensitic dispersion in a soft ferrite matrix, and Transformation Induced Plasticity (TRIP) steels containing retained austenite in addition to martensitic and bainitic dispersions in a soft ferrite matrix. Both DP and TRIP steels develop high strength due to work hardening during stamping. They also have the potential of gaining higher strength during the paint baking cycle after stamping. Another emerging category of high strength steels is called the Ultra High Strength Steels (UHSS), which have strengths ranging upward of 1000 MPa. Although the AHSS and UHSS have much lower ductility and formability than conventional low carbon or high strength steels, they provide much higher crush resistance because of their high strength and are increasingly being selected for the front structure of vehicles.

New developments have also taken place in the manufacturing processes for steel components. Tailored blanking is an example of the new manufacturing processes that did not exist a decade or two ago. Tailored blanking combines different gauge thicknesses and/or different grades of steel in the same blank

Table 1.4 Properties of several steels used for body applications

Material	Yield strength (MPa)	Tensile strength (MPa)	Elongation (%)	n	$\bar{r}$
DQ steel	186	317	42	0.22	1.5
BH 210/340	210	340	34–39	0.18	1.8
IF 300/420	300	420	29–36	0.20	1.6
HSLA 350/450	350	450	23–27	0.14	1.1
DP300/500	300	500	30–34	0.16	1.0
DP400/700	400	700	19–25	0.14	1.0
DP700/1000	700	1000	12–17	0.09	0.9
TRIP 450/800	450	800	26–32	0.24	0.9
Mart 950/1200	950	1200	5–7	0.07	0.9

instead of using the same thickness or the same steel in the entire blank. Tailored blanking allows a more efficient material utilization in the blank and leads to weight saving by eliminating stiffeners or reinforcements. For example, a laser welded two-piece tailor welded door inner panel of a light truck with 1.8 mm thickness in the front and 0.9 mm thickness in the back was 7% lighter than a one-piece design, which was 0.75 mm thick and contained several reinforcements to increase its stiffness. The tailor welded blank not only eliminated the reinforcements, but also reduced the tooling and assembly costs.

There are several advantages of using steel in auto body structures and body panels. The most important of them is steel's high modulus of elasticity, which, at 207 GPa, is the highest among the structural materials considered for automotive applications. The wide variety of strengths available with steel is also an important advantage, since it gives an opportunity to select steel according to the structural need. The availability of AHSS and UHSS allows not only the downsizing of gauge thickness, but improving the load carrying capacity and crashworthiness of the vehicle structure. Furthermore, steel's superior formability compared to aluminum and magnesium alloys, excellent weldability, and recyclability are some of the reasons for steel's predominance in today's automobiles. Indeed, the Ultralight Steel Auto Body (ULSAB) project conducted under the leadership of the American Iron and Steel Institute (AISI) in the mid-1990s clearly demonstrated that significant weight reduction of body structure can be achieved using high strength steels and advanced high strength steels, innovative manufacturing processes, such as tailor welded blanking, tube hydroforming and laser welding, and computer-aided engineering tools (AISI, 1998). A weight reduction of up to 36%, compared with the heaviest benchmarked vehicles of the same class, was possible due to sheet thickness reduction, parts consolidation and design optimization.

Improvements have also taken place in the forging quality steels used for powertrain, suspension and steering components (Yamagata, 2005; Cho *et al.*, 1994). One of these developments is the microalloyed steels containing 0.3 to 0.6% C. The microalloying element is usually a small amount (~ 0.05 to 0.15%) of vanadium, which forms vanadium carbide and nitride precipitates as the forged component is air cooled after hot forging. Tempering after air cooling is not necessary, since the precipitates in the relatively soft ferrite and pearlite matrix strengthen the steel. The microalloyed steels possess a good combination of strength and toughness, which can be further improved by grain size refinement through proper control of the inclusions in the steel as well as the forging conditions. The yield strength and % elongation of microalloyed steels are higher than the conventional forging quality steels of similar carbon content. The fatigue strength is also higher. Furthermore, microalloyed steels do not need quenching and tempering, which not only reduces the cost, but also reduces the possibility of thermal distortion resulting from quenching and tempering. The application of microalloyed forging steels includes connecting rods, wheel hubs and steering knuckles.

1.2.2 Aluminum alloys

Aluminum alloys have the following advantages over steels for automotive applications:

- Lower density (2.7 g/cm^3 for aluminum alloys compared with 7.87 g/cm^3 for steels)
- Higher crash energy absorption per unit weight
- Higher thermal conductivity, which is useful for radiator cores and other heat exchanger applications

Both cast and wrought aluminum alloys are used in numerous applications in automobiles. The cast alloys are mostly the 300-series (Al–Si–Cu or Al–Si–Mg) alloys, such as 319 for intake manifolds and transmission housings, 383 for engine blocks, 356 for cylinder heads, and A356 for wheels. The principal alloying element in these alloys is silicon (Si), which contributes to their high fluidity. They can be cast using a variety of techniques ranging from sand casting and die casting to more intricate permanent mold, lost foam/lost wax casting. They can also be heat treated to produce a range of strength properties.

The major growth of aluminum use in future automobiles is expected to be in the body structures and body panels, such as front rails, roof rails, hoods, deck lids and fenders. The aluminum alloys that are used for these applications are the 5000-series (Al-Mg) alloys, such as AA5754 and AA5182, and the 6000-series (Al-Mg-Si) alloys, such as AA6111 and AA6061. The 5000-series alloys are non-heat treatable, i.e. they cannot be strengthened by heat treatment, whereas the 6000-series alloys are heat treatable, and they are usually strengthened while they are being painted in the paint baking oven. The 5000-series alloys are highly formable, but since stretcher strain marks (Lüder's bands) may appear on their surface as they are strain-hardened during the stamping operation, they are not selected for outer body panels. The 6000-series alloys, used for both inner and outer body panels as well as body structure components, are formed in the relatively soft T4 temper and subsequently age hardened to the T6 temper in the paint baking oven to achieve the final strength.

Many vehicle manufacturers have demonstrated the weight saving potential of aluminum by developing aluminum-intensive concept cars. GM's electric concept car, named 'Impact', had a body structure formed by a combination of stamped, extruded and cast aluminum alloys. The extruded components, such as the front–mid cross member, front fender rails and rear lower link brackets, were made of 6000-series alloys and were integrated into the rest of the structure using either spot welding or mechanical fasteners. The cast components, such as the front shock tower and the rear shock bracket, were made of either A356-T6 or A375-T6 alloy. They were cast using the Vacuum Riserless Casting (VRC) process, which produces high quality castings with wall thickness 4 mm or greater. Chrysler's concept car, 'Neon Lite', was based on a unitized body structure constructed from 6111 and 5454 aluminum alloys. The 6111 alloy was selected for its strength, dent resistance and surface quality, whereas the 5454 alloy was selected for its corrosion resistance and lower cost. Ford built a concept car, named 'Synthesis 2010', which used aluminum in the chassis, body panels, door beams, brake rotors and several other components. The weight of Synthesis 2010 was 35 percent lower than Taurus, which was a similar size car manufactured by Ford in the 1990s.

The leading development of an all-aluminum body structure was due to Audi in their ASF vehicles, which utilized a space frame construction. In the first of these, which went into production in 1994, the space frame members were made of thin-walled extruded tubular beams (both straight and curved) of a 6000-series aluminum alloy, which were joined at the corners or at connecting nodes using vacuum die-cast aluminum components. The skin panels were also made of a 6000-series alloy. About 70% of these panels were punch-riveted (using self-piercing rivets) to the frame members; the rest were either spot welded or weld bonded, which is a combination of spot welding and adhesive bonding. The ASF body was then heat-treated at 210 °C for 30 minutes before applying a zinc, nickel and manganese phosphate coating. The ASF body was about 40 percent lighter than an equivalent steel body of comparable stiffness.

Since aluminum alloys are the main competition to steels for body applications, a comparison between these two types of materials is made in the following paragraph in terms of their design characteristics and formability. Corrosion, joining, cost and recycling issues related to aluminum are also discussed.

(i) *Design characteristics*: The modulus of aluminum is 70 GPa compared with 207 GPa for steel, which means that, for equal bending stiffness, an aluminum component will be 43.5% thicker than a steel component. As a result, the weight reduction achieved by aluminum will not be in the same proportion as the density ratio between the two materials. A simple weight calculation will show that substituting a steel body panel with an aluminum body panel will result in approximately 50% weight saving.

The yield strength of aluminum alloys is in the range of 100 to 400 MPa. With the development of advanced high strength steels, the highest yield strength achieved by steel has increased beyond 500 MPa. The yield strength of steel is more strain rate sensitive than that of aluminum, which means that under impact conditions the yield strength of steel increases much more rapidly than that of aluminum. This is often considered an advantage for steel.

One great advantage of aluminum is that it can be extruded relatively easily compared with steel. Body structural parts, such as roof rails, require multiple stampings and welding when they are made out of steel. With aluminum, a single extruded section can be used. The use of a one-piece extruded section instead of a stamped and welded section can result in tooling and assembly cost reductions.

(ii) *Formability*: The vast majority of the body structure and body panels in both steel and aluminum are manufactured by cold forming operations, such as stamping, roll forming, bending and hemming. In these operations, the two important formability characteristics are the strain hardening exponent (n) and the plastic strain ratio ($\bar{r}$). The n-value of aluminum alloys is similar to that of drawing quality (DQ) steels, which means that both exhibit similar uniform elongations during forming (Table 1.5); however, aluminum has a much lower post-necking elongation than steel. If the cold forming operation is not tightly controlled and the uniform elongation is slightly exceeded, aluminum parts may experience cracking while they are being formed. Aluminum has much a lower $\bar{r}$-value than steel, which means it offers lower resistance to thinning than steel. A comparison of the formability diagram of AA6111-T4 aluminum alloy with that of DQ steel (Fig. 1.1) shows that the formability envelope of the aluminum alloy is lower than that of the DQ steel. This indicates that the aluminum alloy will fail at a much lower strain than the DQ steel under all forming conditions.

In general, formability of medium strength aluminum alloys is about two-thirds that of DQ steels (The Aluminum Association, 1998). Due to lower formability, complex aluminum body panels may require several

Table 1.5 Properties of several aluminum alloys used for body applications

Material	Yield strength (MPa)	Tensile strength (MPa)	Elongation (%)	n	$\bar{r}$
5182-O	130	275	24	0.33	0.80
5454-O	115	250	22	0.30	0.80
5754-O	100	220	26	0.30	0.80
6009-T4	125	220	25	0.22	0.64
6009-T62	260	300	11	–	–
6111-T4	150	280	26	0.28	0.70
6111-T62	320	360	11	–	–
6061-T6	275	310	12	–	–

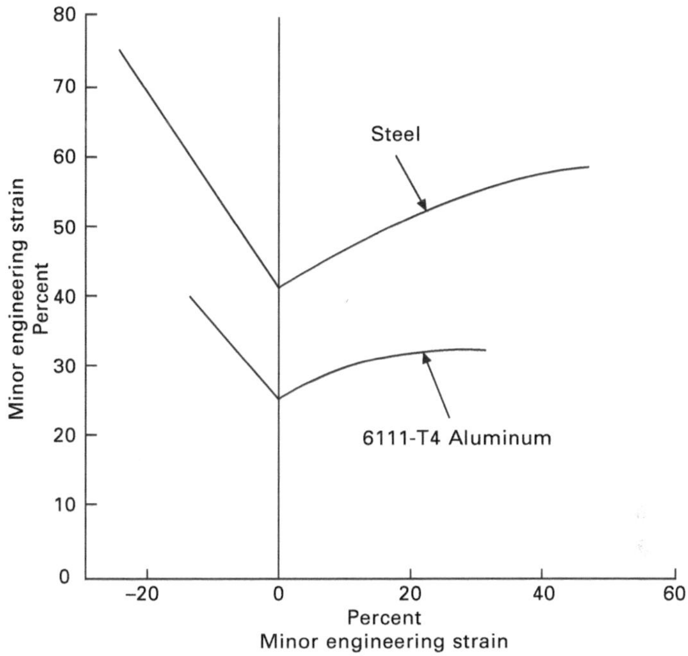

1.1 Formability curves of steel and 6111-T4 aluminum alloy.

additional stamping steps or may have to be produced by assembling several separate stampings. In addition, because of aluminum's lower modulus, aluminum parts exhibit higher elastic springback after forming and therefore poorer shape retention than steel parts. Another important point to note is that aluminum has a greater tendency to gall than steel during the stamping operation and requires larger amounts of lubrication as well as better die surface finish than steel.

(iii) *Corrosion*: Due to the high oxidation potential of aluminum, a thin continuous aluminum oxide film is readily formed on its surface. This oxide film provides a natural protection from corrosion in most environments; however, it creates difficulty for welding and painting, and needs to be removed before performing these two operations.

(iv) *Joining*: Although aluminum alloys can be resistance spot welded like steel, higher welding current is needed for aluminum because of its low electrical resistivity and high thermal conductivity. The welding current for aluminum alloys is 15–30 kA compared to 8–10 kA for steel. This means larger welding machines are needed for spot welding aluminum alloys and energy consumption is also higher. Fusion welding techniques, such as MIG welding, can also be applied to aluminum, but due to aluminum's high thermal conductivity, high heat energy is needed. The two newly developed welding techniques that apply well to aluminum alloys are linear friction stir welding and spot friction stir welding. Other joining techniques that are being used with aluminum alloys are self-piercing riveting, clinching, adhesive bonding and weld-bonding (a combination of spot welding and adhesive bonding).

Prior to the actual joining operation, many panels are folded over each other in a pre-joining bending operation by a process called hemming. For example, the outer door panel is fitted to the inner door panel by bending its peripheral edges 180° over the inner door panel. Because of lower ductility and reduced formability, the hemming limit of aluminum alloys is generally lower than that of steel. Flat hemming, which is common with DQ steels, may not be possible with aluminum alloys. Rope hemming, which is recommended for aluminum alloys, requires a greater thickness allowance in the hemmed area.

(v) *Cost*: The material cost of aluminum alloys per unit weight is three to four times higher than that of steels. In addition, the total tooling cost for manufacturing aluminum body panels may also be higher due to the greater number of stamping tools and better die surface finish required for aluminum alloys (Kelkar *et al.*, 2001). Since slower stamping rates are used to prevent cracking in aluminum, the processing cost of stamped aluminum parts is also higher than that of stamped steel parts. If extruded aluminum profiles are used in body structural components (as in the Audi space frame design), the processing cost is lower than that of stamped aluminum parts, but there may be additional assembly costs that may partially offset the processing cost saving.

(vi) *Recycling*: Aluminum is a highly recyclable material and more than 80% of aluminum used in consumer products, such as aluminum cans, is recycled. A similar recycling rate is also possible with automotive aluminum; however, the shredded mix of aluminum scrap pieces coming from the automobile shredders contains both cast pieces and wrought pieces, and they must be separated before recycling. The reason for this is that cast aluminum contains high amounts of Si (greater than 7%) and Fe (greater than 0.6%). The Si and Fe contents in wrought aluminum are less than 1.2 and 0.4%, respectively. The cast and wrought aluminum alloys are therefore not compatible from the recycling standpoint.

1.2.3 Magnesium alloys

Magnesium alloys, although not much used in today's automobiles, may experience significant growth in the future for the following reasons:

- Lightest structural metal in use (its density is 1.74 g/cm^3 compared to 2.7 g/cm^3 for aluminum alloys)
- Higher strength-to-density ratio than aluminum alloys
- High damping capacity

The modulus of magnesium alloys is 45 GPa, which is significantly lower than that of steel and aluminum alloys; however, because of their low density, the modulus-to-density ratio of magnesium alloys is the same as that of aluminum alloys. Magnesium alloys have low ductility and poor formability; but many magnesium alloys can be cast in thin sections as low as 2 mm in thickness. The common manufacturing method for making automotive magnesium components is die casting, which allows the opportunity for parts consolidation and cost reduction.

Like the aluminum alloys, magnesium alloys can be divided into casting alloys and wrought alloys. Among the casting alloys, AZ91, with aluminum and zinc as the principal alloying elements, is used in many non-structural components, such as brackets, covers and housings, where it provides significant weight saving over aluminum alloy A380. For structural components where higher ductility and crash resistance are important, such as instrument panel beams, steering wheel armatures and seat structures, AM20, AM50 or AM60 is used. The principal alloying elements in the AM-series alloys are aluminum and manganese. Among the wrought magnesium alloys, AZ80 is used for extruded sections and AZ31 is used for sheets. Both of these alloys have yield strength comparable to the 5000 and 6000 series aluminum alloys, but they are less ductile than the aluminum alloys. The room temperature formability of wrought magnesium alloys is also much lower than that of aluminum alloys and steel. Because of this, elevated temperatures, in the range 200–400 °C, are recommended for sheet stamping, bending and other forming operations with AZ31 (Luo, 2005). Elevated temperatures are also used for the extrusion of AZ80.

One major concern with magnesium alloys is their poor corrosion resistance. While the aqueous corrosion resistance of AZ and AM alloys in salt environment is comparable to that of cast aluminum alloys, their galvanic corrosion resistance is very poor. Thus, when a magnesium component is attached to a steel component or two magnesium components are joined together using a steel fastener, magnesium is aggressively corroded. Since the galvanic corrosion of magnesium in the presence of aluminum is much less, aluminum washers, aluminum fasteners or aluminum-coated steel fasteners are often used with magnesium. In the die cast AM50 radiator support assembly in the Ford F150 light truck, galvanic corrosion protection from the attached steel brackets was achieved by a combination of surface coatings and 0.7 mm thick aluminum (AA5052) isolators placed between the magnesium and steel components (Balzer *et al.*, 2003).

Magnesium alloys are currently being considered for several powertrain applications, such as transmission cases and engine blocks. The AZ91 alloy is selected for manual transmission cases where the operating temperature is below 120 °C. The operating temperature of automatic transmission cases and engine blocks can reach up to 200 °C. AZ91 or other conventional casting alloys are not suitable for these applications, since they exhibit significant creep at temperatures higher than 125 °C. Due to creep, the clamping load in the bolted joints of these alloys is reduced, which may cause gas and oil leaks and also increase noise and vibration. Recently, several creep-resistant magnesium alloys containing rare earth elements and alkaline earth elements have been developed, which show promising bolt load retention and are considered better suited for powertrain applications (Pekguleryuz and Kaya, 2003). Some of these creep-resistant alloys are considered good candidates for engine block, oil pan and other engine components, and they are being considered in developing a magnesium-intensive engine with a potential weight saving of at least 15 percent over the conventional aluminum-intensive engine (Hines *et al.*, 2006).

1.2.4 Titanium alloys

The principal advantages of titanium alloys are their low density, high strength-to-density ratio and excellent corrosion resistance. They also maintain high strength at elevated temperatures ranging up to 500 °C. The density of titanium is 4.43 g/cm^3, which is significantly lower than that of steel. The modulus of titanium is 114 GPa, which is nearly half the modulus of steel.

The major drawback of titanium for automotive applications is its high cost compared to steel, aluminum and magnesium. On unit weight basis, the cost of sheet titanium is $18 to $110 per kg compared to only $0.70 to $1.30 per kg for sheet steel and $2.20 to $11 per kg for sheet aluminum (Froes *et al.*, 2004). On the basis of cost, titanium is not expected to compete with steel or aluminum in body panel or body structure applications. However, the potential for saving weight using titanium exists in several other automotive applications. One of these applications is the suspension coil springs, where titanium's relatively low shear modulus and excellent fatigue strength give it an advantage over steel. Since the spring deflection is inversely proportional to the shear modulus, a titanium coil spring can be designed with fewer active coils than a steel coil spring, which contributes not only to weight reduction, but also to increasing its natural frequency of vibration. Titanium coil springs have been used in aircrafts for many years. The first titanium coil spring in the automotive industry appeared in the 2001 Volkswagen Lupo FSI (Faller and Froes, 2001). The titanium alloy for the VW springs was a Ti-4.5 Fe-6.8 Mo-1.5 Al alloy (Timetal LCB), which is 50% lower in cost than the conventional α/β and β-titanium alloys and was specifically developed for automotive applications. The titanium coil springs were about 60% lighter than the steel coil springs they replaced. Titanium's high strength-to-density ratio, fatigue strength and high strength retention at elevated temperatures can be utilized in reducing the weight of engine components that undergo reciprocating motions, such as connecting rods, pistons and piston pins.

Other engine components where titanium has performed well are engine valves, valve retainers and valve springs. Reduced mass of many of these engine components has the secondary effect of reducing friction, which in turn improves the engine efficiency. For example, it was estimated that the use of a titanium valve system can reduce the engine frictional loss by about 10%, which, for a typical driving cycle amounts to 3–4% improvement in fuel economy (Sherman and Allison, 1986). The first titanium connecting rod in a production automobile appeared in the 1992 Acura NSX and the first titanium valves appeared in the 1999 Toyota Altezza. Many of these components can experience a significant amount of wear and since titanium has low wear resistance, the application of titanium for these components required the development of wear-resistant surface treatments, including surface coatings and oxidizing. Another option is to use a titanium matrix composite or titanium aluminide, an intermetallic compound of titanium and aluminum, which not only has a higher wear resistance, but also a higher modulus, desirable in piston pin design.

Another potential application area of titanium is in the exhaust system, since titanium possesses excellent oxidation resistance up to 700 °C. Due to its lower density, considerable weight saving can be achieved over stainless steel, which is currently used for tail pipe, muffler and other components in the exhaust system. Titanium mufflers, offered as an option in the Corvette Z06, were 41% lighter than stainless steel mufflers. Since many of the exhaust system components are cold formed, unalloyed (commercially pure) titanium (grade 1 or 2) is recommended, since it has better strain-to-failure and formability than the α/β or β titanium alloys. However, unalloyed titanium is more suitable for the rear section of the exhaust system, where the temperature is considerably lower than that in the front section.

1.2.5 Stainless steels

The principal use of stainless steels in today's automobiles is in the exhaust system, where their exceptional corrosion and oxidation resistances give them a considerable edge over steel or aluminized steel. Typical choices for the hot end of the exhaust system, which includes the exhaust manifold, down pipe and catalytic converter, are the austenitic grades, such as 309 or 310 (25% Cr, 20% Ni). For the cold end, which includes the resonator, intermediate pipe, silencer and tail pipe, either austenitic grades, such as 304 (18% Cr, 9% Ni) or ferritic grades, such as 409 (12% Cr), are selected. Stainless steel is available in a variety of grades, but the two grades that are used for other automotive applications are the austenitic grade (300 series alloys, containing Cr and Ni as the alloying elements) and the ferritic grade (400 series alloys, containing Cr as the alloying element). The austenitic grade is non-magnetic and has higher yield strength, ductility and corrosion resistance than the ferritic grade.

Neither grade can be strengthened by heat treatment, but both grades can be strengthened by cold work. The austenitic grade has a higher formability than the ferritic grade. A nitrogen-strengthened version of the austenitic grade, called nitronic, is also available, and several nitronic alloys (e.g. Nitronic 19D and Nitronic 30) can be used for automotive applications. Nitronic 19D is a casting alloy and is recommended for suspension components. Nitronic 30 has excellent formability and is recommended for body panels.

The density and modulus of stainless steel are very close to the density and modulus of steel, and therefore, in stiffness-critical applications, direct substitution of steel with stainless steel does not produce any weight reduction. In strength-critical applications, stainless steel can provide weight reduction over steel for the following reasons:

(i) The yield strength-to-density ratio of several stainless steels is higher than that of many high strength steels.
(ii) Stainless steel has a higher work hardening coefficient and formability than steel, which means it can tolerate higher uniform plastic deformation and thickness reduction during forming.
(iii) Stainless steel has a higher strain rate sensitivity than steel, which means it can absorb higher crash energy than steel. Additionally, it also has the capability of collapsing progressively in a controlled manner.

Another great advantage of stainless steel is its corrosion resistance. Anti-corrosion coatings are not needed if stainless steel is used instead of steel. Despite the above advantages, stainless steel has found very little application in automotive structure because of its high cost. A few structural applications where stainless steel has been tried are fuel tanks, knuckle arms and wheels.

1.2.6 Cast iron

With increasing use of high strength steels, light alloys and composites, the role of cast iron in automobiles has decreased considerably. Cast iron, due to its density (as high as that of steel) does not offer any weight saving advantage. Furthermore, cast iron is a low ductility material. The principal advantages of cast iron, for which it continues to be used, are its low cost, high wear resistance, damping and excellent machinability.

The cast irons used in many structural automotive applications are ductile irons, which have high yield strength (275–625 MPa) and relatively high ductility (2–18% elongation). The modulus of ductile irons is 160–170 GPa, which is considerably higher than that of aluminum. Ductile irons are used in steering knuckles, brake calipers, crank shafts, cam shafts and many other powertrain components. Austempered ductile iron (ADI), produced by a heat treatment process called austempering, has a significantly higher yield strength

(400–1200 MPa) and higher fracture toughness than the conventional ductile irons. The yield strength-to-density ratio of ADI is considerably higher than that of cast or forged aluminum, which is the reason for selecting ADI over aluminum alloys in many chassis and suspension components.

Cast iron is also used in many engine applications. One of these engine applications is the cylinder block. Although aluminum is increasingly used for making cylinder blocks for passenger car engines, grey cast iron is still the predominant material for cylinder blocks in diesel engines. However, with the increasing trend toward smaller engines and higher in-cylinder pressures, compacted graphite iron (CGI) is finding increasing use instead of grey cast iron. The graphite particles in CGI appear in vermicular or worm-like form instead of the flaky form observed in grey cast iron or spherical form observed in nodular cast iron. As a result, the properties of CGI fall between those of grey cast iron and nodular cast iron. The tensile strength of CGI is 1.5 to 2 times higher than that of grey cast iron and the modulus of CGI is 150 GPa compared to 105 GPa for grey cast iron. The thermal conductivity of CGI is lower; 38 $Wm^{-1}K^{-1}$ compared to 48 $Wm^{-1}K^{-1}$ for grey cast iron. With higher strength, higher modulus and lower thermal conductivity, CGI cylinder blocks can be designed with lower thickness than the grey cast iron cylinder blocks.

1.2.7 Composites

Polymer matrix composites

Polymer matrix composites (PMC) are produced by combining high strength, high modulus fibers with either a thermoplastic or a thermoset polymer matrix. The fibers used with most polymer matrix composites are glass, carbon or Kevlar fibers (Table1.6). Depending on the processing method selected and the required design need, the fibers can be incorporated in the polymer matrix in continuous lengths or discontinuous lengths, and the fiber orientation can

Table 1.6 Properties of fibers commonly used in polymer matrix composites

Fiber	Density (g/cm^3)	Tensile modulus (GPa)	Tensile strength (MPa)	Elongation (%)
E-glass fiber	2.54	72.5	3445	4.88
S-glass fiber	2.49	85.6	4585	5.7
High strength PAN carbon fiber	1.76	230	3600	1.5
High modulus PAN carbon fiber	1.80	345	2480	0.7
Ultra high modulus PAN carbon fiber	1.96	483	1520	0.38
Pitch carbon fiber	2.20	930	3172	0.25
Aramid fiber (Kevlar 49)	1.45	131	3620	2.8

be unidirectional, bidirectional, random or combinations thereof (Mallick, 2008). With unidirectional fibers (i.e. all fibers oriented in the same direction), the modulus and strength of the composite are the highest in the fiber direction (longitudinal direction), but lowest normal to the fiber direction (transverse direction). For example, the longitudinal modulus of a unidirectional high modulus carbon fiber reinforced epoxy is 207 GPa (which is equivalent to the modulus of steel), whereas the transverse modulus is only 14 GPa. Bi-directional reinforcement (e.g. fabric reinforcement) produces a more balanced set of strength and modulus in the fiber directions; however, they are lower than the longitudinal strength and longitudinal modulus of a unidirectional composite. If the fibers are randomly oriented, the properties are the same in all directions in the plane of the composite; however, they are significantly lower than the properties of composites containing either unidirectional or bidirectional fibers. Thus, unidirectional or bi-directional composites behave as non-isotropic materials, whereas random fiber composites are isotropic.

Most of the polymer matrix composite parts in today's automobiles contain randomly oriented discontinuous glass fibers and they are manufactured using either injection molding or compression molding. E-glass fibers are selected because of their much lower cost than carbon or Kevlar fibers. Because of the discontinuous fiber lengths and random fiber orientation, they do not provide the highest weight saving potential that can be achieved with continuous fiber composites, especially those containing carbon fibers. Continuous fiber composites have lower density, higher strength-to-density ratio and higher modulus-to-density ratio than steel and light alloys (Table 1.7). They also have excellent fatigue strength and fatigue damage tolerance. The possibility of making laminated structures with different fiber orientations in different layers of the laminate or making a sandwich structure with high modulus composite skins and low density foam, balsa wood or aluminum honeycomb in the core provides a tremendous design flexibility that does not exist with metals.

The automotive applications of PMC include both thermoplastic matrix composites and thermoset matrix composites. Thermoplastic matrix composites are used for a variety of interior and body applications, such as instrument panels, seat backs, inner door panels, fender aprons and bumper beams. The thermoplastic polymers in these applications are usually polypropylene, polyamide-6 or polyamide-6,6. The most common thermoset matrix composite is sheet molding compound (SMC), which contains randomly oriented discontinuous E-glass fibers (typically 25 mm long) in a thermoset polymer, such as a polyester or a vinyl ester resin. Examples of SMC parts are hoods, deck lids, fenders, radiator supports, bumper beams, roof frames, door frames, engine valve covers, timing chain covers, and oil pans. These parts are produced by the compression molding process. Another manufacturing

Table 1.7 Material indices and relative costs

Material	Properties (relative values)			Material index for stiffness critical design			Material index for strength critical design		Cost of material
	Density (ρ)	Modulus (E)	Tensile strength (S_t)	Tension $\frac{E}{\rho}$	Buckling $\frac{E^{1/2}}{\rho}$	Bending $\frac{E^{1/3}}{\rho}$	Tension $\frac{E_t}{\rho}$	Bending $\frac{E_t^{1/2}}{\rho}$	(relative values)
DQ Steel	1	1	1	26.3	1.83	0.75	40.3	2.26	1
DP Steel	1	1	2.21	26.3	1.83	0.75	88.9	3.36	1.15
AA6111	0.34	0.34	1.13	25.9	3.10	1.53	133.3	7.03	4–5
Mg AZ91	0.23	0.22	0.76	25	3.73	1.97	133.3	8.6	4–5
SS304	1	0.97	1.94	25.3	1.79	0.74	77.7	3.14	6–8
High strength CFRE	0.2	0.67	4.89	89	7.58	3.33	1000	25.4	15–20
GFRE	0.23	0.19	3.04	21.1	3.37	1.83	521.6	16.8	8
SMC	0.24	0.08	0.52	8.5	2.14	1.35	87.7	6.85	1.5

process used for making thermoset matrix composite parts is called structural reaction injection molding (SRIM). The matrix in composites produced by SRIM is either polyurethane or polyurea.

SMC usage has experienced a large growth in the automotive industry over the last 25 years. Its advantages over steel include not only weight reduction, but also lower tooling cost and parts consolidation. The tooling cost for compression molding SMC parts is 40–60% lower than that for stamping steel parts. An example of parts consolidation can be found in radiator supports in which SMC is used as a substitute for low carbon steel. The composite radiator support will typically be made of two SMC parts bonded together by an adhesive instead of 20 or more stamped steel parts assembled together by a large number of screws. Another example of parts consolidation can be found in the station wagon tailgate assembly, which has significant load-bearing requirements in the open position. The composite tailgate consists of two pieces, an outer SMC shell and an inner reinforcing SMC piece. They are bonded together using a urethane adhesive. In one such application, the SMC tailgate replaced a seven-piece steel tailgate assembly, at about one-third its weight.

Among the chassis components, the first major structural application of polymer matrix composites was the Corvette rear leaf spring, introduced in 1981 (Kirkham *et al.*, 1982). A uni-leaf E-glass fiber reinforced epoxy spring was used with as much as 80% weight reduction as compared to a multi-leaf steel spring. Other structural chassis components, such as drive shafts and road wheels, have been successfully tested in laboratories and proving grounds. They have also been used in limited quantities in production vehicles. They offer opportunities for substantial weight savings, but so far they have not proven to be cost-effective over their steel or aluminum counterparts.

While glass fibers are the primary reinforcing fibers used in today's automotive composites, it is well recognized that much higher weight reduction can be achieved if carbon fibers are used. Carbon fiber reinforced polymers have much higher modulus-to-density and strength-to-density ratios than glass fiber reinforced polymers (Table 1.7). The reason for not using carbon fibers in today's vehicles is that the current carbon fiber price, at $16/kg or higher, is not considered cost-effective for automotive applications. Many development projects in the past have demonstrated the weight saving potential of carbon fiber reinforced polymers; unfortunately, most of these projects did not go beyond the prototyping and structural testing stages due to the high cost of carbon fibers and the lack of manufacturing processes suitable for mass production of composite parts. Recently, several high priced vehicles have started using carbon fiber reinforced polymers in a few selected components. One recent example of this is seen in the BMW M6 roof panel, which was produced by a process called resin transfer molding (RTM). The material is a carbon fiber reinforced epoxy. This panel is twice as thick as a comparable steel panel, but is still 5.5 kg lighter. One added benefit of reducing the weight of the roof panel is that it lowers the center of gravity of the vehicle, which is an important design consideration for vehicle stability.

Carbon fiber reinforced polymers are used extensively in motor sports, where lightweight structure is essential for gaining the competitive advantage of higher speed (Savage, 1991) and cost is not a major material selection decision factor. The first major application of these composites in race cars started in the 1950s, when glass fiber reinforced polyester was introduced as replacement for aluminum body panels. Today, all major body, chassis, interior and suspension components in Formula 1 race cars utilize carbon fiber reinforced epoxy. One major application of carbon fiber reinforced epoxy in Formula 1 cars is the survival cell, which protects the driver in the event of a crash. The nose cone located in front of the survival cell is also made of carbon fiber reinforced epoxy. Its controlled crush behavior is critical to the survival of the driver.

As previously indicated, the major barrier to the application of carbon fiber reinforced polymers is the high material cost, which is solely due to the high cost of carbon fibers. It has been suggested that, if the carbon fiber cost reduces to \$8-\$10/kg, carbon fiber reinforced polymers will become a more viable material option for large-scale automotive applications. The largest contributors to the high cost of carbon fibers are the starting material or precursor cost and the cost of the energy-intensive thermal pyrolysis process used for making carbon fibers. Another current problem with carbon fibers is their availability. Much of the world's production of carbon fibers is consumed by the aerospace and sporting goods industries. New technologies are being developed to produce low-cost carbon fibers and to scale-up the production rate so that it can perhaps meet the automotive industry's need (Warren *et al.*, 2002).

Widespread use of polymer matrix composites, including carbon fiber reinforced polymers, will require the development of processing methods with production cycle time that are competitive with those for steel. The cycle time for the molding processes used today for manufacturing structural automotive composite parts is between 1 to 5 minutes, compared with less than 10 seconds for stamping steel parts. Although the possibility of parts consolidation in composites may reduce the tooling and assembly costs, the processing cost due to higher cycle time causes the total manufacturing cost to be high. For building up confidence in polymer matrix composites for structural automotive applications, long-term durability data, reliable joining techniques and fast non-destructive inspection methods are also needed.

Metal matrix composites

Metal matrix composites (MMCs), by virtue of their low density, high strength-to-weight ratio, high temperature strength retention, and excellent creep, fatigue and wear resistances, have the potential for replacing cast iron and other materials in engines and brakes. Typically, MMCs considered for automotive applications contain either silicon carbide (SiC), aluminum oxide (Al_2O_3) or other ceramic particles or short fibers in a light alloy, such as aluminum, magnesium and titanium. MMCs have been developed for use in diesel engine pistons, cylinder liners, brake drums and brake rotors (Chawla and Chawla, 2006). Other potential applications where MMCs have been tried are connecting rods, piston pins and drive shafts. The major impediment toward their wider use is their high cost. More details on the properties and manufacturing processes of MMCs can be found in Kainer (2006).

1.2.8 Glazing materials

The glazing materials in a vehicle are the laminated glass used for the windshield, and tempered glass used for side windows, rear window and sunroof. The laminated glass is constructed of two 1.8–2.3 mm thick sheets of glass with a very thin layer (typically 0.76 mm thick) of polyvinyl butyrate (PVB) between them. The PVB layer makes the windshield shatter-proof, which is essential for the safety of the driver and the front passengers. Tempered glass is a single sheet of glass (typically 2.4 to 2.6 mm thick), strengthened by heating it above the annealing point of 720 °C followed by rapid cooling. Tempered glass is much easier to penetrate than laminated glass and fractures in a brittle manner when impacted; but it is 3 to 4 times cheaper than laminated glass.

Although the weight of the glazing material is only 2–3% of the total weight of a vehicle, several alternatives are being considered to reduce its weight. One of these alternatives is to reduce the windshield thickness by using thinner glass sheets; however, a large reduction in the thickness may not only raise concern about safety, but also reduce its contribution to the torsional stiffness of the vehicle (which is approximately 10% with the current windshield thickness). Another alternative is to use polycarbonate instead of glass (Mori and Koursova, 2000), which has a density of 1.2 g/cm^3 compared to 2.5 g/cm^3 for glass. Polycarbonate is a transparent thermoplastic with optical properties comparable to glass. It is also a ductile polymer with high impact resistance. However, polycarbonate windshields require a scratch resistant coating on their surface. Since polycarbonate has a lower modulus than glass, polycarbonate windshields are thicker. They are also more expensive.

1.3 Materials selection considerations for lightweight vehicles

1.3.1 Body structure

The structure of most passenger vehicles has changed from the body-on-frame (BOF) construction to body-in-white (BIW) or unibody construction since the 1970s. The BIW construction is not only lighter in weight, but also provides higher stiffness per unit weight than the BOF construction. In the BOF construction, two front-to-back longitudinal frame members connected along their lengths by several cross members provide the required stiffness for the vehicle and serve as the principal load carrying members. The body is built separately and then attached to the frame. The BIW construction consists of an integrated box structure, several skin panels (such as roof, quarter panels and floor pan) and bulkheads (such as floor cross members, roof rails, A-pillars and B-pillars), all of which contribute, to a greater or lesser extent, to the stiffness of the vehicle. Sub-frames of small lengths are used to support the suspension and powertrain components, but they do not provide any structural support to the body. The BIW weight is typically 25% of the vehicle weight. In steel BIW, the box structure and the bulkheads are made by spot welding thin gage steel sheets formed in a variety of shapes and sizes. With increasing use of aluminum and magnesium, there will also be extruded and die-cast components in the BIW. If polymer matrix composites are used, a different approach to designing the body structure may be needed, since they have different material behavior and design characteristics from metals and they require different processing techniques.

The principal design attributes for the vehicle body structure are its static stiffness and frequency response. The static stiffness is determined in both bending and torsion, and depends on the structural configuration, stiffness of the primary components of the body structure, joint design, as well as the joining method. The frequency response is determined using dynamic tests at the engine idling speed, which is typically between 600 and 700 rpm, and depends on both the static stiffness and the mass distribution in the structure. For vibration-free operation, the first natural frequencies for the bending and torsional modes of the BIW must lie within a specific range. Depending on the vehicle class, each vehicle manufacturer sets target values for the static stiffness and frequency response at the initial design stage and determines them using finite element analysis and laboratory tests as the BIW is developed. For the recent BMW 3 series, for example, the target values for the torsional stiffness, first bending frequency and first torsional frequency were 18 100 N-m/degree, 27 Hz and 30 Hz, respectively (Pfestorf and van Rosenburg, 2006).

The stiffness of the primary components in the body structure depends on their shape and dimensions as well as the modulus (E) of the material used.

The selection of material for weight reduction, however, should not only consider the modulus, but also the density (ρ) of the material. For example, if the design objective for a tension linkage is to meet the axial stiffness requirement with minimum weight, the material selection criteria involve not just the modulus, E, of the material, but also the modulus-to-density ratio, E/ρ. The modulus-to-density ratio is considered a material index, and the material that produces the highest value of this material index should be selected for the minimum weight design of the tension link. The material index for minimum weight design depends on the loading condition (e.g. tension vs. bending) and the design requirement (e.g. stiffness vs. strength or maximum load carrying capacity). Expressions for several material indices relevant to automotive structures are listed in Table 1.7. As an example of the use of the material index in preliminary material selection, consider the minimum weight design of a cross member, which can be modeled as a simply supported beam. Since the cross member must be designed with a minimum bending stiffness, the material index to be used is $\frac{E^{1/3}}{\rho}$. If a comparison is made of DP steel, AA6111, unidirectional high strength CFRE and unidirectional GFRE (see Table 1.3 for their modulus and strength values), it is clear that CFRE is the best choice from the standpoint of weight saving. However, there are other considerations that must be taken into account for the final material selection, such as cost, availability, manufacturing constraints, durability, joining issues, environmental factors and life-cycle impact. Another factor in stiffness critical designs is the section shape. For structural components that are mainly subjected to bending loads, closed hollow box sections are preferred, since they provide much higher bending stiffness per unit mass than solid sections. Stiffeners are added in the walls of the box section to improve its stiffness. With steel, such sections are manufactured by stamping two hat sections and then spot welding them along their flanges. With aluminum, extrusion may be a better process for manufacturing such sections. In bending stiffness critical designs, replacing a lower strength steel with a higher strength steel will not cause any weight reduction. Aluminum, on account of its higher $\frac{E^{1/3}}{\rho}$ value, will lead to weight reduction; however, to match the stiffness of the steel section, the aluminum section will be thicker or larger in size. Even with increased thickness, it may be possible to achieve weight saving (Table 1.8); but other considerations, such as formability of the section, may need to be considered. Another approach in increasing bending stiffness without adding significant weight is to use a sandwich section made of thin skins of a high modulus material and a low density core, such as expanded polypropylene (EPP) foam or aluminum foam, a solid polymer, balsa wood and aluminum honeycomb.

Table 1.8 Weight comparisons of various materials for bending applications

Material	Weight comparison (relative values)	
(see Table 1.3 for the property values)	Stiffness critical design	Strength critical design
DQ Steel	1	1
DP Steel	1	0.67
AA6111	0.49	0.32
Mg AZ91	0.38	0.26
SS 304	1	0.72
High strength CFRE	0.23	0.09
GFRE	0.41	0.13
SMC	0.56	0.33

Typically, the modulus of the core is much lower than the modulus of the skin material. For a symmetric sandwich section with a low modulus core (Fig. 1.2), the bending stiffness and the bending stiffness-to-weight are given by

$$\text{Bending stiffness} \approx \frac{btd^2 E_s}{2} \qquad [1.1]$$

$$\text{Bending stiffness-to-weight ratio} \approx \left(\frac{E_s}{\rho_s}\right) d \qquad [1.2]$$

where, E_s = modulus of the skin material, b = width of the section, t = thickness of the skins and d = distance between the skins. Equations [1.1] and [1.2] show that increasing the core thickness of a sandwich section increases its bending stiffness as well as its bending stiffness-to-weight ratio. Steel–polymer–steel (Dunand and Gacel, 2006) and aluminum–polymer–aluminum (Burchitz *et al.*, 2005) are two commercially available sandwich materials developed for body structures as well as for body panels. Sandwich panels with carbon fiber reinforced polymer in the skins and a polymer foam core can provide much higher weight saving opportunity for body applications.

The joining method also has a large influence on structural stiffness. Continuous adhesive bonding instead of spot welding significantly increases the torsional stiffness. To improve the durability of adhesive joints, a combination of adhesive bonding and welding or riveting can be used. Joint design can also be used to increase structural stiffness. For example, using a larger corner radius in a T-joint between the B-pillar and the roof rail has been shown to not only reduce stress concentration at the joint, but also to increase the stiffness of the joint (Patton and Li, 2002). In the case of spot welded joints, the spacing between the welds, their closeness to the panel edges and the amount of panel overlap are some of the factors that influence both stiffness and strength of the joint. Locally filling the hollow

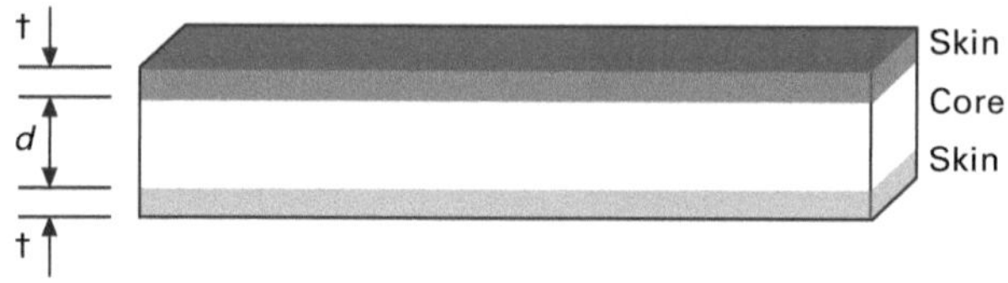

1.2 Sandwich construction.

body structure at the joints and other locations with expandable structural epoxy foam (Casey and Weber, 2004), polyurethane foam (Barpanda *et al.*, 2004) or aluminum foam (Claar *et al.*, 2000) is also an effective method of increasing stiffness, NVH characteristics and crush resistance without adding any significant weight.

Many structural elements in the body structure, as well as many chassis components, experience random fatigue loading. Fatigue analysis methods for steel and aluminum alloys for random fatigue loading are available in the literature. In general, steel has higher fatigue life than aluminum alloys. In the high cycle fatigue area ($>10^6$ cycles), steel shows an endurance limit, whereas for aluminum alloys, fatigue strength continues to decrease with an increasing number of fatigue cycles. Some carbon fiber reinforced polymers show fatigue performances that are better than steel's. They also have significantly better fatigue damage tolerance than steel.

One major concern with a lightweight body structure is its crashworthiness. To improve crashworthiness, the front and rear end structures of a vehicle are designed to absorb energy through progressive deformation while maintaining force levels commensurate with tolerable deceleration rates. The occupant compartment structure, which includes A-pillar, B-pillar, roof rails and door frame, is designed to provide protection for the occupants in all modes of collision and, in particular, for side impact and roll-over accidents. The front end structure is designed with a primary crush zone and a secondary crush zone. The primary crush zone is located in the fore section of the powertrain compartment and is the main energy absorbing element. The secondary crush zone includes the fire wall and the toe-board areas, and acts as the structural interface between the primary crush zone and the passenger compartment. The rear end structure similarly includes a primary crush zone, which is immediately in front of the rear bumper, and a secondary crush zone, which is the structural interface between the shock absorber and the passenger compartment.

For improved crashworthiness, the primary crush zone (also called the crumple zone) is designed to crush progressively with an accordion-type folding (Fig. 1.3) and controlled energy absorption. For a full frontal barrier impact at 35 mph, the length of this crush zone is usually between 500 and 900 mm. To initiate a progressive crush, crush initiators (Fig. 1.4) are

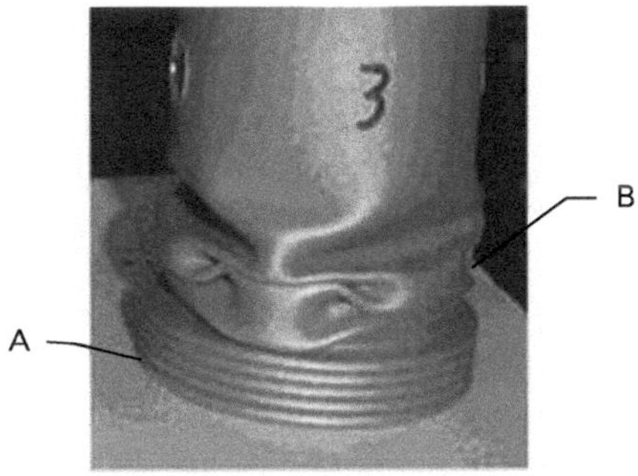

1.3 Accordion folding pattern (A: symmetric, B: asymmetric) in a round tube in quasi-static crush testing.

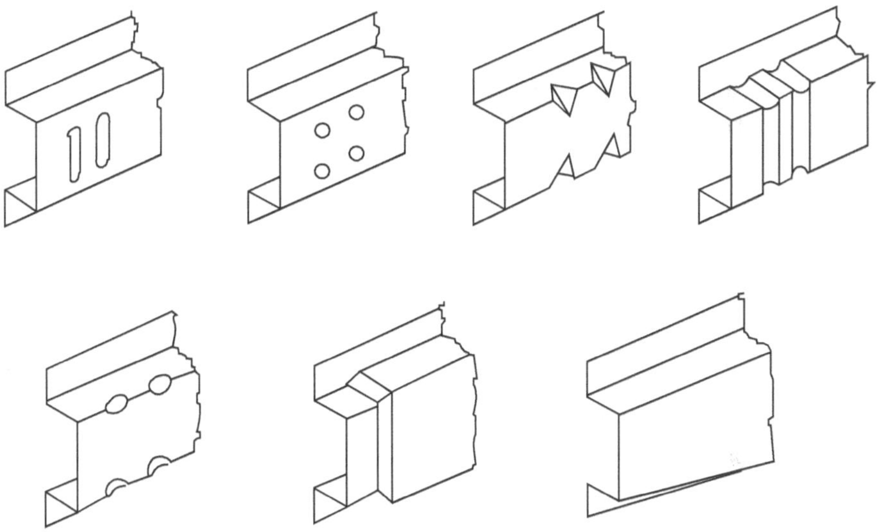

1.4 Different crush initiators used in front rails to promote progressive folding.

designed in the front section of the front rails. Controlled energy absorption is obtained through progressive folding of the front rails and plastic bending or folding of various connected beam sections. The secondary crush zone is designed not to fail or collapse prematurely so that the primary crush zone can perform its energy absorbing role.

Material selection and shape design play key roles in the controlled energy absorption of the primary crush zone. Both steel and aluminum alloy front rails can be designed to crush with accordion-type folding. On a unit weight basis, aluminum shows higher energy absorption than steel of equivalent yield strength. Energy absorption can be increased by increasing the gauge thickness; but since that will result in increasing weight, a better approach is to select a material with a higher yield strength. This is one reason for the increasing application of advanced high strength steels, such as DP steels and TRIP steels, in the front rails and other crush resisting structural elements. Section shape also has an influence on energy absorption. In general, a section with multiple corners, such as a hexagonal section, shows higher energy absorption than a hat section.

Polymer matrix composites can also be designed to exhibit high energy absorption under axial crushing load. On a unit weight basis, the energy absorption of polymer matrix composites is two to three times higher than that of either steel or aluminum. However, the failure mode of polymer matrix composites may not be of a progressive folding type; instead, depending on the fiber type, fiber architecture and the fabrication method, it may include delamination, fiber micro-buckling, fiber splaying and shear failure.

1.3.2 Body panels

There are two types of body panels: (i) outer body panels, which include horizontal panels (hoods, deck lids and roof) and vertical panels (door and quarter panels) and (ii) interior body panels, such as wheel house and rear floor. Closure panels, such as hoods and deck lids, usually consist of an inner panel and an outer panel that are joined together along their perimeters.

The material selection requirements for body panels and body structures are listed in Table 1.9. As with the body structure, the bending stiffness of body panels is the first major design consideration, since it controls the panel deflection under load. For a bending stiffness-critical design, the initial material selection is based on the material index $\frac{E^{1/3}}{\rho}$. The bending stiffness can also be increased by increasing the panel thickness. However, since increasing the panel thickness also increases weight, the bending stiffness of interior body

Table 1.9 Body structure and body panel material requirements in a BIW construction

Body structure	Body panel
• Bending stiffness	• Bending stiffness
• Torsional stiffness	• Oil canning (outer)
• First bending frequency	• Dent resistance (outer)
• First torsional frequency	• First bending frequency
• Crashworthiness	• Surface finish and smoothness (outer)
• Fatigue	• Painting characteristics (outer)
• Corrosion resistance	• Corrosion resistance
• Formability	• Formability
• Joining	• Attachment
• Recycling	• Recycling

panels is usually increased by adding stiffeners, reinforcements, curvature, intermediate supports, corrugation, ribs, etc. For outer body panels, stiffness increase is achieved through double curvatures, adding reinforcements on the non-visible side and, in some cases, forming character lines along their lengths. Two other design requirements for outer body panels are resistance against oil canning and resistance against denting. Oil canning is local elastic buckling which is manifested by a sudden reversed curvature on the surface when the panel is pressed by hand. Since it is an elastic deformation, it does not leave any permanent depression. Resistance to oil canning requires high elastic modulus, but high yield strength is also desirable, since it delays the plastic deformation. Denting, on the other hand, is a permanent deformation and it occurs when the local stresses in the dented region exceed the yield strength of the material. Denting can occur in a number of ways; for example, by stone impact, low-speed accidents or an unintentional hit by the opening door of a neighboring car. The denting energy of steel panels, which is a measure of their dent resistance, has been found to follow the empirical equation (DiCello and George, 1974):

$$\text{Denting energy} \propto \frac{S_{yd}^2 t^4}{K} \qquad [1.3]$$

where, S_{yd} is the dynamic yield strength of the panel material, t is the panel thickness and K is the panel stiffness, which depends on the modulus, panel thickness, curvature and geometry as well as its support conditions. Equation [1.3] shows the importance of high dynamic yield strength in providing dent resistance of an outer body panel. It should be noted that polymer matrix composite panels will not dent in the way steel or aluminum panels dent. Instead, under low impact conditions, there will be local damages. Some of these damages may be visible on the surface; but there may also be sub-surface damages, such as delaminations or internal cracks, that can only be detected using non-destructive testing. These local damages may not pose any immediate structural problems, but they may have to be repaired to reduce their spread over time. The damage repairing methods for composites are different from the dent repairing method used for steel panels and are not yet fully developed.

1.4 Conclusion

To achieve the much needed fuel economy and emission control, it is expected that future vehicles will be much lighter than the vehicles of the past. Low carbon steel has been the principal material for body and chassis construction of a typical vehicle till the middle of 1990s. To improve collision safety, the material of the front end structure of many vehicles has been slowly changing to high strength steels and some advanced high strength steels. For significant weight reduction, future vehicles will use a mix of materials, such as advanced high strength steels, aluminum and magnesium alloys, and carbon fiber reinforced composites. These materials have either higher strength, lower density or both, and can be utilized to design components with lower weight. Many current vehicles are already adopting some of these materials in increasing quantities. However, there are several technical and non-technical issues that still remain to be addressed for large-scale material substitution to occur. The technical issues related to using a mix of materials with different mechanical behavior and design characteristics include safety and reliability, joining and assembly, surface interfacing, recycling, environmental impact and life cycle values. Cost is a primary consideration in the automotive industry and it includes not only the material cost, but also the manufacturing and assembly costs, as well as the product design and development costs. Most of the lightweight materials are more expensive than steels and their processing requires dies and tools that are different from the ones used with steels. Thus, affordability is one of the major barriers in the adoption some of the materials discussed in this chapter, particularly the carbon fiber reinforced polymers which have the greatest weight saving potential among all of them. This was exemplified in the US Department of Energy initiated Partnership for a New Generation of Vehicles (PNVG) concept vehicles, which achieved 20 to 30 percent weight reduction through extensive use of aluminum and/or carbon fiber reinforced polymer, but at significant price premium (Anon., 2001). Another barrier is the lack of familiarity and design experience. The vehicle manufacturers have long experience in designing and manufacturing with low carbon steels for body and chassis constructions. They are much less familiar and experienced with the emerging materials and it may take many years to build the confidence that will lead to using them in lieu of low carbon steels. Significant changes in the materials scenario will also require a shift in design philosophy, manufacturing strategies and business plans by the vehicle manufacturers.

1.5 References

AISI (1998), *Ultralight Steel Auto Body: Final Report*, Washington, D.C., American Iron and Steel Institute.

Anon. (2001), *Review of the Research Program of the Partnership for a New Generation of Vehicles – 7th Report*, National Research Council, National Academy Press.

Balzer J S, Dellock P K, Maj M H, Cole G S, Reed D, Davis T, Lawson T and Simonds G (2003), 'Structural magnesium front end support assembly', *2003 SAE World Congress*, Paper No. 2003-01-0186, Warrendale, Soc. of Automotive Engineers.

Barpanda D, Boven M L, Allen M P and Bilotto F V (2004), 'Polyurethane foam inserts for NVH and structural applications', *2004 SAE World Congress*, Paper No. 2004-01-0461, Warrendale, Soc. of Automotive Engineers.

Brooke L and Evans H (2009), 'Lighten up!', *Automotive Engineering*, 117, 16–22.
Burchitz I, Boesenkool R, van der Zwaag S and Tassoul M (2005), 'Highlights of designing with Hylite – a new materials concept', *Materials and Design*, 26, 271–179.
Casey K and Weber P E (2004), 'Expandable epoxy foam: a systematic approach to improve vehicle performance', *2004 SAE World Congress*, Paper No. 2004-01-0243, Warrendale, Soc. of Automotive Engineers.
Cho W-S, Kim K-S, Jo E-K and Oh S-T (1994), 'Development of medium carbon microalloyed steel forgings for automotive components', *1994 SAE World Congress*, Paper No. 940784, Warrendale, Soc. of Automotive Engineers.
Chawla N and Chawla K K (2006), 'Metal–matrix composites in ground transportation', *JOM*, November, 67–70.
Claar T D, Yu C-J, Hall I, Banhart J, Baumeister J and Wolfgang S (2000), 'Ultra-lightweight aluminum foam materials for automotive applications', *2000 SAE World Congress*, Paper No. 2000-01-0335, Warrendale, Soc. of Automotive Engineers.
DiCello J A and George R A (1974), 'Design criteria for the dent resistance of auto body panels', *1974 SAE World Congress*, Paper No. 740081, Warrendale, Soc. of Automotive Engineers.
Dunand M and Gacel J-N (2006), 'USILIGHT: A cost-effective solution to lighten cars', *2006 SAE World Congress*, Paper No. 2006-01-1216, Warrendale, Soc. of Automotive Engineers.
Faller K and Froes F H (2001), 'The use of titanium in family automobiles: Current trends', *JOM*, 27–28.
Froes F H, Friedrich H, Kiese J and Bergoint D (2004), 'Titanium in the family automobile: The cost challenge', *JOM*, 40–44.
Hines J A, McCune R C, Allison J E, Powell B R, Ouimet L J, Miller W L, Beals R. Kopka L and Reid P P (2006), 'The USAMP magnesium powertrain cast components projects', *2006 SAE World Congress*, Paper No. 2006-01-0522, Warrendale, Soc. of Automotive Engineers.
Kainer K U (ed.) (2006), *Metal Matrix Composites: Custom-made Materials for Automotive and Aerospace Engineering*, Wiley-VCH.
Kelkar A, Roth R and Clark J (2001), 'Automobile bodies: Can aluminum be an economical alternative to steel?', *JOM*, 53, 8, 28–32.
Kirkham B E, Sullivan L S and Bauerie R E (1982), 'Development of the Liteflex suspension leaf spring', *1982 SAE World Congress*, Paper No. 820160, Warrendale, Soc. of Automotive Engineers.
Luo A A (2005), 'Wrought magnesium alloys and manufacturing processes for automotive applications', *2005 SAE World Congress*, Paper No. 2005-01-0734, Warrendale, Soc. of Automotive Engineers.
Mallick P K (2008), *Fiber Reinforced Composites*, Boca Raton, CRC Press.
Mori H and Koursova L (2000), 'Automotive glazing: Issues and trends', *2000 SAE World Congress*, Paper No. 2000-01-2691, Warrendale, Soc. of Automotive Engineers.
Patton R and Li F (2002), 'Causes of weight reduction effects of material substitution on constant stiffness components', *2002 SAE World Congress*, Paper No. 2002-01-1291, Warrendale, Soc. of Automotive Engineers.
Pekguleryuz M O and Kaya A A (2003), 'Creep resistant magnesium alloys for powertrain applications', *Advanced Engineering Materials*, 5, 12, 866–878.
Pfestorf M and van Rosenburg J (2006), 'Improving the functional properties of the body-in-white with lightweight solutions applying multiphase steels, aluminum and composites', *2006 SAE World Congress*, Paper No. 2006-01-1405, Warrendale, Soc. of Automotive Engineers.

Powers W F (2000), 'Automotive materials in the 21st century', *Advanced Materials and Processes*, 38–41.

Savage G (1991), 'Composite materials in Formula I racing', *Metals Mater.* 7, 617–624

Sherman A M and Allison J E (1986), 'Potential for automotive applications of titanium alloys', *1991 SAE World Congress*, Paper No. 860608, Warrendale, Soc. of Automotive Engineers.

The Aluminum Association (1998), *Aluminum for Automotive Body Sheet Panels*, Publication AT3, Washington DC, The Aluminum Association, Inc.

Warren C D, Shaffer J T, Paulauskas F L and Abdullah M G (2002), 'Low cost carbon fiber for the next generation of vehicles: Novel technologies', *2002 SAE World Congress*, Paper No. 2002-01-1906, Warrendale, Soc. of Automotive Engineers.

Yamagata, Y (2005), *The Science and Technology of Materials in Automotive Engines*, Cambridge, UK, Woodhead Publishing.

Yang, B, Nunez S W, Welch T E and Schwaegler J R (2001), 'Laminate dash Ford Taurus noise and vibration performance', *2001 World Congress*, Paper No. 2001-01-1535, Warrendale, Soc. of Automotive Engineers.

Part I

Materials for lightweight automotive structures

2
Advanced steels for lightweight automotive structures

C. D. HORVATH, General Motors, USA

Abstract: High strength steels are commonly used in the automotive industry to reduce mass and improve structural performance. This chapter discusses the history of steel in the manufacture of automobiles, the types of steels currently being used and the new advances in the types of steels that can be used for lightweighting automotive structures. The chapter also reviews the manufacturing and forming aspects of these steels along with some design considerations for the selection of these materials.

Key words: steel, advanced high strength steel, dual phase, trip, complex phase, martensite, boron steel, press hardening, spot welding, adhesive bonding, spot welding, welding, MIG welding, vehicle design.

2.1 History of steel in automobiles

The use of steels in the manufacture of automobiles has a rich and varied history. By most accounts, the era of the automobile is considered to have started in the late 19th century. It was in 1885 that the German engineer Karl Benz built what is generally considered to be the first automobile powered by an internal combustion engine. By today's standards, these first internally combustion driven automobiles were little different from the horse-drawn carriages common in the 18th and early 19th centuries. Both were largely constructed from wooden frames and body panels and even used wooden wheels.

Sometime in the late 19th and early 20th centuries, after the advent of sheet metals and the manufacturing processes to form them into complex shapes, the largely wooden structures of early automobiles were gradually being replaced by sheet metals. It was sometime in the early 1900s, that bodies manufactured from a combination of metal sheets of steel and aluminum, although still built largely on wooden frames, were being driven around the streets of North America. With the introduction of the Model T Ford in 1908, the era of mass produced, largely steel automotive bodies was up and running, and the automobile was well on its way to becoming what many consider to be the invention that most significantly impacted the lives of common working people and, just as importantly, led to the development of today's mobile and modern society.

The use of steel virtually dominated automotive body design for the next 100 years. The primary grades used in these first automobiles were mild steels, or as they are more accurately referred to, low carbon steels. For the existing demands of the time, these steels provided the nearly perfect balance of strength, formability, cost and design flexibility that the industry needed to expand throughout the developed regions of the world. It was not until the first oil shock in the early 1970s and the adoption of fuel economy standards in North America that the industry began to seriously look towards higher strength steels to lower the mass of their vehicles and to improve fuel economy. Since that time, the structural designs used for automobiles and the grades of steel used to manufacture them have been in a constant state of evolution in response to not only consumer demands, but also to the complex array of regulations for fuel economy, emissions and crashworthiness that the modern automotive industry must now account for in the designs of their products.

It was common in the early development of the automobile, and even through the late 1930s and early 1940s, that most vehicles were built on frames that carried the body and skin panels – hence it was commonly referred to as body on frame construction or BOF. This type of construction not only allowed automotive designers to easily change the visual styling of the body to meet the ever changing demands of consumers, but also provided an easy and cost effective way to maintain styling differentiation between manufacturers. However, as vehicles became smaller during the 1960s, and especially after the oil shock in the early 1970s, a new type of body costruction became popular. This new design, called body frame integral (BFI), gained favor and started a steady transition to this construction philosophy. Largely unknown to the industry at the time, the increase use of this integrated vehicle structural design would eventually become a key enabler for the future widespread use of high strength steels

In contrast to the BOF designs, this new design was considered to be more mass and fuel efficient, since the body and frame are designed as one system. This is in contrast to the BOF design where the body and frame are independent from each other, resulting in part function redundancy and increased mass. With the increasing focus on mass and fuel economy in the 1970s and 1980s, the more efficient BFI structural designs became more popular and eventually began to dominate automotive designs for passenger cars. In contrast, light trucks and heavy duty vehicles, where high load capacity and durability were considered more important than fuel economy, largely kept the BOF design that was considered superior for the needs of those vehicles.

With the automotive industry's large scale conversion to BFI designs that did not require heavy frames, the stage was set for an increased use of high strength steels. The increased number of stampings used in this design, and the ability to tailor the strength and shape of these panels, gave designers and manufacturing engineers new freedom to adapt part shape and forming technologies to the characteristics of high strength steels. As a result, a new age in automotive steel development would be ushered in to improve product safety and meet the increasing demands of government regulators and consumers for safer, more fuel efficient vehicles: the face of automotive design and engineering would be changed forever.

However, unknown to the automotive industry at the time, the road they were about to embark on was steep and full of potholes. Not surprisingly, the industry's initial attempts at using high strength steels, which largely consisted of high strength low alloy and solid solution strengthened steels, were largely unsuccessful. The relatively low ductility of these steels compared with the low carbon steels in general use at the time presented many challenges that designers and manufacturing engineers were unprepared to handle. The part designs and manufacturing processes used on these initial applications were the same, or at best little changed, from those used with low carbon steels and did not suit the specific needs of high strength steels. Compounding matters, steel companies had not yet perfected their processes to make high strength steels consistently, and the resulting variability in mechanical properties essentially sealed the fate of these first attempts to use them.

Initial attempts to form high strength steels routinely resulted in severe forming cracks and springback that could not be effectively resolved with techniques that were available at the time. The lessons learned during these early experiences were both difficult and costly. Unfortunately, the lessons did not come fast enough to prevent the industry from returning to the low carbon steels that they understood much better and that had contributed to the early successes of the industry. These first attempts at using high strength steels were, by nearly all accounts, a virtual disaster. Even so, the benefits of these steels were recognized across the industry and manufacturers continued to use them in a select number of the most critical structural parts in the automobile. This foresight and perseverance would end up teaching the entire industry valuable lessons that would be used in later years to effectively incorporate these steels into future vehicle designs.

2.2 Types of high strength steels

There are many types of steels used by today's automotive designer to reduce mass and improve energy absorption in crash events. These steels can easily be divided into three general categories that roughly separate materials by their microstructures. The first category consists of both low carbon and conventional types of high strength steels (HSS) that have been the backbone of automotive design for a number of decades. The second category is generally referred to as the first generation of advanced high strength steels (AHSS) and the third category consists of the second generation of AHSS. Each of these categories has a number of different materials that, while different from each other, share some unique physical and metallurgical properties that distinguish them from materials in the other categories. These categories and the materials included in them are shown in Table 2.1.

Low carbon steels with carbon levels of 0.13% or less have been the most commonly used steels in automotive body panels and structural components for many years. The combination of low cost, excellent formability, weldability, high quality appearance after painting and overall ease of manufacturing has contributed to their widespread use. There are many grades of low carbon steel used in the automotive industry, but the most common types are hot rolled and cold rolled grades with much lower carbon contents than the maximum generally allowed by industry specifications. The lower carbon content grades such as SAE 1005 and the highly formable, ultra-low carbon interstitial free (IF) steels, are still very commonly used on most of today's vehicles. However, the amount of these steels used in structural components has been steadily declining in the last ten years as more high strength steels are designed into critical stampings to reduce weight and improve strength and overall vehicle performance.

2.2.1 Interstitial free and bake hardenable steels

With the advent and proliferation of continuous casting technologies, homogeneous alloys of very formable steels are now commonly available to the industry. In particular, interstitial free (IF) steels have had a major impact on how steels are stamped and the use of the complex features commonly incorporated in today's vehicle designs. These steels are produced with very low carbon contents, typically in the range of 30–50 ppm, and frequently

Table 2.1 Types of automotive steels

Low carbon and conventional high strength steels (HSSs)
Low carbon steels (LC)
Solid solution strengthened (SSS)
Bake hardenable (BH)
High strength low alloy (HSLA)
First generation advanced high strength steels (AHSSs)
Dual phase (DP) – ferrite/martensite
High hole expansion (HHE) – ferrite/bainite
Stretch flangeable (SF)
Transformation induced plasticity (TRIP)
Complex phase (CP)
Fully martensitic (MS)
Boron heat treatable steels
Second generation advanced high strength steels
Twinning induced plasticity (TWIP)
Lightweight steels with induced plasticity (L-IP)

have manganese levels of less than 0.25%. They offer very high elongations that can approach and sometimes even exceed 50%, and very high strain hardening coefficients that are advantageous when stamping very difficult-to-form parts.

Perhaps more importantly, at least from the aspect of steels used for lightweighting, the use of vacuum degassing technology to produce ultra-low carbon (ULC) steels has served as a base for the development of new classes of highly formable high strength steels. These steels owe their combination of high formability and high strength to the use of solid solution strengthening (SSS) and grain size control. The ultra-low carbon rephosphorized steels and some bake hardenable steels are good examples of how ULC technology and SSS have produced new grades of steel that meet many of the needs of modern automotive designs. In some cases, multiple strengthening mechanisms are used in conjunction with ULC technology. The addition of micro alloying elements such as niobium, in conjunction with the SSS elements such as phosphorus, manganese and silicon, along with grain sizes in the 7 to 8 μm range, have also produced some steels with very high strength and good formability. In this specific case, niobium is added in excess of what is needed to stabilize carbon. The resulting steel is not only interstitial free, but also contains well-dispersed fine carbo-nitride precipitates that contribute to higher tensile strengths through a refined grain structure as well as precipitation hardening (Toshiaki *et al.*, 2003). Some of these steels are shown in Table 2.2.

In addition to these grades of steel, micro-alloyed high strength low alloy (HSLA) grades of steel are commonly used throughout the body structure, especially for structural parts that carry significant loads and where very good spot weldability is required. These steels combine very good formability, high strength and modest cost, and are therefore commonly used by many automotive manufacturers. Parts such as motor compartment rails, rocker side panels and rear longitudinal rails are prime applications for these steels. Some of the more common HSLA steels are shown in Table 2.3.

Bake hardenable steels are commonly used in the automotive industry on exposed body panels and even structural stampings. The effect of bake hardening on the improvement in dent resistance is well established and bake hardenable steels have been used for many years to downgauge exterior panels without reducing denting performance (Schaeffler *et al.*, 1996; Okamoto, 1982; Campbell, 1989). These low carbon steels are thermally treated such that small amounts of solute carbon are held in solution which, upon heating, allows carbon to diffuse and pin dislocations. Mechanistically, solute carbon migrates to the dislocations until the concentration of carbon increases to the point that it precipitates and pins the dislocation (Wuebbels *et al.*, 2002). Typically, hardening from this bake hardening effect causes the yield strength to increase by 30–40 MPa in steels of this type. The key for

Table 2.2 Mechanical properties and chemistries of selected solid solution strengthened and bake hardenable steels (source: Toshiaki, 2003 and General Motors Corp., 2008)

	Mechanical properties					Chemistry				
	YS (MPa)	TS (MPa)	El (%)*	*n*-value	*r*-value	C (%)	Mn (%)	Si (%)	P (%)	Nb (%)
180 MPa SSS	186	340	37	0.22	1.6	.005 max	0.4	n/a	0.05	–
210 MPa SSS	223	335	35	0.20	1.5	.005 max	0.4	n/a	0.07	–
240 MPa SSS	252	363	34	0.19	1.4	.005 max	0.4	n/a	0.09	–
180 MPa BH	193	342	37	0.21	1.5	0.005 max	0.5	n/a	0.04	–
210 MPa BH	218	337	35	0.18	1.3	0.005 max	0.6	n/a	0.05	–
240 MPa BH	245	344	34	0.18	1.3	0.005 max	0.6	n/a	0.06	–
440 IF HSS	290	446	34	n/a	1.9	0.005	0.62	0.01	0.04	0.068

*Total elongations are in 50 mm with an ISO #1 tensile bar.

Table 2.3 Mechanical properties of common HSLA steels (source: SAE, 1999 and General Motors Corp., 2007)

Grade designation	Yield strength, (MPa) minimum)	Tensile strength (MPa) minimum)	%Total elongation, minimum, cold reduced*	%Total elongation, minimum, hot rolled
300	300	400	24	26
340	340	410	22	25
380	380	450	20	23
420	420	490	18	22
550	550	620	–	18

*Total elongations are in 50 mm with an ISO #1 tensile bar.

producing bake hardenable steels is to stabilize the carbon in solution such that they do not harden over time at room temperature, but rather harden after exposure to elevated temperatures such as those that occur in automotive paint bake ovens. By convention, the bake hardening index is measured after the steel is strained 2% in the direction of rolling, as shown in Fig. 2.1. The procedure consists of testing a standard tensile bar to determine its yield strength. Another tensile bar is then strained by 2% and is baked at a standard temperature, typically 170 °C for 20 minutes, at which time the tensile bar is pulled again until it breaks. The bake hardening index is determined by the difference in the yield strength of the original tensile bar and the bar that was strained and then baked.

It is important to note that the hardening effect from baking and from ambient age hardening will occur with less than 2% strain; tests show that strains of 0.5% are sufficient to activate the hardening mechanism. It is also interesting to note that there is strong strain path dependence for bake hardening. Several papers have reported that both batch annealed and continuously annealed bake hardenable steels do not show a bake hardening effect if the steels are biaxially stretched prior to baking. The change in strain path between a uniaxial tensile bar taken from a biaxially strained sample has been proposed as a reason for this effect, although mechanistic explanations for this observation have not yet been proposed (McCormick *et al.*, 1998; DiCostanzo and Matlock, 1996).

2.2.2 Dual phase steels

The term Advanced High Strength Steels (AHSS) generally refers to a wide range of steel grades that are part of the first generation of AHSS. While conventional HSS are principally composed of ferritic structures, the AHSS are characterized by their multiphase structures where ferrite is accompanied by other phases that have a significant effect on the mechanical, forming and even energy absorbing properties. In contrast to the use of precipitation

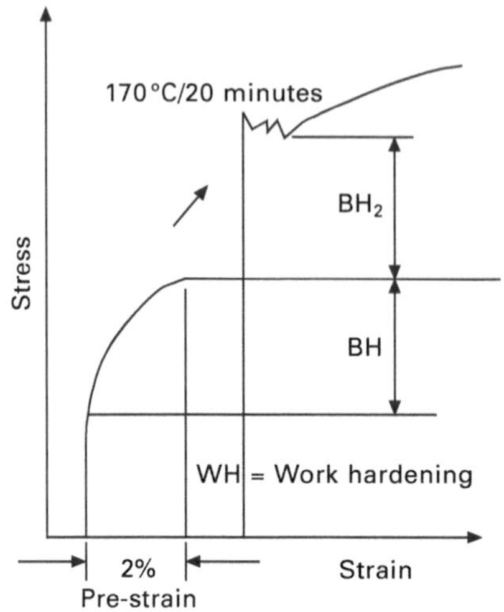

2.1 Bake hardening measurement (source: General Motors Corp., 2007).

hardening or solid solution strengthening common in many of the early HSS, the AHSS are noteworthy not only by their use of microstructural hardening, but, increasingly, steel manufacturers are incorporating other hardening mechanisms and multiple metallurgical phases in their products. Of all the types of AHSS that are currently available, the most commonly used in automotive applications are the dual phase steels. The difference in the microstructures of dual phase and the precipitation hardened HSLA steels is shown in Fig. 2.2.

The appearance of dark and irregular shaped martensitic islands in a ferrite matrix is clearly evident in the dual phase steel. In contrast, HSLA steel is mostly ferrite with evidence of fine and well dispersed niobium, titanium or vanadium carbides. Dual phase steels generally contain 10–70% volume fraction martensite, where increasing martensite content corresponds to increases in tensile strength (Shaw and Zuidema, 2001). Dual phase steels are also notable for their high tensile strengths compared to conventional HSLA steels at similar yield strengths, their high work hardening rates, and higher energy absorbing capabilities.

Dual phase steels are produced by controlled cooling from the austenite or austenite plus ferrite phase such that some of the austenite is converted into martensite. This is followed by rapid cooling that transforms the remaining austenite to martensite. With the addition of appropriate alloying elements, hardenability of the alloy is adjusted such that the austenite to martensite transformation occurs at cooling rates that are achievable in modern continuous annealing lines for cold rolled products, or the run-out table of a hot mill in the case of hot rolled products. A typical cooling curve for a dual phase steel is shown in Fig. 2.3.

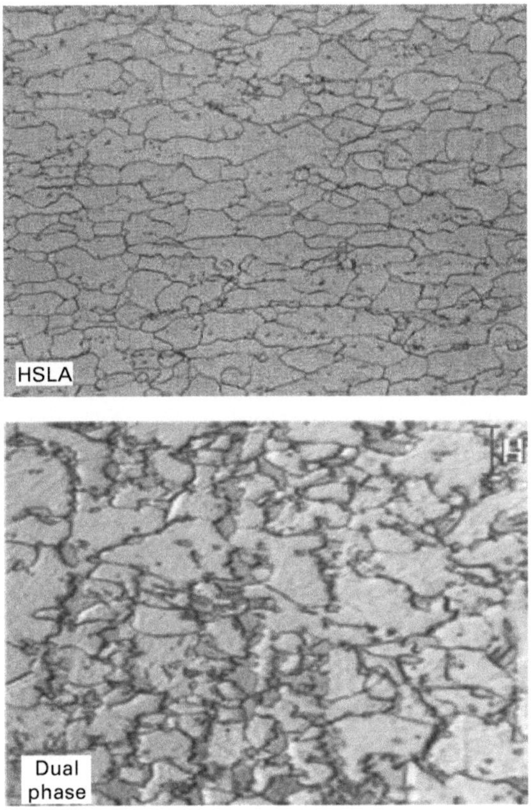

2.2 Typical microstructures of HSLA and dual phase steels.

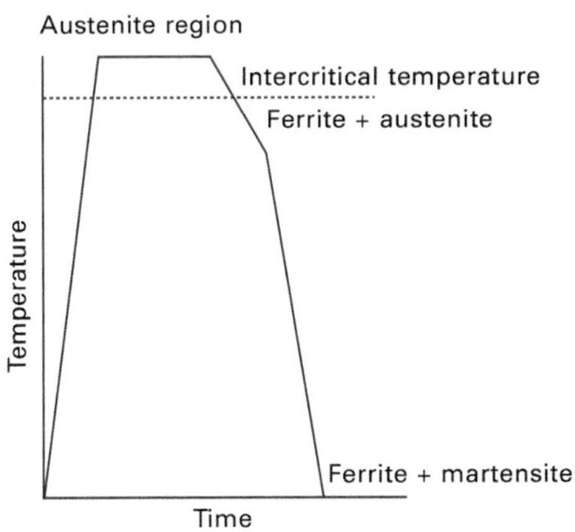

2.3 Typical dual phase steel cooling curve (Shaw, 2001).

The alloying elements commonly used to achieve the required hardenability include carbon, manganese, chromium, molybdenum, vanadium or nickel. While proper alloy content is required, it is important to balance hardenability and cooling rate to minimize alloy additions (Shaw and Zuidema, 2001). Alloy additions are not only expensive, but high alloy contents increase carbon equivalent and result in reduced spot weldability. Even with finely tuned welding processes, interfacial fracturing of weld buttons can occur, especially with higher tensile strength grades. This may cause traditional weld inspection techniques, such as chisel impact evaluations, to be less reliable with AHSS than with conventional steels that routinely result in weld button pull-out from the surrounding steel. To some extent, increasing the size of the spot weld fusion zone has been shown to have a positive effect on the fracture mode of dual phase steels (Sun *et al.*, 2006) and is one of the ways automotive engineers commonly modify design requirements to adapt to specific material characteristics.

Some of the standard dual phase steels and their mechanical properties are shown in Table 2.4 along with a number of common precipitation hardened steels. The higher tensile strength characteristics of dual phase steels is clearly evident in this table and in the engineering stress–strain curves shown in Fig. 2.4. For the selected dual phase steels, the yield strength to tensile strength ratios are much lower than for precipitation hardened steels. This is a characteristic that was common during the early stages of dual phase steel development. In later stages, metallurgists began to modify the yield to tensile strength ratios, sometimes by the use of multiple strengthening mechanisms, in an attempt to improve the performance of these products and to address specific formability needs. The incorporation of multiple strengthening mechanisms in the ferrite–martensite dual phase steels eventually resulted in a new classification of advanced high strength steels commonly referred to as multiphase steels. These steels will be discussed later in this chapter.

Table 2.4 Comparison of dual phase and precipitation hardened steels (source: SAE, 1999 and General Motors Corp., 2007)

Grade designation	Yield strength (MPa, minimum)	Tensile strength (MPa, minimum)	%Total elongation, minimum, cold reduced*	YS/TS ratio
DP600	340	590	21	0.58
DP800	420	780	14	0.54
DP1000	550	980	8	0.54
340 HSLA	340	410	22	0.83
420 HSLA	420	480	18	0.88
550 HSLA	550	610	18[†]	0.9

*For 550 HSLA, the reported elongation value is for hot rolled substrate.
[†]Total elongations are in 50 mm with an ISO #1 tensile bar.

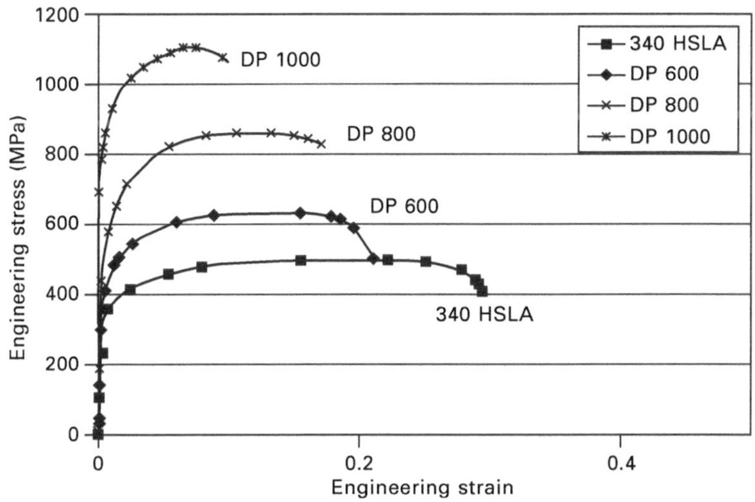

2.4 Engineering stress–strain curves of selected HSLA and dual phase steels (source: General Motors Corp., 2008).

Numerous studies of the energy absorbing behavior of these steels have shown the significant benefit of dual phase steels for absorbing energy during crash events (Yan *et al.*, 2005; Chen *et al.*, 2005). Furthermore, the potential for absorbing energy has also been shown to be highly correlated to tensile strength. This feature makes it convenient to make a general estimate of the improvement in energy absorption with different tensile strength grades.

Figure 2.5 shows a plot of the crush height of a U channel section with flat plates welded to the flanges, which is a common design for load absorbing rails in the motor compartment or in the rear compartment. The plot shows a highly linear relationship between crush distance and the 'as produced' tensile strength of the steel. This effect is evident with the impact speeds that are commonly used in testing steels for automotive use.

Another type of dual phase steel, the ferrite–bainite steels, are commonly referred to as stretch flangeable or high hole expansion steels. While both products are technically dual phase steels, it is convention in the automotive industry to distinguish these materials by referring to the ferrite–martensite products as 'dual phase' and the ferrite–bainite products as simply 'ferrite–bainite' or more conventionally, hole expansion steels. These steels have a higher resistance to edge cracking compared to conventional ferrite–martensite dual phase steels and precipitation hardened steels, making them well suited in applications where flanged holes are required. They are generally available only as hot rolled products, and they are generally limited to heavier thicknesses, usually greater than 2.0 mm. Given their thickness restrictions, the ferrite–bainite steels are most valuable in applications where there are flanged holes that experience high loads which demand high strength and

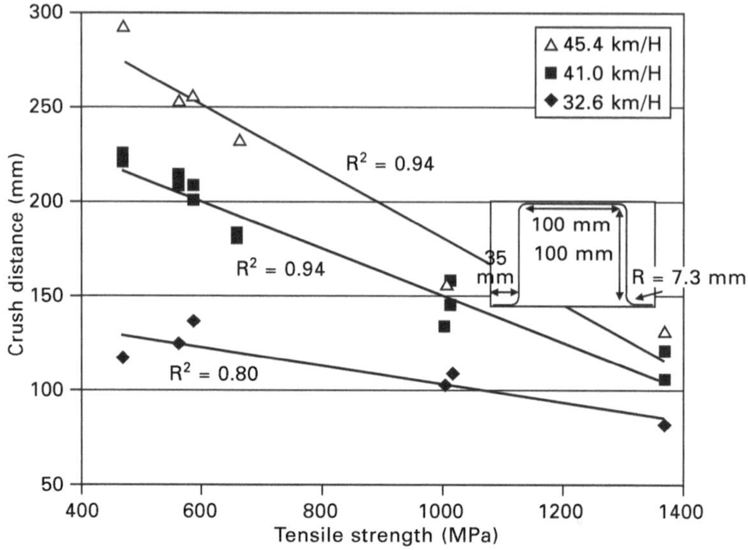

2.5 Crush distance of axially loaded longitudinal sections as a function of tensile strength of the steel (source: Yan, 2005).

heavy gauges. These characteristics make them especially suited to a number of applications for chassis and suspension parts, especially control arms.

The most common grade of ferrite–bainite steel currently used in automotive applications has a tensile strength of 580 MPa, with some manufacturers relying almost exclusively on this type of product for control arms. However, development of a 780 MPa tensile strength grade with adequate ductility to form control arms holds promise in the future to challenge aluminum in mass savings potential while providing a much lower component cost. One of the limitations of these steels is that they are not available with hot dip applied zinc or zinc–iron coatings. The thermal treatment caused by the molten zinc bath and the post zinc coating furnace required for galvanneal are not generally compatible with this microstructure. In addition, suppliers generally will not electroplate corrosion protective coatings on hot rolled substrates. This lack of corrosion preventative coatings somewhat limits the application of these products.

2.2.3 Advanced high strength steels – multiphase

Multiphase steels were developed as a modification of the early dual phase steels and, in their purest form, incorporate multiple strengthening mechanisms to improve mechanical properties without the addition of high levels of alloying elements that not only reduce weldability, but also significantly increase costs. The specific strengthening mechanisms used in these steels are selected based upon a number of factors, including the desired properties and the types of heat treatments that are available to the steel maker.

This wide definition for these steels has led to an increasing number of steels that are available at high tensile strengths. In particular, in the 980 MPa tensile strength category, there are many versions available with various combinations of yield strengths, elongations and hole expansion capability. In a recent review of available products (General Motors Corporation, 2008), the several dozen different grades that are being offered were identified by their manufacturers as drawing types, high elongation, low silicon, bending, high hole expansion, high yield and intermediate types. Adding to the confusion, there seems to be little relationship between mechanical properties and these product descriptions. As is likely becoming evident by now, there is little standardization in this tensile strength category. The term 'multiphase' is therefore a somewhat wide definition and it is cautioned not to take the term too literally. Over time, some materials could be re-classified, or additional classes may be developed to better describe them. Work is ongoing in the industry in attempting to evaluate the different variations of this grade to determine if the number of variations can be reduced. This is important since the large number of currently available grades in this tensile strength range is resulting in lower volumes than steelmakers generally want to produce for any individual product. As a result, higher prices and availability problems are not uncommon and are beginning to present significant issues to automotive manufacturers, especially if one of these materials is required in multiple regions around the world.

While global commonization of the 980 MPa tensile strength grade has not yet occurred, there are a number of strength levels that are loosely becoming standards in the industry. Some grades that are becoming relatively standard are shown in Table 2.5.

Applications for high strength multiphase steels include structural parts where very high strength is desirable to restrict part deformation, such as in and around the passenger compartment. Typical parts for these steels include

Table 2.5 Selected multiphase steels at 980 MPa tensile strength (source: General Motors Corp., 2007)

Grade designation	Yield strength (MPa, minimum)	Tensile strength (MPa, minimum)	%Total elongation, minimum, cold reduced*	YS/TS ratio
980T/550Y	550	980	~ 8	0.56
980T/650Y	650	980	~ 8	0.66
980T/700Y†	700	980	~ 8	0.71

*Total elongation is in 50 mm with an ISO 1# tensile bar.
†Grade designation has not been standardized.

the lower side members (rocker panels), the center pillar outer reinforcement and associated parts, and in upper body roof structures that are critical in limiting roof crush in roll over events. These very high strength grades have also gained widespread application in seat tracks and seat structure parts that previously were dominated by low carbon, SSS and HSLA steels. The use of these steels has provided significant opportunities for mass reduction and improvements in strength and performance of these structures.

2.2.4 Cold rolled martensitic and heat treated boron steels

Martensitic steels, which consist of a matrix rich in martensite with smaller amounts of bainite and/or ferrite, are produced by quenching at very high rates from the austenite–ferrite region. In the automotive industry, martensitic steels are produced by several different methods; the microstructure is developed while the steel coil is being produced or after the part is stamped from steel with high hardenability. In coil form, cold rolled products are produced on water quenched continuous annealing lines with high cooling rates while hot rolled products are produced by water quenching on the run-out table in the hot mill. A typical cooling curve for water quenched martensitic steel is shown in Fig. 2.6.

In the water quenching process, steels of varying carbon content are used to produce martensitic steels with different strength levels. The use of lower carbon content steels of around 0.09% carbon and 0.50% manganese will result in martensitic steels with tensile strengths around 900–1000 MPa. If higher strengths are desired, the carbon content of the steel can be increased. When carbon contents are increased to about 0.23% carbon, the martensitic structure is strengthened such that the tensile strength is increased

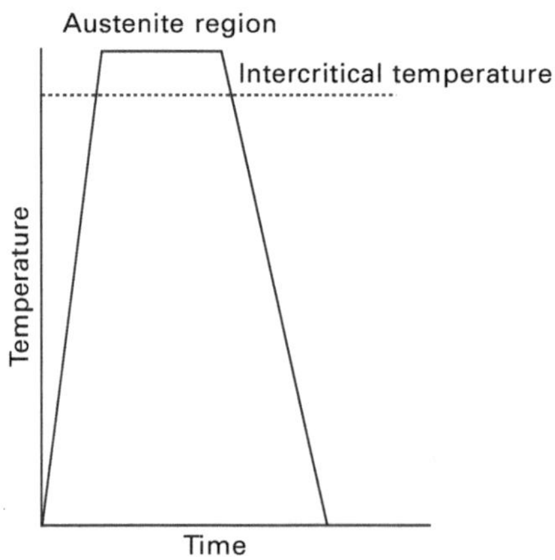

2.6 Martensitic steel cooling curve. Source: Shaw, 2001.

to 1300–1400 MPa. All of this strength comes at a cost, however, in that the ductility of the steel is reduced such that total elongations are now 5% or less. Even so, there is enough ductility for this steel to be roll formed or bent into the necessary shape. Martensitic steels are very useful in automotive applications for door beams, bumpers, very lightweight and high strength lower side members (rocker panels), and cross car bars and beams that are designed to prevent intrusion into the passenger compartment.

There are other methods used to make martensitic steel structural parts, and all of them form martensite '*in situ*' after the parts have been formed. One technique is to employ water cooled dies to stamp steel blanks that have been pre-heated above the intercritical temperature. Cooling in the die is designed to provide high enough cooling rates to form the martensite in the formed part. Quite logically, this process is commonly referred to by the industry as '*hot stamping*' or '*die quenching*'.

Producing martensitic structures in water cooled dies has been increasing for a number of automotive applications, and likely will continue to increase as fuel economy and safety regulations increase. Since the cooling rate in water cooled dies is lower than that which can be achieved in direct water quenched continuous annealing lines, the alloys used in this process must have a higher hardenability rate. Increased hardenability is achieved by the use of steels with moderately high carbon, generally in the range of 0.20 to 0.25%, with manganese at approximately 1.2% along with the addition of 0.0005–0.001% boron. With chemistries such as these, tensile strengths of 1400 MPa can routinely be achieved. In the industry, steels used in the hot stamping process are commonly known as boron steels.

The advantage of hot stamped boron steels over fully martensitic cold rolled or hot rolled steels is that they are stamped at high temperatures where ductility is much higher than at room temperature. This allows for the formation of much more complicated stampings than could be formed from conventionally stamped, cold rolled martensitic steel. The downside is that the cycle time required to stamp and cool the steel in the die is long and the tooling is quite expensive. Compared with stamping conventional steels, where it is not uncommon to produce twelve or more parts per minute, hot stamping generally is restricted to about one per minute, although this production rate can be increased somewhat by designing dies that stamp more than one part.

Other techniques for making martensitic steel parts include direct water quenching after forming, and heating tubes produced from steels very similar to those used in hot stamping. For years, tubes made in this manner were extremely common in applications such as door beams. Another method is to fixture the cold formed part in a rigid fixture, and then selectively induction heat the part followed by an immediate water quench. This process is commonly referred to as post-form induction heat treatment. It has the advantage of being able to selectively heat treat areas of a part and not heat treat others. In this way, strength is added where it is needed, but is not added were it is not desirable. This allows the potential to 'tune' how a part will crush or bend.

2.2.5 Transformation induced plasticity steels

Transformation Induced Plasticity, or TRIP steels, are produced in a similar manner to dual phase steels. Both require continuous annealing lines except that TRIP steels require a somewhat slower cooling rate and an isothermal hold at a temperature below the intercritical temperature. It is this isothermal hold, along with the higher carbon and silicon or aluminum contents of these steels that leads to a microstructure with appreciable amounts of metastable retained austenite. A typical cooling curve is shown in Fig. 2.7.

The microstructure of TRIP steels consists of ferrite with interdispersed hard phases of martensite and bainite, along with volume fractions of austenite that are greater than 5%. Typical chemistry for a TRIP steel with a tensile strength of approximately 700–800 MPa is shown in Table 2.6.

The primary benefit of TRIP steels is their increased formability as compared with dual phase steels of similar strength. This high level of formability is attributed to their extremely high work hardening rate and the presence of austenite in the matrix that, upon straining, changes into martensite. This takes advantage of the soft ferrite matrix for enhanced formability with a very high strain hardening rate provided by the transformation of austenite to martensite, which also strengthens the matrix. A comparison of the strength of a TRIP steel to selected other steels is shown in Table 2.7.

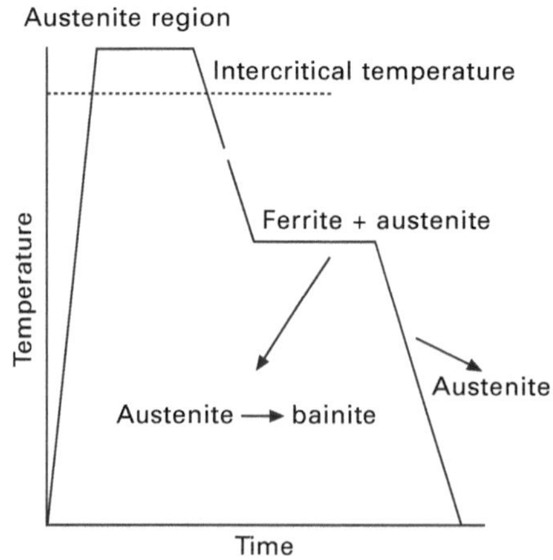

2.7 TRIP and complex phase cooling curve. Source: Cornette, 2001.

Table 2.6 Chemistry of a typical TRIP steel (source: General Motors Corp., 2008)

%C	%Mn	%Si	%Al	%P	%Cr	%Ni	%N	%Ti
0.18	1.5	0.05	2	0.01	0.02	0.002	0.003	0.003

Table 2.7 Comparison of TRIP mechanical properties with other steels (source: Cornetle, 2001 and General Motors Corp., 2008)

	Tensile strength (MPa)	Yield strength (MPa)	(%)	Total elongation* (uniform, %)	n
Mild steel	181	296	42.4	21.4	0.208
340 HSLA	375	480	24	13	0.15
Dual phase 600	355	623	26	19	0.18
Dual phase 800	440	825	17	18	0.13
TRIP800	503	831	27.6	21.5	0.236

*Elongation is in 50 mm with an ISO #1 tensile bar.

It is evident from these mechanical properties that the formability of TRIP steels is much like that of low carbon steels. Compared to the other high strength steels shown in this table, the high strain hardening capabilities (n-value) along with high elongations are very favorable for the TRIP steels. As a result, these steels are very applicable to stampings that are relatively complex, but which also need to be of very high strength. Until the development of TRIP steels, these attributes were not available in any commonly produced steel. However, there are downsides to these steels. In particular, the very high alloy contents and resultant high carbon equivalents are detrimental to spot welding. As a result, these steels have not been as widely applied as other grades of AHSS by automotive manufacturers as of this time (2009). As welding practices advance or other joining methods become more popular, this trend could change and may allow the properties of this steel to be more widely utilized.

2.2.6 Complex phase steels

Complex phase steels are similar to TRIP steels in composition except that they are adjusted to contain less retained austenite. A typical cooling curve is similar to that used to produce TRIP steel and is shown in Fig. 2.7. In place of the large volume fraction of retained austenite as in TRIP steels, the complex phase steels contain fine precipitates along with a fine microstructure of ferrite and a high volume fraction of hard phases. Precipitation hardening of these steels is accomplished with the use of the addition of niobium, titanium and/or vanadium in addition to the normal alloying elements that dual phase and TRIP steels utilize (Cornette *et al.*, 2001). The resultant materials are of very high strength and are becoming a more common strategy for achieving strength levels of 800 MPa and above. While the formability does not approach that of TRIP steels, the complex phase steels are capable of achieving higher strengths with lower alloy contents, which improves weldability at lower cost.

These steels and the performance improvement they provide have ushered in a new era in the types of steels that are used in the automotive industry. It has been said that most of the steels being specified on new vehicle architectures were not available 10 or 15 years ago. While that claim might be hard to prove, it would be equally difficult to find anyone involved in specifying steels for the automotive industry who would take much of an issue with that statement. The automotive and steel industries are in the midst of a revolution in the types of steels being manufactured for use on today's automobiles, and there appears to be no end in sight to that rate of change.

2.2.7 TWIP steels: the second generation of advanced high strength steels

Twinning Induced Plasticity (TWIP) steels and Lightweight Steels with Induced Plasticity (L-IP) are the latest generation of Advanced High Strength Steels that are being developed for the automotive market. These steels are designed to provide the combination of very high tensile strengths and exceptional ductility, far above what is available with the current generation of AHSS.

In contrast to the ferrite based structure which characterizes the first generation AHSS, this latest generation consists of a face centered cubic austenite structure that is achieved through the use of very high levels of manganese. In the case of the recently introduced TWIP steels, the resultant material has a tensile strength of 1000 MPa with a total elongation well in excess of 50% (Cornette *et al.*, 2005). However, this combination of high strength and exceptional ductility comes at a significant cost. With manganese levels in excess of 20%, the cost of these steels is expected to approach that of some grades of stainless steel and aluminum, which presents a significant challenge for this material in the very cost sensitive automotive industry, and may ultimately be one of the limiting factors in its future success.

This latest generation of AHSS, like the dual phase steels that preceded it, is not a recent invention. A high manganese austenitic alloy similar to TWIP steels was discovered by Hadfield, perhaps as long ago as the early 1880s. Like many steel grades, there has been a long time between its discovery and its 're-discovery' and development into a product that may provide benefit to the automotive and other industries that could use a material with its uniquely high tensile strength and superior ductility. The unique properties of this steel are tied to the fact that it remains fully austenitic at all the temperatures that it has been evaluated at, and that it not only deforms by the more conventional processes of layers slipping between each other and the movement of dislocations, but also by twinning inside of its grains. It has been reported that the boundaries from twinning function in a similar manner to grain boundaries, with respect to their ability to restrict the movement of dislocations (Cornette *et al.*, 2005). The seemingly magical mechanical properties of this steel and those of other common steels are shown in Table 2.8.

2.2.8 Overview

There is a wide range of steels available to the automotive designer and materials engineer for use in designing modern vehicle architectures. These steels provide a variety of strength levels and ductility that can meet most of the requirements for automotive structures and, when properly utilized, enable one to significantly reduce mass from what has been typically available in today's market. In order to make some sense of all these grades, a chart comparing these steels has commonly been used that graphically shows the relationship between strength and ductility. This graph, commonly called the 'banana chart' or 'ductility ladder' in the industry, is shown in Fig. 2.8 and has evolved over the years to the point that it is difficult to determine who came up with the original concept.

A chart such as Fig. 2.8 very effectively shows the positive relationship of tensile and elongation for the AHSS as compared with the conventional low carbon, bake hardenable, solid solution strengthened and precipitation hardened (HSLA) steels. In comparison, the first generation AHSS, consisting of TRIP, DP and CP steels, have higher elongation values for a given tensile strength than conventional HSLA steels. In stark contrast, materials that compose the second generation of AHSS, such as TWIP steels, greatly increase the ductility to tensile strength relationship to previously unheard of levels in commonly used carbon steels.

Table 2.8 TWIP steel properties compared to other HSSS Source: General Motors Corp., 2008

Grade	Yield strength (MPa, typical)	Tensile strength (MPa, typical)	Total elongation (typical, %)*	*n*-value
Low carbon IF steel	140	270	42	0.22
DP600	360	610	23	0.14
DP800	440	820	16	0.11
DP1000	650	1020	9	n/a
340 HSLA	360	450	24	0.13
420 HSLA	640	550	21	0.11
FeMn TWIP 1000	464	1009	>50	0.42

*Total elongation in 50 mm with an ISO #1 tensile bar.

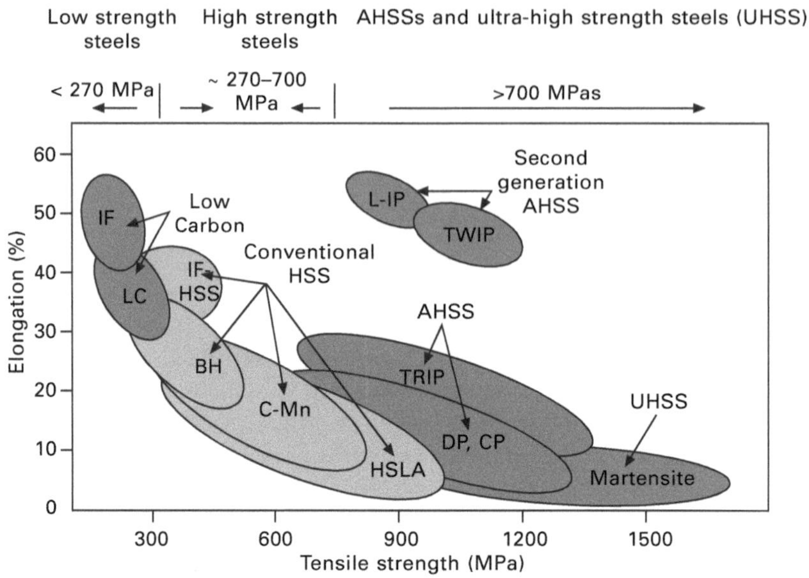

2.8 Automotive steel strength/ductility ladder.

2.3 Third generation advanced high strength steels

The American Iron and Steel Institute (AISI) initiated a program to investigate the types of steels that could be developed to fill in the gap in the ductility ladder between the first generation and second generation AHSS. The potential for this new, third generation of AHSS is being studied in the hope that it will be a more cost effective alternative to the second generation while maintaining a much higher level of ductility at similar strength levels to steels currently available with the first generation of AHSS. In terms of the 'infamous' banana chart, these steels are expected to fill the gap between the first and second generation AHSS and are shown in Fig. 2.9. There are three phases to the research: the first is a survey and evaluation of the latest trends, the second is the application of known strengthening mechanisms and investigation of the potential for modified base structures or novel or non-conventional strengthening mechanisms, and the third phase will consist of an evaluation of the potential processes to produce the structures and properties that were identified in the second phase. Some of the strengthening mechanisms under study, in addition to the more common mechanisms, include grain twinning, shear band formation, strain induced transformations such as found in the TRIP effect, and dislocation pinning.

In conjunction with these strengthening mechanisms, the AISI will also examine new base structure combinations including, but not limited to, duplex ferrite with austenite and new alloy approaches to produce leaner austenitic

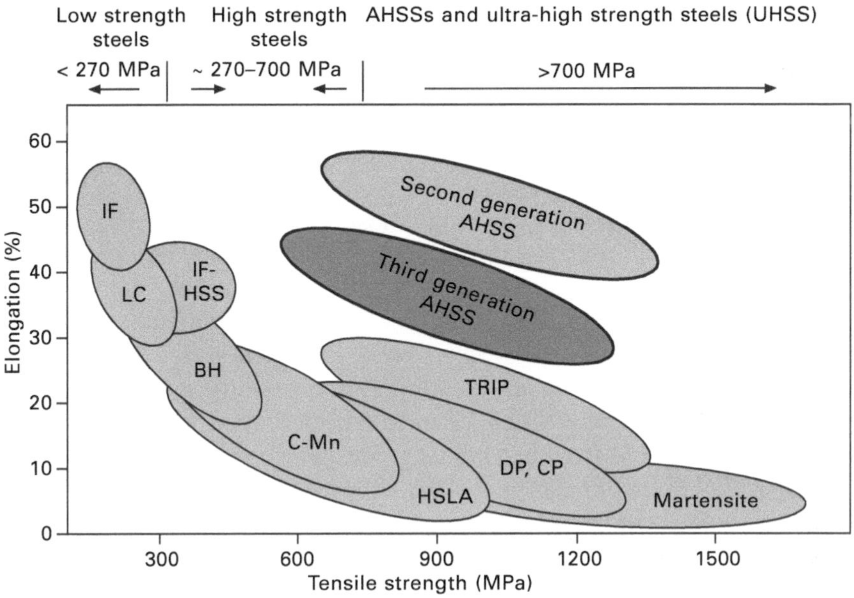

2.9 Third generation of advanced high strength steels.

structures, ultra-rapid cooling, and also controlling the crystallographic texture of steel.

The goal of all this investigation and research is to develop new steels that will provide additional and very cost effective means of achieving the very challenging goals that automotive manufacturers have in meeting future fuel economy, emission and safety requirements for the next generation of automotive body structures in a cost effective manner compared to lightweight metals and composites (AISI, 2006).

2.4 Manufacturing and forming high strength steels

The major processes used to form and manufacture vehicles from high strength steels are similar to those used with conventional steels except that special techniques and guidelines have been developed by each automotive manufacturer to adapt to the special needs of these steels. The typical issues with forming high strength steels, such as springback, are well known but not entirely predictable, and there still exists some 'art' and 'trial and error' built into the process. Even so, much has been learned about how to manufacture parts from these steels and much has still to be learned. In addition to forming, joining high strength steels also presents its own set of challenges. While the most common process for joining steel in the automotive industry is spot welding, other joining techniques such as adhesive bonding, laser assembly welding and MIG welding are also used in automotive manufacturing.

2.4.1 Forming high strength steels

Of all the processes used by automotive manufacturers to form steel sheets into parts, conventional stamping is the most commonly employed process. To a large extent, the reliance on stamping is related to the complex nature of the parts, the readily available infrastructure that supports stamping and its high volume capability that suits this industry. While stamping is the most common forming process, other processes such as roll forming, hot forming (or hot stamping as it is commonly referred to) and hydroforming are all used at various times. The following discussions will give the reader some background as to why the various processes are selected, some of their advantages and disadvantages and some examples of the types of parts that are formed with these technologies.

Stamping

Stamping has been the preferred method of forming parts in the automotive industry for decades. The industry's manufacturing infrastructure is largely based on stamping and the technologies necessary to make that process successful. There are several primary reasons why stamping has been so successful; stamping processes are able to make the complex shapes that are commonly required in the industry and stamping is compatible with the high volume that the industry requires.

Conventional stamping requires a set of dies that form the shape of the part, create the flanges that are necessary to join parts together, as well as trim, flange and punch any holes or openings that may be required. As a result, stamping processes generally require at least three dies; the forming die, the flanging and trim die, and the piercing die that adds any necessary holes. Depending on the complexity of the part, additional forming steps may be added such as pre-forming or doming the part to move metal to where it is needed before the forming die creates the shape of the part. This process can be used when forming very deep drawn parts and is most commonly employed with very formable low carbon steels. However, the technique is not commonly used with high strength steels, especially the advanced high strength steels. As mentioned in previous sections, the advanced high strength steels work harden very quickly. Pre-forming reduces the elongation and ductility of the steel and makes it difficult to move the steel into the desired shape in a second forming operation. As a result, it is common practice to form parts into the final shape, or as close to the final shape as possible, in the first stamping operation.

In its simplest form, a traditional set of line dies consists of a number of presses that are set together in a line such that steel blanks are fed into the first die, the part is stamped, and then the part is transferred to the second die, which is in a press next to the first, and the next process is performed in that press. This process continues through all the dies until the part is finished. Since there are a number of expensive dies involved in stamping, in order to be cost effective, this process must produce a large number of parts. In most cases, a line of stamping dies will produce between 10 and 14 parts per minute. This translates into the potential to produce more than 6000 parts in an 8-hour shift.

Another commonly used stamping process uses a progressive die. A progressive die is a single die that incorporates multiple forming operations in successive steps as the steel is progressively moved through each step between strokes. This process is most suited to small parts that do not have very deep forming operations but require multiple stages to make the desired shape. In a progressive die process, steel is fed directly from a coil into the tool, thereby eliminating the costs of blanking and transferring of blanks from a stack to the die. Another advantage is that many operations can be incorporated into that one process and there is no need to transfer parts between dies. The number of parts produced per minute can also be significantly higher than with traditional line dies, especially with small parts that are optimum for this process. Although the die sets can be expensive, this process can be less expensive in total cost compared to a conventional set of line dies.

Global formability

In order for stamping to be successful, a die must be designed with the knowledge of how metal flows and how the steel will react to applied strains. Today, that job is largely fulfilled by computer forming programs that evaluate part shape and formability of the steel, the die design, friction between the surface of the die and the steel, as well as how the metal flows in the die. These programs predict where the strains will be highest and where there is danger of excessive metal thinning that requires modifications to either die geometry or part shape. Metal flow in a die is influenced by many factors; strength of the steel, friction differences between the steel surface and the die surface, ductility of the steel, the ability of the steel to resist thinning, lubrication, strain rate sensitivity of the steel and how the metal is restrained from flowing into the die are but a few of the factors affecting formability.

A relatively simple way to determine if the selected steel grade has enough formability to make a part is to use a forming limit diagram. The forming limit diagram, also known as the Keeler–Goodwin diagram after its founders, was developed as a method to plot strains on stamped parts and predict how close the part is to splitting (see Fig. 2.10).

The forming limit diagram is a very useful, robust and relatively easy-to-apply tool. Since the shape of the curve is effectively constant with most sheet metals, the forming limit diagram is modified for the material of interest simply by moving the diagram up and down the *y*-axis using a very simple formula to calculate the *y*-intercept, which is know as the FLD_0. That formula is based on the thickness of the steel and the strain hardening exponent, or *n*-value, which is the exponent on the following Power Law equation:

$$\sigma = K\,\varepsilon^n \qquad [2.1]$$

where σ = true stress
ε = true strain
K = strength coefficient at $\varepsilon = 1.0$
n = the strain hardening exponent

The FLD_0 is then calculated using the following formula:

$$FLD_0 = (23.3 + 14.1\,t) \times n/0.21 \quad \text{where } t = \text{thickness (mm)} \qquad [2.2]$$

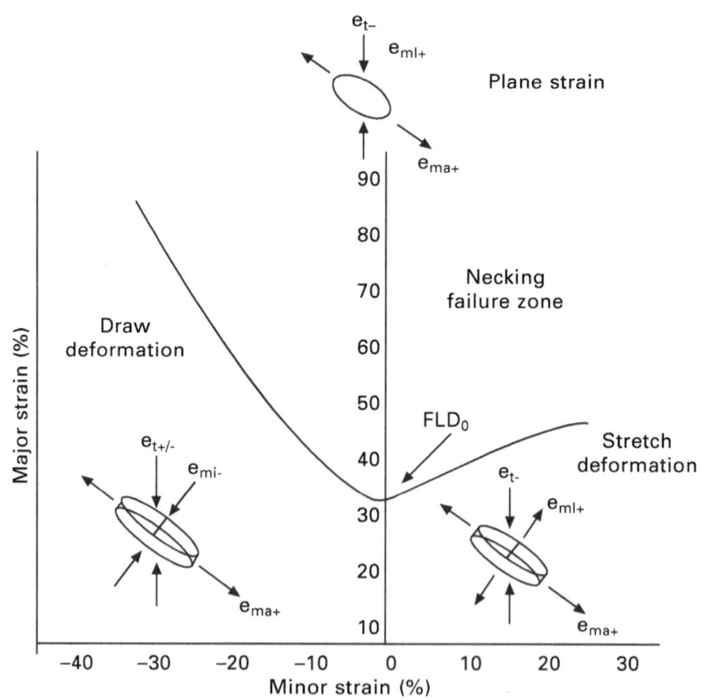

2.10 Forming limit diagram.

For most steels, the *n*-value ranges from 0 to about 0.3, with higher values being indicative of higher formability and improved ability to distribute strains and avoid necking.

This diagram is very effective when measuring global formability of sheet metal stampings. The strains present on a panel are determined by etching circles of known diameter on a steel blank before it is stamped. The circles are formed using templates made of thin films that have circles of known diameter etched imprinted over their entire surface (5 mm circles being common). This blank is placed into the forming die and then, under the same conditions as in production – i.e. same blank holding pressures, same ram speed, same lubrication, etc. – is stamped only in the forming die. The formed panel is then removed and the technician then can examine how the circles changed shape and calculate the forming strains at that location on the part. A diagram of the way the steel surface and the circles look before and after stamping is shown in Fig. 2.11.

These strains, which represent the in-plane major and minor forming strains, are plotted on the forming limit diagram such as is shown in Fig. 2.10 after the FLD_0 is calculated for the steel being tested. If strains at any location are above the solid line forming the forming limit diagram, that location is considered to be at risk for necking or even splits. Strains that are below this line are considered safe. There are many good books and papers written on sheet metal formability that can provide a working background for the interested engineer.

Local formabilty

Local formability, in contrast to global formability, generally relates to issues that are not predicted by the forming limit diagram. Examples of

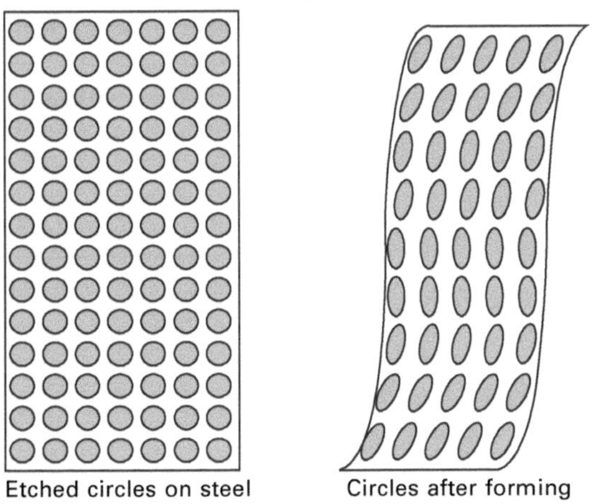

2.11 Examples of etched steel surface.

local formability issues include cracking on the outside radius of a bent flange, cracks on the edges of extruded holes, and cracking that occurs in areas of lower strain than would be predicted to cause cracking. Unlike conventional HSS such as HSLA and carbon–manganese grades, some of the AHSS, especially dual phase steels of 780 MPa and above, have shown a susceptibility to cracks in areas of low strain, especially at formed radii. These types of fractures have been called shear fractures and are currently the subject of extensive evaluation. Other AHSS, such as complex phase and TRIP steels, have shown greater resistance to stress fractures than have dual phase steels (Walp, 2006).

Springback

While all steels experience some degree of springback, the phenomenon is most severe in high strengths steels. In many cases, springback is more of an issue than cracking and necking when forming high strength steels because it is very difficult to predict springback. An example of springback of a flange is shown in Fig. 2.12. There are a number of ways to compensate for springback. One method is to design the tool such that the flange is over bent during the stamping process with the intent that, after forming, the flanges will 'springback' to the desired location. This is good in theory, except that the absolute amount of springback is affected by residual stresses that may be in the part and other factors such as the radii of the die at the inside of the bend. In practice, it is easy to either under- or over-predict the amount of springback, resulting in a flange that is not in the proper design location. Worse yet, when springback causes the entire part to twist out of shape, it can be very difficult to determine the cause. All of these issues can result in a very long and drawn-out process of re-cutting the tool face to adjust springback in an attempt to achieve the proper final location of a flange or shape of a part.

The amount of springback can be improved in a simple flange by reducing the radii of the die, and hence the inside radii of the part flange. However,

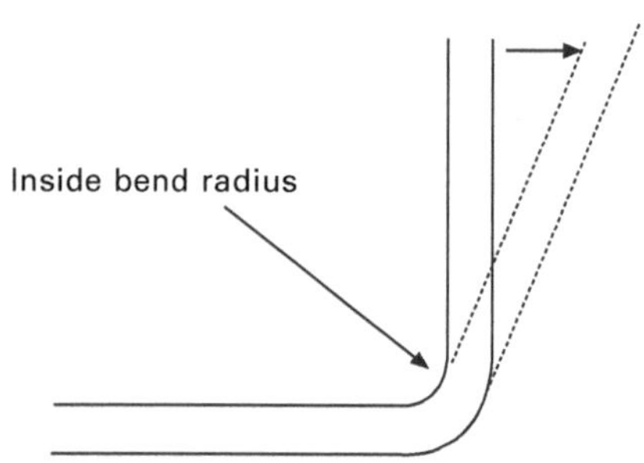

2.12 Springback in high strength steels.

this must be done with the proper knowledge of the minimum bend radii that the material of choice can withstand before fracture. In addition, for many materials, flange direction in relation to the rolling direction of the steel can be critical. It is not uncommon for the minimum bend radii to be four times as large parallel to the rolling direction (i.e. longitudinal) as it is perpendicular (i.e. transverse) to the rolling direction. By convention, bend radii are reported as a function of not only the radius, but also the thickness of the steel and are expressed in the following formula:

$$\text{Bend radius} = r/t \qquad [2.3]$$

where: r = radius in mm
t = thickness in mm

Prediction of springback is an area where a significant amount of study is being dedicated within the automotive industry. Steel companies, the Auto Steel Partnership and a number of research laboratories are studying springback prediction and modeling, and a detailed discussion would take a book entirely dedicated to this topic.

Roll forming

Roll forming is a process that is very applicable to high strength steels (especially steels of very high strength and low ductility such as cold rolled martensite), although almost any material can be roll formed. Roll forming utilizes a series of rolls that gradually form the steel into the required shape. Since this process is a continuous, coil fed, straight line forming process, part shape needs to be constant throughout the length of the part. Changes in section shape are not consistent with this process; however, there are a number of post roll forming processes that can bend a part and change its sweep. There are some superior examples where extremely high strength steels have been fabricated into some very complex sections using this process. Some example parts include bumper bars made out of DP1000 or cold rolled martensite, and rocker panels that have also been made of cold rolled martensite. Both are good examples of parts that are generally intended to resist deformation under all but the most severe loads. Figures 2.13 and 2.14 show some typical designs for roll formed parts.

Press hardening

Press hardening, or die quenching as it is also known, is a manufacturing method that generally provides the highest strength and lowest mass option for parts that are relatively complex in shape. In press hardening, a steel grade with sufficient hardenability, usually a carbon, manganese, boron

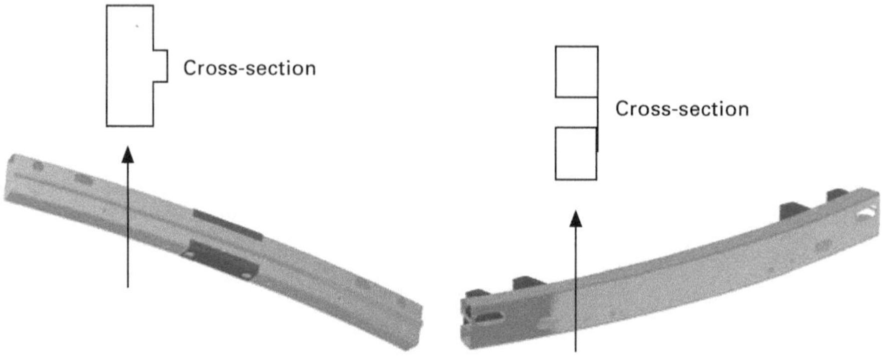

2.13 Examples of ultra high strength steel roll formed bumpers.

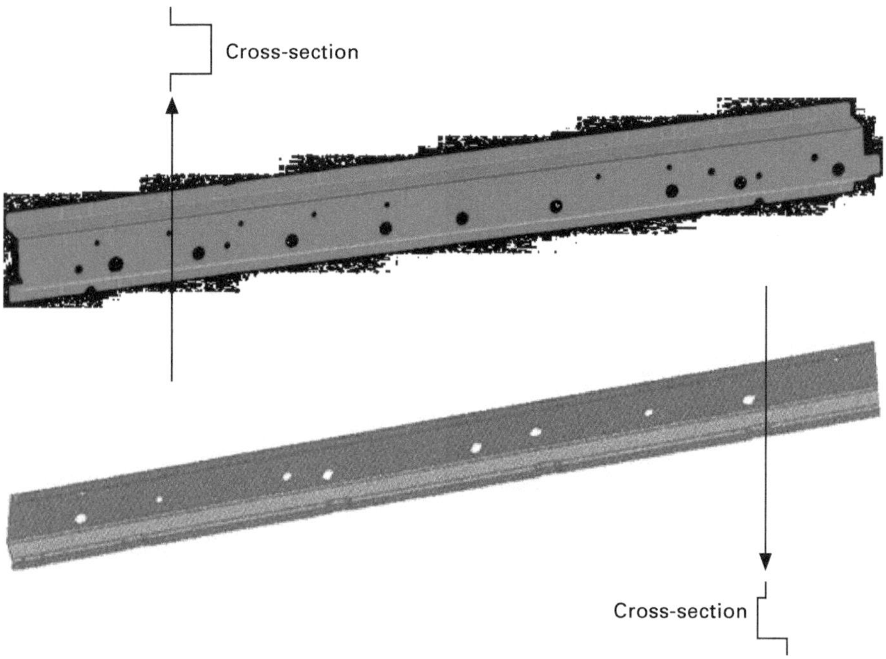

2.14 Examples of ultra high strength steel roll formed rockers.

steel alloy, is heated into the intercritical region, usually around 950 °C, and then the part is transferred to a water-cooled die and immediately stamped and quenched into the final shape. The resultant microstructure is almost purely martensitic. This process results in parts that have tensile strengths of 1400–1500 MPa and has the advantage of being tolerant of complex shapes that can otherwise be formed only with much lower strength steels using conventional stamping processes.

Press hardening has greatly expanded in the number of its applications in the automotive industry since early 2000, and is now a process that is almost routinely specified on a number of automotive applications such as door beams and center pillar reinforcements. The technique is used on parts that are critical to safety and frequently includes parts essential for the protection of passengers in side impact and/or roll over events. The parts produced by press hardening provide the highest strength among carbon steels currently in production for body structure stampings in the automotive industry. Some typical parts are shown in Fig. 2.15.

This process does have disadvantages that the automotive engineer must be aware of. Stampings from this process are covered with a heavy oxide scale and must be cleaned to facilitate welding and to provide a surface that is capable of providing suitable adhesion for automotive primers and paint. Removal of the oxide coating is usually accomplished by shot blasting, which not only adds additional cost, but also stress relieves the part and can contribute to wider dimensional variation. In order to avoid shot blasting, the steel industry is offering a product with an aluminum/silicon alloy coating that is highly resistant to oxidation at high temperatures. This coating eliminates the need for post-form cleaning, although it has its own set of challenges. The presence of this coating inhibits spot welding, which is already problematic due to the high alloy content of the steel. In addition, the coating alloys with

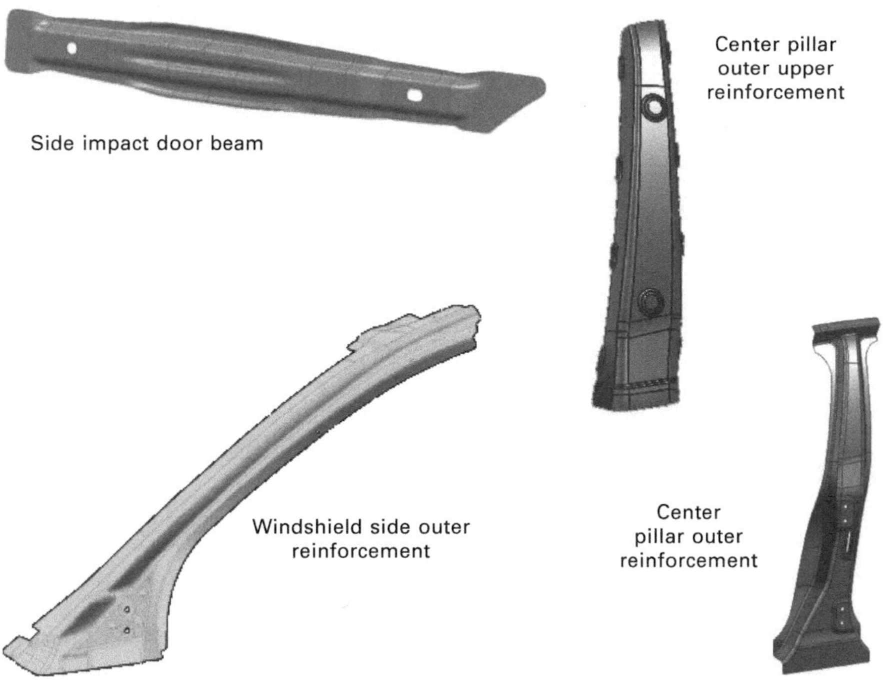

2.15 Examples of press handened safety critical stampings.

iron in the steel to such an extent that the corrosion resistance that would be afforded by the coating is largely rendered ineffective.

As previously mentioned, steels used in this process must have sufficient hardenability such that they can achieve the high strength, martensitic microstructure with cooling rates that are practical to achieve in a water cooled die. With the required high alloy contents, weld cracks can be a concern and will be discussed at length in the next section on joining.

2.4.2 Joining high strength steels

Most of the methods commonly used to join lower carbon steels are also used to join high strength steels, although each of these methods possesses its own set of advantages and disadvantages that the automotive designer must understand for successful execution. The most common joining method for high strength steels is spot welding, just as it is for other steel grades that are commonly used in automotive body structures. Other joining methods such as arc welding, adhesive bonding and laser welding are also used, although arc welding is frequently limited because of the wide heat affected zone which exhibits significantly lower strength than the surrounding metal. Each of these methods is discussed briefly in the following paragraphs.

Spot welding

Automotive manufacturing plants are primarily facilitated for assembling vehicles with spot welds, and changes to high strength steel grades have had little effect on changing the industry's joining preferences. However, while the basic technology used for spot welding is little changed, improvement in weld controls and process enhancements has extended the ability to successfully spot weld steel grades with higher carbon equivalents in high volume processes common in the industry. Robotic welding, automatic weld feedback, sophisticated weld controllers and weld electrode tip dressing are but a few of the improvements in this joining method. Furthermore, the willingness to adapt weld cycles to changing materials and the detailed testing and evaluation of weld stacks have greatly extended the types and grades of steels that can be routinely welded with the very high quality levels that are required.

One of the most critical analysis tools available to the weld engineer when evaluating spot welds is to prepare and examine a metallurgical cross section. This technique will reveal critical information about the weld, the presence of weld cracks or microcracks, depth of weld penetration and the heat affected zone (HAZ). With the addition of other micro-analytical analysis techniques, the investigator can also determine hardness of the weld and adjacent areas, and the segregation of elements that may be detrimental to the integrity of the weld. This type of information is invaluable, and weld development without access to this type of testing is very limited in what type of work can be done.

A cross-section of a weld between two DP600 steel sheets is shown in Fig. 2.16. The cross section has been mounted, polished and etched to show the grain structure of the weld and HAZ. In addition, a microhardness traverse across the weld clearly shows the hardness of the base metal, the HAZ, and the weld metal. There are a number of observations that can be made from this weld. The first is that the weld fusion zone is consistent and has excellent penetration into both steel sheets. Partial penetration or uneven penetration can suggest potential problems with welding conditions. Also, the hardness is consistent across the weld nugget and, as is clearly evident in the photomicrograph, the amount of weld electrode deformation of the weld is minimal. The HAZ is evident in both the photomicrograph and in the microhardness values, which show a sharp reduction in the hardness across this zone until the hardness of the base metal is reached. A weld such as is

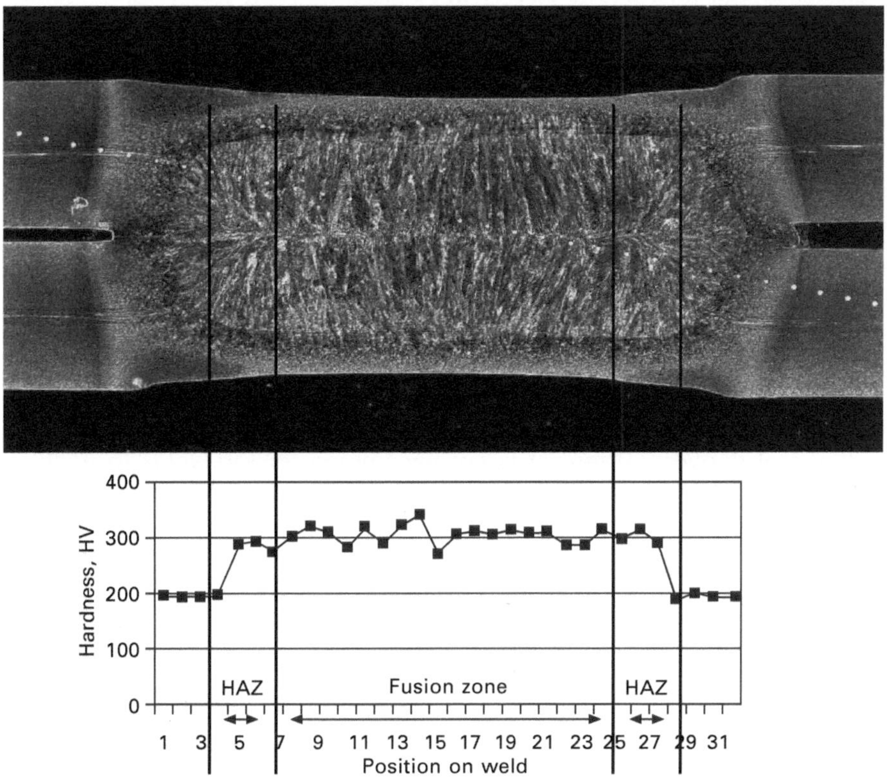

2.16 Fusion zone and hardness of DP600 to DP600 spot welds. Source: general Motors Corp., 2008.

shown in Fig. 2.16 would be considered a good weld and would be predicted to perform well.

The performance of a weld is generally considered acceptable if the design load can be carried by the weld without fracture. However, this acceptance criterion is difficult to apply during an accident investigation since actual loads are difficult to estimate after the accident has occurred. From a practical perspective, it is preferred to have the weld button, or fusion zone of the weld, stay intact and force the fracture outside the weld. When this occurs, one can be sure that the weld did not fail. This type of fracture mode is called button pull-out and is shown in Fig. 2.17.

Another possible fracture mode is interfacial fracture. Interfacial fracture is said to occur when the fracture occurs through the fusion zone of the weld. This type of fracture mode can occur if the HAZ is small or does not exhibit significant softening, or the materials being welded are of very high strength. There may also be instances where there is a partially fused weld, or the weld conditions or materials being welded causes the weld metal to become brittle. This type of fracture can occur at lower strengths than the design load and is clearly not desirable. Fortunately, examination of the fracture surfaces can easily reveal the fracture mode and eliminate any question about the performance of the weld. The important point is that interfacial fracture, while not considered the preferred fracture mode, does not necessarily indicate that the weld did not meet the loads that it was designed to meet. This is an important point since interfacial fractures are more common with very high strength steels than they are with lower strength steels. However, it is salient to note that if the steel is thick enough, even low carbon steels can fracture interfacially.

In many automotive assembly operations, weld quality is evaluated by chisel impact of the weld which, with common steels, results in full perimeter weld metal pull-out. Under these conditions, the weld is known to have been stronger than the parent steel and, by definition, was an acceptable weld. However, with some AHSS, and in particular when the thickness of the steel is over 2.0 mm, it is not uncommon for the welds to fracture across the weld nugget itself. This creates inspection issues for the assembly

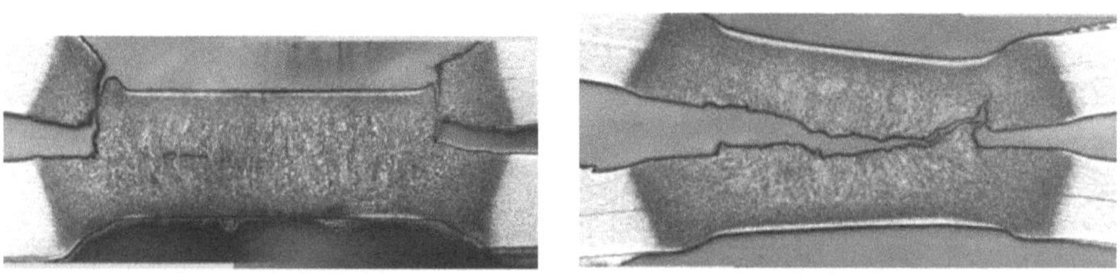

Button pull-out   Interfacial fracture

2.16 Weld fracture modes.

plant since it may not be clearly evident if the weld metal was properly fused without running a time-consuming metallurgical mount of the weld. However, in order to avoid potential legal responsibility, it is imperative that the manufacturer have proof that the weld met all applicable requirements and that the weld carried the necessary loads before it fractured. This may motivate manufacturers to utilize other methods to verify the condition of welds of these types of steels, such as the use of ultrasonic evaluation.

In real world applications, welds can also fracture in mixed modes, where there may be varying degrees of interfacial fracture and button pull-out. In the final analysis, either fracture mode can be acceptable in terms of the load that was carried, and in the end, engineering specifications should consider these fracture modes and specify what is acceptable. This is especially important when spot welding AHSS.

There are a number of formulae developed to assess cracking susceptibility (Matsuda, 1990; Olson *et al.*, 1993), with each one containing different elements and contributions to cracking. For many steels, such as HSLA and leaner dual phase steels, the carbon equivalent (CE) formula (Saito, 1983), developed at Nippon Steel, has been found by some to have a good relationship between chemical composition and weld-fracture from cross-tension tests. Using this formula, with normal levels of residual phosphorus, sulfur and silicon commonly found in boron steels, a carbon equivalent of 0.36 is not unusual. Testing has shown that, when the carbon equivalent of steel was higher than 0.24, weld interfacial fractures were generally observed.

$$\text{Carbon equivalent} = \%\text{C} + \%\text{Mn}/20 + \%\text{Si}/30 + 2\,(\%\text{P}) + 4\,(\%\text{S}) \qquad [2.4]$$

More recent evaluations of higher strength dual phase steels indicate that this formula, which does not include a number of alloying elements used in some of these grades, may not be the most accurate in its predictions for all alloy compositions and strength levels. The investigator would be prudent to evaluate a number of different alloys with different chemistries before applying any formula for weldability and, even then, to follow accepted practices to evaluate weld quality such as microstructural evaluations.

Other welding methods

While spot welding is the most common method used to join automotive body structures, other techniques such as metal inert gas welding (MIG), laser assembly welding, structural adhesive bonding and laser blank welding are also employed. MIG welding is a type of gas metal arc welding (GMAW) that utilizes a consumable electrode that makes a weld under an inert or semi-inert gas shield. The process is commonly used in automotive manufacturing, but has its limitations when joining many grades of advanced and ultra, high strength steels. The relatively high heat input required to fuse the electrode material and the work piece can result in a large heat affected zone as compared to resistance spot welding. As a result, this technique needs to be carefully studied with respect to the alloys being joined and to the amount of softening that is occurring in the HAZ. Therefore, techniques that limit heat input such as localized rather than continuous welds, weld speed, shielding gas, and the size of the weld need to be considered along with the structural requirements of the joint. The sensitivity of many of the newer advanced and ultra high strength steels to HAZ softening should lead one to perform careful evaluations of the alloy, weld stack and welding practice before this process is selected. It should be noted, however, that this technique has a long and positive performance history in welding relatively heavy gauge low carbon, HSLA and solid solution strengthened steels.

Adhesive bonding

Adhesive bonding of automotive structural joints is a technique that is receiving a lot of attention. The development of new generations of adhesives that combine high strength without the brittleness of conventional structural adhesives has enabled the use of increased structural bonding and has opened up new opportunities for the automotive designer. These new adhesives, commonly referred to as crash toughened adhesives, use synergistic rubber toughening to create particles in the adhesive that exhibit high fracture toughness. This reduces the tendency for fracture during impact events and therefore improves energy absorption. Another critical improvement with this new generation of adhesives is the higher working temperature range that better suits the operating temperature extremes in which automobiles need to operate.

The use of these adhesives creates the opportunity for significant weight savings through the reduction of weld flange size and improvement in the structural efficiency of joints. At this time, most crash toughened adhesives are used to augment conventional spot welds; however, the opportunities for these adhesives to reduce or even replace spot welding is a major area of interest. An entirely adhesive bonded structure may be years away from reality, but there are many talented engineers working on this challenge and the use of structural adhesives in joining structural components is likely to continue to grow.

2.5 Designing with steels for lightweighting automotive structures

In an environment where fuel economy, safety and emissions are becoming so critically important, the ultimate goal for the automotive engineer is to develop an efficient design that incorporates the right material on each part such that the overall material and design work together to enable a highly efficient vehicle architecture. Given the overall critical nature of design, no discussion of the materials used to lightweight automotive designs would be complete without at least some discussions on how those materials are effectively incorporated into a vehicle structure.

In order to design efficient vehicle body structures with the lowest potential mass, the design of each part of the vehicle, how these parts function as a subsystem and ultimately, as an overall vehicle, are extremely critical to overall efficiency. The complex interactions between individual parts, these parts and their subsystems and between each subsystem will ultimately determine how successful the entire design is in terms of structural and mass efficiency.

The role a material plays in overall vehicle performance, while being critical to an efficient design, is not as important as the part and system designs that make up the vehicle. In other words, overall vehicle design and structural load path efficiency are more critical than the materials that are selected for each of the parts in that vehicle. To look at this in another way, an inefficient structural design cannot be transformed into an efficient design through the use of the latest and greatest materials that are available. However, excellent material selection can make a really good design exceptional, and that should be the goal of every automotive design.

One of the first things automotive engineers need to accomplish when designing new vehicle architectures is to develop a philosophy of how vehicle loads and crash energy will be managed and how all the major components and subsystems will be 'packaged' within the allotted space. Frequently, it is the amount of space available for a load carrying part or subsystem that will ultimately determine how efficient that part and/or subsystem is. For the automotive designer, the strength of components and their ability to carry and distribute load is critical. In order to better understand how this works, let us look at a simplified vehicle design such as in Fig. 2.18 to see how loads could be transferred through the architecture.

The body structure in Fig. 2.18 represents most of the major components in a vehicle structure. The number of critical body structure parts is somewhat difficult to identify with great precision since it is common practice to break up many of the larger parts into smaller parts that are more manufacturable. While this number can vary between manufacturers depending upon their design and manufacturing philosophies, of the 400–450 parts that typically comprise a body structure, there are in the order of 40 to 50 major components. This figure does not include the closure panels: hoods (bonnet), fenders, doors, and rear decklid (boot). These panels are generally not considered critical to the vehicle structure. The remaining parts are comprised mostly of miscellaneous reinforcements and brackets that may be critical to vehicle performance, but their discussion is beyond the scope of this chapter.

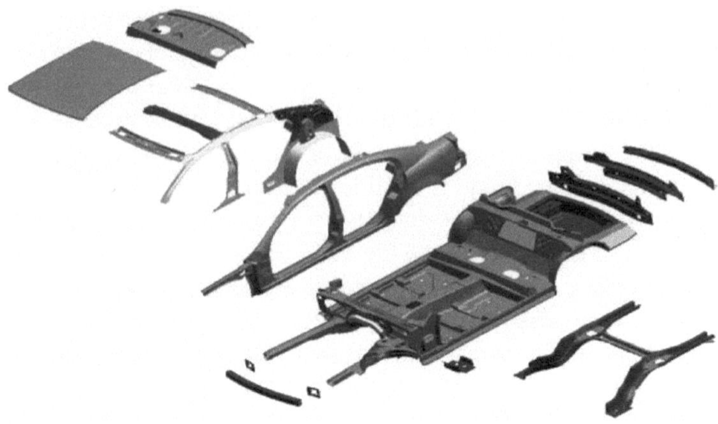

2.18 Simplified vehicle structure.

A typical front end structure is shown in Fig. 2.19. This structure consists of the primary load carrying members: bumper, motor compartment rails and underbody side rails. Additional parts consist of the wheelhouse, dash, plenum and front body hinge pillar. While these parts are important to the overall vehicle structure and do contribute to energy absorption, the primary load path for energy from a front end collision is through the bumper, to the motor compartment rails, the engine cradle (not shown) and underbody side rails as shown in Fig. 2.20.

In this simplified view, the bumper bar experiences the initial load of impact and transfers load to the motor compartment rails; any remaining energy is transferred to the underbody side rails. The function of the bumper bar is to take the high loads from impact and, to the extent possible, transfer those loads to the motor compartment rails. The bumper bar is designed to resist deformation, and therefore is made from very high strength steel. Steels with high tensile strength are preferred for the bumper, so the highest strength dual phase steels and martensitic steels are generally preferred. Roll forming these steels to the proper shape is an effective and low cost manufacturing process.

In contrast, the role of the motor compartment rails is to absorb energy and reduce the overall amount of energy transferred to the occupants. In order to absorb significant amounts of energy, the rails are designed to undergo significant, but controlled, deformation. Traditionally, motor compartment rails have been manufactured from conventional high strength steels, with precipitation hardened and carbon–manganese steels of around 300–350 MPa yield strength being very common. The ability for dual phase steels to absorb significantly higher energies than conventional steels at the same yield strength has led to these steels becoming more common in this application. Dual phase steel with a tensile strength of 590 MPa and a yield strength of

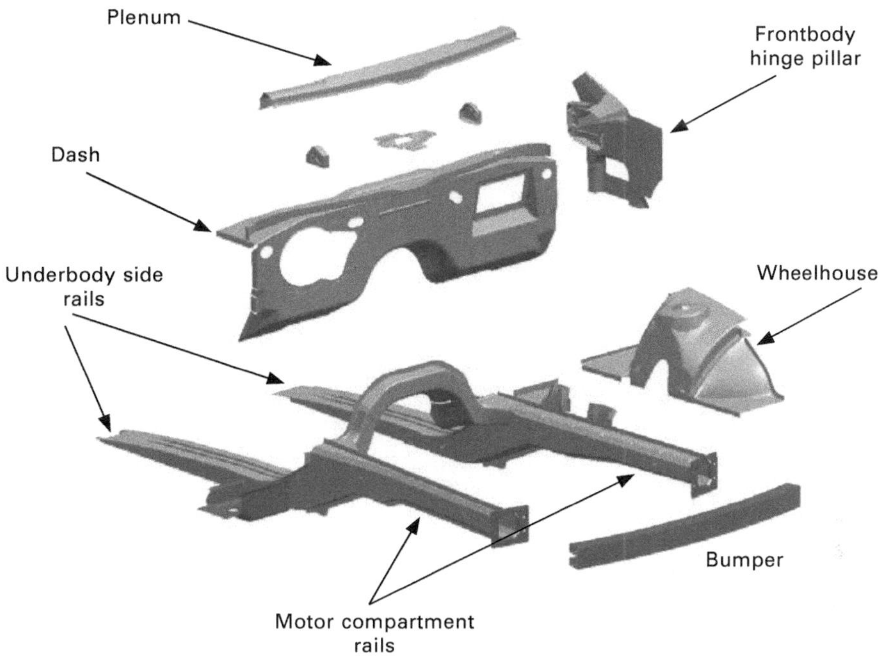

2.19 Front end structure.

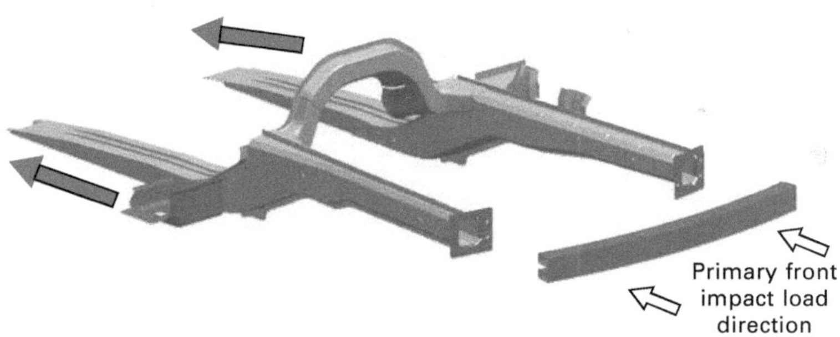

2.20 Front end structure load direction.

340 MPa, commonly referred to as dual phase 600, or just DP600, has been successfully used in the manufacturing of motor compartment rails and is now in full production by a number of vehicle manufacturers from Europe, Asia and North America. With proper designs, DP800 grades have demonstrated their ability to carry even higher loads and also show great promise in these types of applications. In order to absorb the most energy, engineers commonly try to design these rails to crush like an accordion. That is, they want the rails to form folds as they crush axially, which maximizes the amount of absorbed energy. In Fig. 2.21, a rail section before and after axial crush is

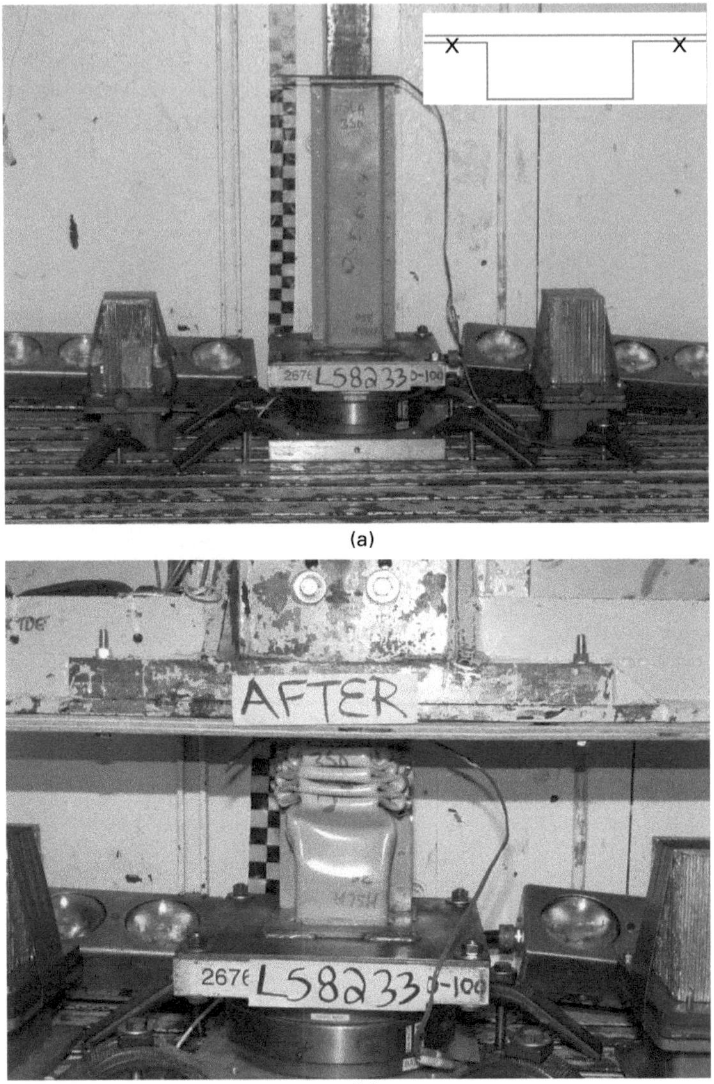

(a)

(b)

2.21 Axial crush of a motor compartment rail section (a) motor compartment section before crush testing (b) motor compartment section after crush testing.

shown. In the lower photograph, the folds in the rail after testing are clearly evident in the circled area. These folds absorb significant amounts of energy from bending and unbending during deformation and can be effectively computer modeled. As previously mentioned, simulations and testing have shown that dual phase steels absorb more energy than conventional high strength steels. For equivalent yield strengths at approximately 340 MPa, the corresponding dual phase steel at 590 MPa tensile strength will absorb

10–15% more energy than a corresponding microalloyed steel at 410 MPa yield tensile strength (Yan *et al.*, 2005). Advantages such as these are key enablers for lightweighting automotive structures.

The underbody side rails in Figs. 2.19 and 2.20, in contrast to the motor compartment rails, are intended to resist buckling. This is important because these rails are directly under the passenger compartment where deformation is not desirable. Maintaining the passenger space with as little deformation as possible is highly desirable and is a key strategy to prevent significant injury. This concept is referred to in the industry as the 'safety cage', or an area around the passengers in which most of the structure is designed for protecting the interior volume of the vehicle and reducing the potential for serious injury.

Nowhere in the vehicle is the concept of the 'safety cage' more evident than in the side structure of the vehicle. In the side structure, as shown in Fig. 2.22, there is very little or no available crush space to absorb energy without intruding on the passenger compartment. In side impact, significant intrusion into this safety zone for the passenger is something that automotive designers try to limit. This area is so critical for passenger safety that in many regions around the world, regulators and safety groups have adopted standards to measure vehicle performance against side intrusion.

The complexity of the body side structure and the number of parts and subsystems that need to function together as a complete system are evident from the diagram. The process of designing the side structure, materials selection and tuning the performance of the subsystem such that it provides the proper performance is extremely difficult and requires a team of very experienced and talented engineers. While there is no way to perform an adequate review of side structure design and load management in a book such as this, a high level review of the major loads and potential material selection philosophies is possible.

The load direction in side impact is principally through the center pillar of the vehicle, or B-pillar as it is commonly called. In contrast, the outer 'skin' of a vehicle, which is shown in Fig. 2.23, is primarily comprised of very thin, low carbon and bake hardenable steels and, in most cases, is not a major contributor to the overall structure of the vehicle.

However, the parts that comprise the inner structure of the vehicle: the B-pillar outer reinforcement, the rocker outer and the roof rails, are the major load carrying members. As mentioned earlier, it is the function of the side structure to oppose the crash force and to not significantly deform such that the passenger space inside the vehicle is compromised. As a result, these types of parts should be made of the highest strength steels that are manufacturable into the shapes that are required. In particular, the B-pillar outer reinforcement and the rocker outer are the two members that carry the highest loads and are arguably the most important members for side

2.22 Bodyside structure.

impact. Since these parts are designed to resist crushing or to crush only in pre-determined areas, for all the reasons previously discussed, they are made of the highest strength that can be formed into the necessary shapes. For the B-pillar outer reinforcement, the highest strength steel is martensite, but forming martensitic steels at room temperature would be impossible for the shapes that are common to these types of parts. As a result, these parts are more frequently being made from hot stamped boron steel. With this process, the tensile strength of the final part is commonly around 1400 MPa, with a corresponding yield strength close to 1000 MPa. From a performance perspective, when high loads need to be managed and deformation must be

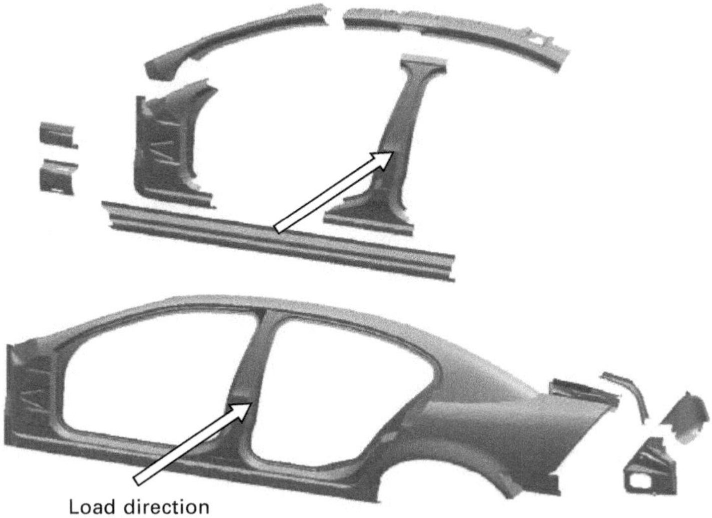

2.23 Load direction in side impact.

limited, hot stamped parts are generally considered the best option from a mass and performance perspective.

The roof rails also play an important role in side impact. As force is transferred from the B-pillar on impact, the loads are transferred to the rocker and to the roof side rails. Significant intrusion into the upper passenger compartment would compromise headspace and cause significant potential for occupant injury. As a result, press hardened steels need to be considered as possible materials in this type of application as well. This area is so critical that automotive manufacturers have been incorporating side curtain airbags in a number of vehicles to help protect the passenger from severe trauma.

However, there are many more nuances to side structure design than the simple analysis provided above. There are strategies designed to modify materials and section strength in these areas such that more load is transferred into the underbody and away from the upper side structure. This lowers the potential force that can be transferred to an occupant's head and thorax, and transfers that load into the auto's underbody, where it can be harmlessly absorbed by crushing other less critical components. In such a strategy, the designs and materials for these components will be adjusted to move the loads to areas where they present less danger to vehicle occupants.

In addition to the side structure, the roof structure, as previously shown in Fig. 2.22. also is utilized in side impacts where the force is high enough to travel into the upper vehicle structure. Here, forces are transferred from the B-pillar reinforcement to the roof side rails, and ultimately, if loads are high enough, the center roof bow can come under load. In some designs, this roof bow is being designed with high strength steels in order to support the roof side rails and to ultimately transfer load to the other side of the vehicle and away from the primary load path.

There are additional load paths that can be important in some crash events. For rear crashes, the rear bumper and longitudinal rails perform much the same function as the front motor compartment rails. These rear rails perform double duty in that they not only protect the passengers, but they also protect the fuel tank from high loads, deformation, and ultimately, from the potential loss of fuel. In the case of a rollover, the upper and side structures provide protection just as they do in side impacts, except that the loading directions are very different from the side impact events. The high strength roof side rails and B-pillars prevent collapse of the roof and protect interior passenger space.

2.6 Conclusion

The complexities facing an automotive engineer are daunting. The difficulties of designing the parts of a vehicle to function together to meet the myriad of requirements that modern vehicles must meet is extremely complex. Because of this, the overall potential for success of a vehicle architecture is determined far before the first parts are made and the first prototype vehicle is assembled. Architectural design, energy management and efficient part designs are the cornerstones of lightweight, efficient designs. However, to optimize a design, proper material selection is necessary for the part and the vehicle that it is a component of, in order for it to function to its optimum design capability. It has been estimated that the weight reduction potential through material substitution, based on some work in 1980 (Hsia and Kidd, 1980), was in the range of 10–30%. These numbers are probably still valid given the continual advancements in steel development.

While the mass savings potential by switching from steel to aluminum used to be in the area of 50%, with optimum designs and properly using the first generation of steels, that number is now far less than that with the currently available AHSS. There are a number of design examples where an efficient steel design nearly matches an aluminum design for the same part function, and this can be accomplished with much lower cost than the 'lightweight metal' alternatives. In the end, all materials – steel, aluminum, magnesium, cast iron and composites – have attributes that can, and should, be utilized to their fullest. To do less is to produce products that are not optimized and may not use our precious resources as effectively as they should. To that end, the latest generation of advanced high strength steels, and the new materials that are yet to be developed, will take automotive design to the next level and beyond.

The process of designing vehicle architecture is very complex and time consuming, and as such is way beyond the scope of this chapter. However, combining the aspects of material selection, energy management in vehicle architecture, and joining and manufacturing considerations such as have been presented, gives the aspiring engineer a good introduction into the world of engineering for optimum lightweight automotive structures.

2.7 References

1 AISI (2006), *Talking Points: Solicitation of Research Proposals and Department of Energy Sciences Funding to Develop Third Generation Advanced High Strength Steels,* American Iron and Steel Institute, Washington D.C., pp 1–3.

2 Campbell, M P (1989), 'Evaluation of a Continuous Annealed Bake Hardenable Steel for Improved Dent Resistance', Society of Automotive Engineers, *SAE Publication 890711.*

3 Chen X, Shi M, Chen G, Kamura M, Watanabe K, Omiya Y (2005), 'Crash Performances of Advanced High Strength Steels of DP780, TRIP780 and DP980', Society of Automotive Engineers, *SAE Publication 2005-01-0354*, pp 6–7.

4 Cornette D, Hourman T, Hudin O, Laurnet J P and Reynaert A (2001), 'High Strength Steels for Automotive Safety Parts', Society of Automotive Engineers, *SAE Publication 2001-01-0078.*

5 Cornette D, Cugy P, Heldenbrand A, Lavato G and Bouzekri M (2005), 'Ultra High Strength FeMnTWIP Steels for Automotive Safety Parts', Society of Automotive Engineers, *SAE Publication 2005-01-1327*, pp 2–3.

6 DiCostanzo, G P, Matlock, D K (1996), 'Effect of Tensile Properties on Dent Resistance of Sheet Steels', Society of Automotive Engineers, *SAE Publication 960024*, p 46.

7 General Motors Corporation (2007), 'High Strength Sheet Steel, 180 MPa through 700 MPa Yield Strengths: GMW3032', Englewood, *IHS Engineering Documents.*

8 General Motors Corporation (2008), unpublished study.

9 Hsia, H-S and Kidd, J (1980), 'Weight Reduction Potential of Passenger Cars and Light Trucks by Material Substitution', Society of Automotive Engineers, *SAE World Congress 1980*, Paper 800803, p 1.

10 Matsuda, F (1990), '*Proceedings of 2nd International Conference, Gatlinburg, TN, May 1989*' (1990), ASM International, pp 127–136.

11 McCormick, M A, Fekete J R, Muleman D J and Shi M F (1998), 'Effect of Steel Strengthening Mechanisms on Dent Resistance of Automotive Body Panels', Society of Automotive Engineers, *SAE Publication 980960*, pp 7–8.

12 Okamoto A, Takeuchi K, and Takagi M, A Mechanism of Paint Bake-hardening, *The Sumitomo Search No. 39*, 1989, pp 183–194.

13 Olson, D L, Edwards, G R, Liu, S, Frost, R (1993), *ASM Handbook Volume 06 – Welding Brazing and Soldering*, American Society for Metals.

14 SAE (1999), '*Categorization and Properties of Dent Resistant, High Strength, and Ultra High Strength Automotive Sheet Steel: SAE J2340*', *Society of Automotive Engineers*, Warrendale, PA.

15 Saito, T (1983), *Welding Technique*, 31 (4), p 27.

16 Schaeffler, D J, Stoddard, P A, Horvath, C D (1996), 'Quasi-static Dent Resistance Evaluations and Formed Panel Properties of Door Assemblies', *International Body Engineering Conference* '96', Detroit.

17 Shaw J and Zuidema B (2001), 'New High Strength Steels Help Automakers

Reach Future Goals for Safety, Affordability, Fuel Efficiency and Environmental Responsibility', Society of Automotive Engineers, *SAE Publication 2001-01-3041*, pp 3–4.

18 Sun X, Stephens E, Khaleel M (2006), 'Effects of Fusion Zone Size on Failure Modes and Performance of Advanced High Strength Steel Spot Welds', Society of Automotive Engineers, *SAE Publication 2006-01-0531*, p 8.

19 Toshiaki U, Fujita T, Ono Y, Yamasaki Y and Hosoya Y (2003), 'Development of IF High Strength Steel with Fine Grain Structure for Exposure Panels, *SAE Publication 2003-01-2769*, pp 1–4.

20 Walp M, Wurm S, Siekirk J, Desai A, 'Shear Fracture in Advanced High Strength Steels', Society of Automotive Engineers, *SAE Publication 2006-01-1433*, 2006, pp 3–7.

21 Wuebbels T, Matlock D K and Speer J G (2002), 'The Effects of Room Temperature Aging on Subsequent Bake Hardening of Automotive Sheet Steels', Society of Automotive Engineers, *SAE Publication 2002-01-0041*, p 1.

22 Yan B, Kanter C, Zhu H, Nadkarni G, Horvath C (2005), 'Evaluation of Crush Performance of a Hat Section Component Using Dual Phase and Martensitic Steels', Society of Automotive Engineers, *SAE Publication 2005-01-0837*, pp 3–6.

3 Aluminum alloys for lightweight automotive structures

J. C. BENEDYK, Illinois Institute of Technology, USA

Abstract: An overview of automotive aluminum alloys and their applications as castings, sheet and plate, and extrusions is presented. To assist engineers and metallurgists unfamiliar with aluminum alloys, the classification systems for aluminum alloy designations and tempers are described, and comparative guidelines are presented for lightweight materials selection by automotive engineers. A review is made of progress by US, European, and Japanese automobile companies in lightweighting automotive vehicles utilizing innovative technologies for processing aluminum alloys in all forms.

Key words: aluminum alloys, automotive lightweighting, casting, mill products, extrusions.

3.1 Introduction

The critical property of aluminum that makes it so attractive is its low density (2.69 g/cm^3), which is 1/3 that of steel. Aluminum can be alloyed and strengthened by cold working and/or heat treatment to achieve high strength and so can achieve a high strength to weight ratio. However, alloying of aluminum does not change its modulus of elasticity (69 GPa) to any significant degree, and this property is about 1/3 that of steel (210 GPa). At the same time, aluminum alloys are amenable to a variety of production processes and are therefore available in many forms: castings, extrusions, stampings, forgings, impacts, and machined components. Compared with stampings that need to be consolidated by welding or other joining process, aluminum castings and extrusions make it possible to consolidate product forms and functions, thereby reducing the number of components necessary to produce a vehicle and improving productivity. With proper design and taking full advantage of the low density of aluminum, it is generally accepted that one pound of aluminum can replace two pounds of commonly used steel or iron in an automotive vehicle.

In 2006, aluminum surpassed cast iron as the second most-used material in a North American vehicle, behind steel. North Americans consume most aluminum in transportation (32%), followed by containers and packaging (21%), and building and construction (13%); this market breakdown for aluminum is typical of major consuming countries as well. *Per capita* consumption of aluminum in North America is about 35 kg (77 lb) annually, although it is lower in Europe and Japan. Over the last 34 years, aluminum increased its share of metal content in North American automotive vehicles from 36 kg (81 lb) in 1973 to 147 kg (324 lb) in 2007 (Fig. 3.1).[1–3] In North America, more than 60% of engine blocks, almost 100% of transmission cases, and about 60% of wheels are made of aluminum alloys as castings. Several applications for wrought aluminum alloys exist as well. With the need to increase fuel efficiency and the current 35 miles-per-gallon fuel economy standard mandated by 2020 for North American vehicles, the edge in materials choice for automotive vehicles will undoubtedly go to the lightweight material that is most cost effective while promising a reduced carbon footprint.

The growth trend for automotive aluminum continues not only in North America but also in Europe and Asia. In Europe, the world's largest producer of automobiles, average aluminum content in automobiles has increased from 37 kg (82 lb) in 1973 to 126 kg (278 lb) in 2007. This trend has no doubt been prompted by the higher fuel prices in Europe compared with North America and, although the average content of aluminum today is less in Europe than in North America, automotive vehicles in Europe are smaller. In Asia, again where automotive vehicles are smaller, average automotive aluminum content has also grown in a trend similar to that of Europe. In 2000, worldwide automotive aluminum usage amounted to 5.8 million tonnes and this is expected to double by 2020, given conventional projections of growth of automotive vehicle production, although due to the current recession aluminum will be in surplus until 2012.[3] Average automotive US usage belies the fact that in 2007 there were some 50 types of vehicle representing two million units that had over 227 kg (500 lb) of aluminum per vehicle, and there were a number of models with complete aluminum body structures (aluminum intensive vehicles or AIVs).

Today, aluminum in vehicle structures, power trains, and accessories makes it possible to produce vehicles that are lighter, safer, and have higher performance in terms of acceleration, handling, and braking distance. These lighter vehicles, in turn, lower fuel consumption and therefore reduce greenhouse gas (GHG) emissions and general pollution levels. In an extensive study of worldwide aluminum production, the energy balance for the replacement of steel or iron with aluminum shows that, on average, each tonne of aluminum is estimated to save 18 tonnes of carbon dioxide emissions over the average lifetime of a mid-size sedan, based on the generally accepted estimate that a 10% reduction in vehicle weight translates, on average, to a 5–6% fuel saving.[1]

At the same time, it is well known that it takes more energy to win aluminum metal from its ore than it does for iron; however, recycling aluminum requires a specific remelting energy of only 5% of that needed to produce primary aluminum. And, with 95% of aluminum in end-of-life

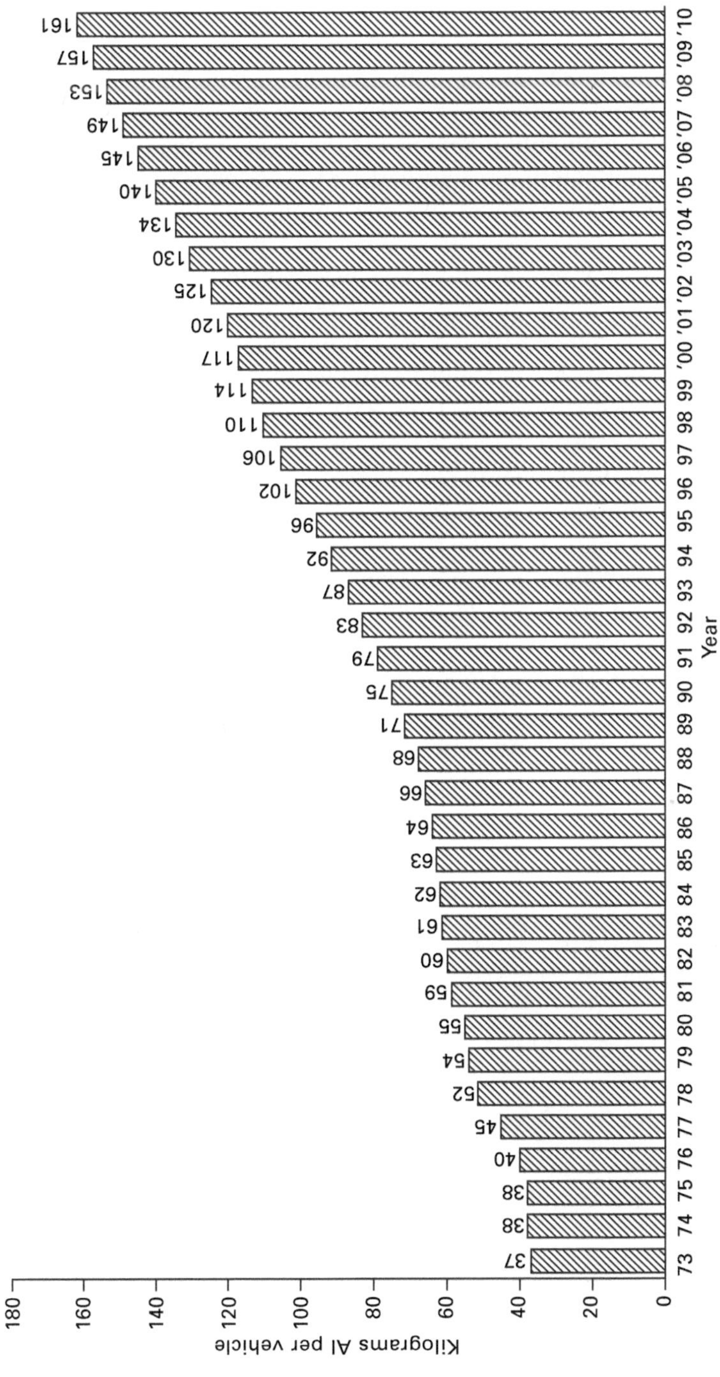

3.1 North American light vehicle aluminum content increasing over three decades.[1,2] © Light Metal Age, reproduced with permission.

vehicles being recycled, this secondary aluminum offers a significant reduction in life cycle greenhouse gas emissions.[1,4–6]

The issue of vehicle crashworthiness and its relation to lightweighting has been extensively studied by auto engineers, with general agreement that high strength and lightweight materials such as aluminum can improve crashworthiness in automotive vehicles if their size is maintained. In fact, crashworthiness and safety are hallmarks of aluminum intensive vehicles such as the Audi A8, which was the first automobile that Audi built with their patented all-aluminum spaceframe design (Audi Spaceframe or ASF); the A8 had a five star safety rating and was considered one of the safest vehicles on the road in the 1990s, as it is today.

Performance is another hallmark of aluminum intensive vehicles. After the aluminum intensive A8, A2, and TT, the latest ASF vehicle that Audi has built to date is their R8 sports car (Fig. 3.2), which has an entire aluminum body shell weighing just 210 kg (463 lb), giving it one of the top ratings of torsional rigidity to body weight for a sports car.[7]

In designing automotive vehicles and components from aluminum alloys, automotive engineers have a wide selection of aluminum wrought or casting alloys from which to choose. They also need to select a temper with required mechanical properties of tensile yield, tensile strength and percent elongation. There has been much effort among aluminum associations in different countries to standardize aluminum alloy designations, chemistries, and mechanical properties, and it is important for automotive engineers to become acquainted with these standards.[8–10]

3.2 The Audi R8 sports car built with an aluminum intensive body shell (courtesy of Audi). © Light Metal Age, reproduced with permission.

3.2 International designation systems for aluminum alloys

Compared with the number of ferrous alloys used in automotive applications, the number of aluminum alloys is far more numerous. For the automotive engineer, it is important to recognize the significant differences between wrought and cast aluminum alloys and, in the wrought category, between heat treatable and non-heat treatable alloys. Aluminum alloys are classified according to a standard designation system recognized by either international standards agencies or by standards agencies of countries that prescribe their own designation system. The Aluminum Association aluminum alloy designation system for alloys and tempers is recognized by the American National Standards Institute (ANSI) as the national standard in the US and it is incorporated into ANSI Standards H35.1.and H35.2.[8]

The maintenance of the system is managed under ANSI charter by the Aluminum Association, Inc. In addition, there is an international accord recognizing the Aluminum Association wrought alloy designation system as the international standard, ratified and accepted by almost all of the world's aluminum producing countries. Table 3.1 briefly describes the internationally recognized Aluminum Association designation system for wrought aluminum alloys.

The Association has also established a four digit designation system for aluminum casting alloys (Table 3.2) that is approved by the Society of Automotive Engineers, although this aluminum casting alloy designation system is not internationally ratified. Thus, the CEN (European Committee for Standardization) has established a five digit designation system for casting alloys which differs in many respects from the four digit Aluminum Association system, as do standard designation systems for casting alloys in Japan, China, and most other non-CEN member countries.

Table 3.1 Aluminum Association designation system for wrought aluminum alloys.[8–11] © Light Metal Age, reproduced with permission

1xxx – Pure Al (99.00% or greater)
2xxx – Al–Cu alloys
3xxx – Al–Mn alloys
4xxx – Al–Si alloys
5xxx – Al–Mg alloys
6xxx – Al–Mg–Si alloys
7xxx – Al–Zn alloys
8xxx – Al+other elements
9xxx – Unused series

First digit – principal alloying constituent(s).
Second digit – variations of initial alloy.
Third and fourth digits – individual alloy variations (number has no significance but is unique).

Table 3.2 Aluminum Association designation system for aluminum casting alloys.[9,11] © Light Metal Age, reproduced with permission

lxx.x	Pure Al (99.00% or greater)
2xx.x	Al–Cu alloys
3xx.x	Al–Si + Cu and/or Mg
4xx.x	Al–Si
5xx.x	Al–Mg
7xx.x	Al–Zn
8xx.x	Al–Sn
9xx.x	Al+other elements
6xx.x	Unused series

First digit – principal alloying constituent(s).
Second and third digits – specific alloy designation (number has no significance but is unique).
Fourth digit – casting (0) or ingot (1,2) designation.
Variations indicated by preceding letter (A, B, C).

International reference guides, such as Aluminium Schlüssel/Key to Aluminium Alloys,[9] can identify an aluminum alloy by designation, country of origin, brand name, and reference or material number (all cross indexed) along with its specified chemistry, mechanical and physical properties, as well as technological characteristics. It should be noted, however, that aluminum alloys developed for special automotive applications by aluminum companies, such as Alcoa's C-446 aluminum casting alloy (described in Section 3.4.1), and those developed by automotive companies, will typically bear a unique designation that has proprietary significance.

3.3 International temper designations for aluminum alloys

Temper designations in the CEN member countries (EN standards) are substantially the same as those proposed by the Aluminum Association in the US for wrought and cast aluminum alloys (Table 3.3). However, different temper designations are used in other countries and until recently by the International Standards Organization (ISO). Mechanical properties such as yield, tensile strength, and percent elongation are specified for each particular temper by the governing standards organization, e.g. the Aluminum Association in the US and the CEN in European Union countries. Thus, it is important to examine appropriate mechanical property standards for a given aluminum alloy and temper before product certification.[8–10]

Table 3.3 Aluminum Association and European Union (EN) aluminum alloy temper designation system.[8–11] © Light Metal Age, reproduced with permission

F – as fabricated
0 – annealed
H – strain-hardened (wrought products only)
W – solution heat-treated
T – thermally treated to produce tempers other than F,O,H (usually solution heat-treated, quenched, and precipitation hardened)
Numeric additions indicate specific variations:
H14 – strain hardened, ½ hard
T4 – solution treated and naturally aged
T6 – solution heat treated and artificially aged
T65 – solution heat treated, stress relieved by stretching a specified amount, and artificially aged

3.4 Aluminum alloys used in lightweight automotive vehicles

Figure 3.3 shows a composite of some applications of aluminum alloys in a typical sedan, without consideration of body design, i.e. unibody or body-on-frame. Of some seven million tonnes of aluminum used worldwide in automotive manufacture, the majority (~80%) is in the form of castings used for powertrain components, although most automotive vehicles also contain radiator and condenser components of ten made of AA1200 and AA3005 aluminum alloys, the latter in the form of brazing sheet (AA3005 alloy metallurgically bonded with an Al-Si braze alloy). The range of aluminum alloys shown in Fig. 3.3 is exemplary but not universal. In fact, the universe of aluminum alloys in automotive applications is very large. These alloys are modified in many cases by minor alloying additions but within chemical specifications required by the respective standards, and their processing parameters (casting, rolling, extrusion, forging, heat treatment, welding, surface finishing, etc.) are controlled to attain optimum properties in the finished component tailored to automotive performance requirements.

As mentioned above, aluminum alloys used in automotive applications are far more numerous than classes of competitive materials in order to match the alloy to the performance of a given automotive component and because aluminum alloys can be formulated as wrought or cast, with either amenable to all types of processing. Thus, wrought aluminum alloys are available as sheet or plate, extrusions, forgings, impacts, and even semisolid shapes; aluminum casting alloys can be cast by a wide variety of casting processes into complex near net or net shapes or formed by semisolid processing. Tables 3.4 and 3.5 list some cast and wrought aluminum alloys and their potential automotive applications. Chemical composition and mechanical

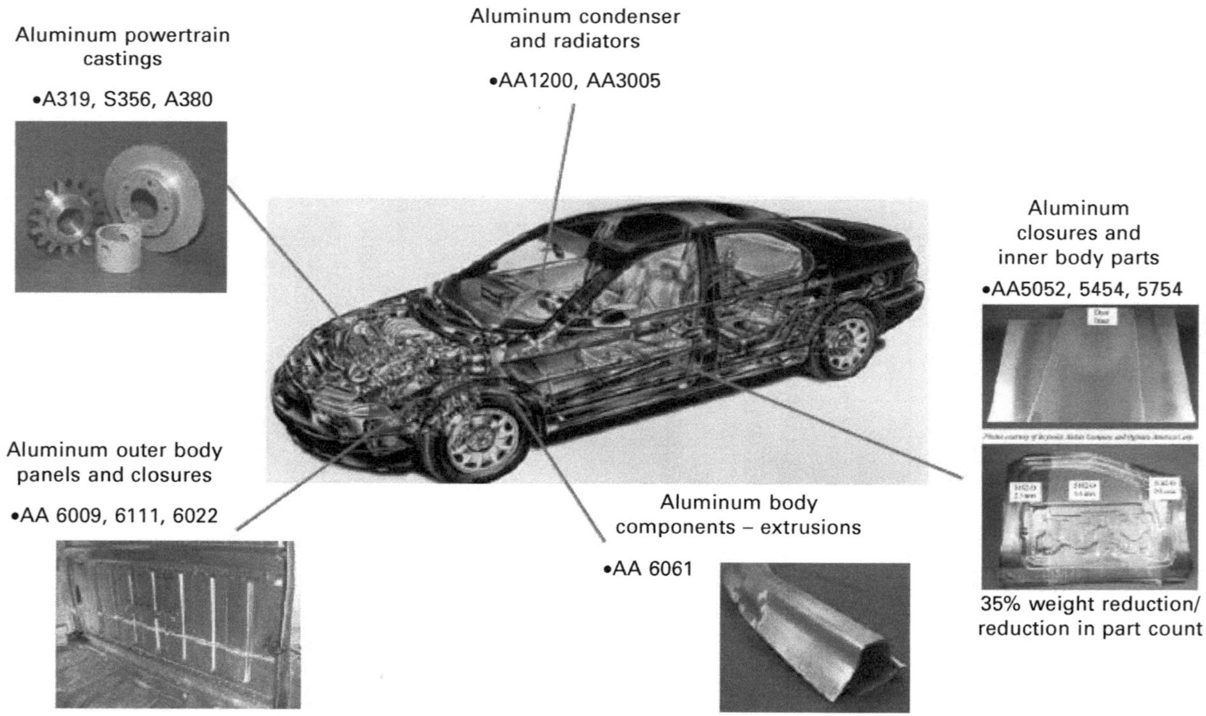

3.3 Some typical automotive aluminum alloy applications and product forms. © Light Metal Age, reproduced with permission.

property specifications for these alloys are given in published standards and are readily available from aluminum suppliers.[8,9]

3.4.1 Aluminum alloy castings in lightweight automotive vehicles

Although cast aluminum engines were used in many early automobiles, e.g. the 1931 Marmon V-16 convertible sedan and the 1932 Packard sedan built in the US, the trend toward use of aluminum alloy castings in many automotive powertrain applications in North America really began in the 1950s and has grown substantially since then. The casting processes used to produce automotive components from aluminum alloys are many and include sand casting, permanent mold casting, low and high pressure die casting, lost foam casting, pressurized lost foam casting, high integrity die casting processes (described below), and the recently developed ablation casting process. A distinction between North America and Europe is that much of the aluminum content in North American vehicles is in castings for engines and powertrains, whereas Europe, with its smaller vehicles, uses more aluminum sheet and extrusions in non-powertrain applications such as body structures and closures.

Table 3.4 List of cast aluminum alloys and automotive applications[1]. © Light Metal Age, reproduced with permission

2xx.x	**Series**
201.0	Structural members, cylinder heads, pistons, connecting rods, rocker arms
204.0	Structural members, brake calipers, powertrain castings
206.0	Structural members, gear housings, cylinder heads for gasoline and diesel engines, supercharger impellors
208.0	Manifolds, valve bodies
242.0	Diesel pistons, air-cooled cylinder heads
295.0	Rear axle housings, crankcases
3xx.x	**Series**
319.0	Cylinder heads, crankcases, internal engine parts
332.0	Gasoline and diesel engine pistons, pulleys, sheaves
335.0	Liquid-cooled cylinder heads and blocks
339.0	Pistons
356.0	Transmission cases, liquid-cooled cylinder blocks, manifolds
A356.0	Wheels, chassis castings
357.0	Chassis castings
A380.0	Brackets, housings, internal engine parts, steering gears
383.0	Liquid-cooled engine blocks, transmission housings/parts, fuel metering devices
390.0/A390.0/B390.0	Engine blocks, high-wear applications such as ring gears and internal transmission parts, disc brakes, brake shoes, compressor scrolls

Table 3.5 List of wrought aluminum alloys and automotive applications. © Light Metal Age, reproduced with permission

1000	**Series**
1100	Trim, nameplates, appliqués
1200	Extruded condenser tubes and fins
2000	**Series (often clad with 6000 or 7000 alloys for body panels)**
2008	Outer and inner body panels or structural applications
2010	Outer and inner body panels or structural applications
2011	Screw machine parts
2017	Mechanical fasteners
2024	Mechanical fasteners
2036	Outer and inner body panels, load floors, seat shells
2111	Applications requiring good machinability
2117	Mechanical fasteners
3000	**Series**
3002	Trim, nameplates, appliqués
3003	Braze clad welded radiator tubes, heater cores, radiator, heater and outlet tubes, oil coolers, air conditioning liquid lines
3004	Interior panels and components
3005	Braze clad fins for radiators, heaters, and evaporators
3102	Extruded condenser fins
4000	**Series**
4032	Pistons and high temperature service parts
4043	Weld filler wire for MIG/TIG welding of all wrought and cast aluminum alloys
5000	**Series**
5005	Trim, nameplates, appliqués
5052	Interior panels and components, bumpers, some body panels
5252	Trim
5182	Inner body panels, reinforcement members, brackets
5356	Weld filler wire for MIG/TIG welding of aluminum alloys with a high magnesium content (>3% Mg)
5456	Armor plate
5454	Wheels, engine brackets and mounts, various welded structures
5457	Trim
5657	Trim
5754	Inner body panels, splash guards, heat shields, air cleaner trays and covers, structural parts, load floors
6000	**Series**
6009	Outer and inner body panels, load floors, bumper face bars, bumper reinforcements, structural and welded parts, seat shells
6005	Seat frame components (extruded)
6009	Outer and inner body panels
6010	Outer and inner body panels, bumper reinforcements, seat shells, seat tracks
6016	Outer and inner body panels
6020	Lead-free replacement for 6262 in high machinability applications, ABS manifolds, brake housings, brake pistons, transmission valves and sleeves, impacts
6053	Mechanical fasteners
6061	Body components (extruded), brackets (extruded and sheet), suspension parts

Table 3.5 Continued

	(forgings), driveshafts (tubes), driveshaft yokes (impacts and forgings), spare tire carrier parts (extruded), bumper reinforcements, mechanical fasteners, brake cylinders (extruded), wheels (sheet), ABS manifolds (extruded), fuel delivery systems
6063	Body components (extruded)
6070	Structural components (extruded)
6082	General structural parts (extruded or forged), brake housings
6111	Body panels
6262	Brake housings, brake pistons, screw machine products
6463	Bright anodized automotive components
7000	**Series**
7003	Seat tracks, bumper reinforcements
7004	Seat tracks, bumper reinforcements
7021	Bumper face bars, brackets (sheet), bright anodized bumper face bars, bumper reinforcements
7033	Forged suspension and control arms

The variety of aluminum casting alloys and the modification of existing aluminum casting alloys by minor alloying constituents used in automotive applications are constantly growing and changing. Basically, aluminum casting alloys can be classified as hypoeutectic, eutectic, or hypereutectic, depending on whether they are below, at, or above the eutectic composition (lowest melting point) of the alloy system. For the Al–Si system, the eutectic composition is at 12.5% Si. The highly castable hypoeutectic 356 aluminum alloy (7% Si–0.3% Mg nominal composition), besides being used for powertrain components, is also widely used in automotive chassis and suspension components as well as competing with wrought 6061 aluminum alloy in ABS manifolds due to its superior machinability. The near eutectic 413 aluminum alloy (12% Si nominal composition) has excellent castability, especially in thin walled and intricate castings. The hypereutectic 390 aluminum alloy (17%Si–4.5%Cu–0.6%Mg nominal composition) developed in the late 1950s for the automobile industry for unlined bores in engine blocks and other wear resistant applications has low thermal expansion and high elevated temperature strength.

Although suitable for cylinder heads and blocks in conventional gasoline engines operating at pressures of up to 130 bar and temperatures up to 150 °C, conventional 3xx.x alloys such as 356 do not have the desired high temperature strength required in diesel or advanced supercharged gasoline engines, which operate at >240 bar and >200 °C; thus, special Al–Si–Cu–X alloys have been developed for this purpose. At the same time, the need for higher strength in chassis and suspension components in some cases has resulted in the substitution of high strength aluminum alloys such as A206 aluminum alloy to achieve weight reductions of as much as 55% in a near net shape over a component assembled from various stampings of high strength low alloy (HSLA) steel.[12]

Aluminum alloy castings can achieve near net or even net shape in large thin wall shapes by means of pressure die casting or by advance casting processes such as squeeze casting and vacuum die casting. Special aluminum alloys may be required to achieve castability and properties in such components. The recently introduced Nissan GT-R sports sedan features inner door panels, 1300 × 700 mm in plan area with walls of 2–3 mm thickness, cast from Alcoa's new C-446 aluminum alloy by means of Alcoa's vacuum die casting process, resulting in a 25–35% weight reduction over conventional designs and materials (Fig. 3.4).[7]

Vacuum die casting, squeeze casting, and semisolid molding are considered high integrity die casting processes that have significantly advanced the state of the art in terms of reducing porosity, shrinkage cracks, and other casting defects, while allowing aluminum alloy castings to achieve high performance levels and greater weight reductions than with conventional casting processes.[13]

Squeeze casting in the US was originally developed at what was then the Illinois Institute of Technology Research Institute (IITRI) in the early 1970s, and the first automotive applications in pistons, wheels, hubs, and other automotive components were done to improve performance over

3.4 Inner door panels for the Nissan GT-R sports sedan produced by Alcoa from their C-446 aluminum alloy at the Alcoa Soest, Germany plant[1]. © Light Metal Age, reproduced with permission.

conventionally cast components or to replace forgings rather than to achieve major lightweighting advantages.[14–16] In squeeze casting, porosity and solidification shrinkage are eliminated or reduced by control of molten metal flow and the application of high pressure (100–1000 bar) to solidifying metal. Various cast and wrought aluminum alloys, as well as aluminum based composites, were squeeze cast successfully in prototype production in the 1970s to establish process and product feasibility. Today, squeeze casting is used worldwide to produce a variety of automotive components that take advantage of the sound metallurgical structure and improved properties (Fig. 3.5). Porsche, for example, has used the squeeze casting process to produce its Boxter engine block, with integral aluminum matrix composite cylinder liners, from a European version of 380 aluminum alloy.[13]

Vacuum die casting of aluminum alloys can control gas porosity by coupling a vacuum system to a die casting machine, but does not offer any improvement over conventional die casting with regard to shrinkage porosity during solidification. Also, like typical pore-free squeeze cast components,

3.5 Squeeze cast automotive components produced by SPX Contech Corporation (upper – wheels, lower left – valve housing, lower right – steering column housing). © Light Metal Age, reproduced with permission.

but unlike those cast by conventional die casting, vacuum die cast aluminum alloys can be heat treated to increase strength without blistering.[17]

Semisolid molding – a process in which a finely divided solid/liquid metal mixture or slurry is injected into a die cavity – not only reduces gas porosity and solidification shrinkage but also significantly modifies the microstructure of the alloy. Alloys with a broad solidification range, such as 356 aluminum alloy, work best in creating slurry suitable for die casting or squeeze casting. Typically, the shrinkage that occurs in die casting a 50:50 solid:liquid slurry of 356 aluminum alloy is less than 3% as compared with 6% in casting the 100% liquid. Semisolid molding of 356 aluminum alloy was highly promoted in the 1990s, especially by the Alumax Engineered Metal Products (AEMP) Division of Alumax Aluminum, Inc., and used successfully for fuel rails, control arms, and other automotive components.[17,18] However, the high cost and narrow processing window required for casting the solid/liquid mixture into billet and subsequent reheating it to slurry for forming made the AEMP process untenable, resulting in the bankruptcy and sale of the plant real estate and all the AEMP equipment in early 2003.[19] Subsequent efforts in developing what is called a 'slurry-on-demand' semisolid molding process, which creates an aluminum alloy slurry by in-line mechanical or electromagnetic stirring right before molding, have resulted in a more cost effective process for manufacturing automotive components, including engine blocks.[20]

As the mechanical properties of aluminum alloy castings depend not only on composition and temper, but also on solidification rate (metallurgical structure control), shape, and degree of soundness, specifications of mechanical properties depend on the type of casting process used and other factors. In most cases, the cast aluminum alloy component is rated on the basis of functional and performance criteria as well as weight reduction relative to the component made of assembled steel stampings or made of cast iron.

In aluminum intensive vehicles, a significant amount of aluminum alloy castings is used in combination with aluminum alloy sheet and extrusions in body structures. The spaceframes for the Audi A8 and A2, for example, contain about equal parts castings, extrusions, and sheet components, joined together by MIG and laser welding and fasteners, including self-piercing rivets. A measure of this evolution at Audi is the reduction in the number of parts in the A2 aluminum spaceframe to 238 from 334 in the A8, thanks to the extensive use of aluminum castings in integrating functions and providing clever rib designs to precisely control crash energy absorption.[21]

3.4.2 Aluminum alloy sheet and plate in lightweight automotive vehicles

Aluminum alloy sheet for automotive and light truck applications is predominantly made of work hardenable 5xxx (Al–Mg) aluminum alloys such as 5182, 5454, and 5754, which are supplied in the annealed O temper condition, and of age hardenable 6xxx (Al–Mg–Si) alloys such as 6009, 6022, and 6111, which are supplied in the solution annealed and stabilized T4 or T41 temper condition. Typical mechanical properties of some aluminum body sheet alloys are given in Table 3.6.

The 6xxx alloy sheet is commonly used in external body panels due to the resistance of these alloys to 'ludering' or strain marking in stamping, and their higher strength, some of which is achieved in paint baking. The lower cost 5xxx alloy sheet is more common in internal body panels and body structure components. Relevant information on design considerations, forming, joining and finishing of aluminum alloy automotive sheet as well as on crash energy management and repair of aluminum automotive sheet is available in various monographs published by The Aluminum Association.[22–24]

In 1999, The Aluminum Association, with the support of the US Department of Energy, provided an *Aluminum Industry Roadmap*[25] for the automotive market to address challenges posed by the automotive industry for expanded application of aluminum and its alloys in all forms in vehicles through enabling technologies, which for aluminum sheet were defined as follows:

- Improved product forms (direct-chill and continuous belt or twin roll casting)

Table 3.6 Typical mechanical properties and formability parameters of some aluminum body sheet alloys.[22,23] © Light Metal Age, reproduced with permission

Alloy and temper	Ultimate tensile strength (MPa)	0.2% Yield strength (MPa)	Elongation in 50 mm or 2 in (%)	Modulus of elasticity tension/ compression (GPa)	Strain hardening exponent *n*	Plastic strain ratio *r*	Hem type
5182-O	275	130	24	71	0.33	0.80	Standard
5454-O	250	115	22	70	0.30	0.80	Standard
5754-O	220	95	26	71	0.30	0.80	Standard
6009-T4	220	125	25	69	0.22	0.64	Standard
-T62	295	260	11	69	–	–	–
6022-T4	255	150	26	69	0.25	0.67	Roped
-T62	325	290	12	69	–	–	–
6111-T4	290	150	26	69	0.28	0.70	Roped
-T62	360	315	11	69	–	–	–

- Improved manufacturing (modeling, sliver control, hemming, and aluminum handling)
- New/improved forming processes (stamping, tailored blanks, hydroforming, electromagnetic forming, warm forming, and superplastic forming)

In 1999, of the 293 024 tonnes of aluminum mill products produced in North America, 21% or some 62 000 tonnes represented automotive shipments used for heat exchangers, heat shields, closures,structural sheet for complete or partial body assemblies, and body panels. Of the 62 000 tonnes, about 46 000 tonnes represented aluminum body panels and structural sheets. Aluminum alloy body panel and body structure sheet consumption for automobiles and light trucks in North America, however, was expected to grow to about 204 000 tonnes by 2009.[26]

Novelis Aluminum Company, spun off from Alcan in 2005 as an independent company and acquired in 2007 by Hindalco Industries Ltd, is the world's largest producer of aluminum automotive body sheet, with their sheet used in over 70 models of passenger automobiles and light trucks, representing some 4.5 million vehicles.[27] In the last three years, Novelis has launched a major breakthrough in the production of clad aluminum alloy sheet products utilizing their patented Fusion™ process that simultaneously direct chill (DC) casts single or double alloy layers into a single aluminum rolling ingot (Figs. 3.6 and 3.7).[28] Automotive applications of clad aluminum alloy sheet produced by rolling Fusion ingot are numerous but are especially promising

3.6 Novelis Fusion ingot ready for rolling into clad aluminum alloy sheet. © Light Metal Age, reproduced with permission.

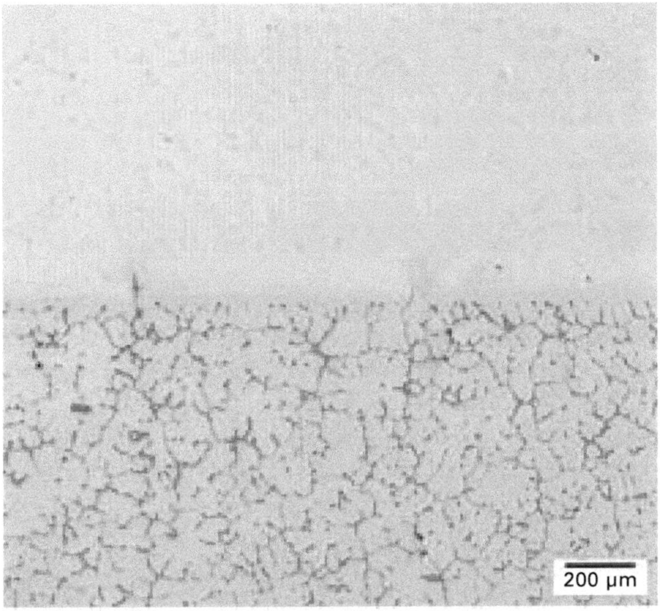

3.7 Clad/core interface microstructure of 1200/2124 aluminum alloy ingot cast by Novelis Fusion technology. © Light Metal Age, reproduced with permission.

in brazing sheet. Aluminum brazing sheet, commonly used in the automotive industry for making radiators and condensers, is made of an Al–Mn alloy core with a cladding of an Al–Si braze alloy, typically AA4147 (nominally 12% Si, 0.19% Mg, balance Al and impurities). Other Fusion applications involving several dozen clad/core aluminum alloy options include automotive sheet with both strength and improved formability to enable new design options and sheet with superior anodizing characteristics.

Europe, the world's largest automotive vehicle producer, is also the worldwide leader in the use of aluminum for new, innovative applications in the automobile, particularly for sheet applications. Among the many applications of innovative use of aluminum alloy sheet in European automobiles is the Jaguar XJ2200 sport sedan (Fig. 3.8), which represents the seventh generation in the Jaguar model program. The XJ is the first in the Jaguar series of automobiles with a monocoque body built out of aluminum alloy stampings, extrusions, and castings. These various aluminum alloy components are connected by adhesive bonding and about 3200 self-piercing rivets that give the body extreme stiffness and robustness. The epoxy adhesive used hardens during the paint baking cycle. The XJ body is 60% stiffer, 40% lighter, and roomier than its predecessor.[29]

The launching of Ford's Jaguar XJ, with its aluminum alloy monocoque

3.8 Schematic of the aluminum intensive Jaguar XJ2200 sport sedan with a monocoque body-in-white containing adhesively bonded and riveted stamped aluminum alloy sheet components. © Light Metal Age, reproduced with permission.

body, was preceded by the experience gained in its P2000 Prodigy body structure development, which was Ford's contribution to the Automotive Lightweighting Materials (ALM) R&D initiative established by the US Department of Energy (DOE) in 1992 during the Clinton administration and brought under the umbrella of the PNGV (Partnership for a New Generation of Vehicles) program. The P2000 Prodigy contained 333 kg (733 lb) of aluminum alloy or 37% of the vehicle weight; by comparison, the production model 1997 Taurus GL contained 129 kg (284 lb) of aluminum alloy or 8.6% of vehicle weight. In comparison to a steel body-in-white, the aluminum alloy body-in-white was 47.4% lighter.[30]

The ALM/PNGV initiative of the DOE was incorporated into the US FreedomCAR program in 2002 by the Bush administration in recognition of the fact that the passenger vehicle mix in the US had changed significantly since 1994. In the most recent peer review of DOE ALM projects that are part of the FreedomCAR program, an economic analysis of the Ford P2000 vehicle concluded that the total manufacturing cost increase was just over US$100 or about US$145 at the retail level based on the price of primary aluminum at the time, while fuel savings over the P2000 lifetime would net a total ownership savings of about US$1200.[31,32]

With aluminum intensive spaceframe construction, exemplified by the Audi A8 and A2 series, again stamped aluminum alloy sheet components are joined to aluminum alloy extrusions and castings. Audi began employing aluminum alloy sheet in the early 1980s for door units in the Audi 80 and Audi 100 series automobiles, and again later in 1997 for door units and front hoods for the Audi A6. Their first generation Audi spaceframe or ASF®, used in the Audi A8 that was introduced in 1994, was followed by the second generation ASF in the Audi A2, both having a body-in-white constructed of aluminum alloy sheet stampings joined to aluminum alloy castings and extrusions. Obviously, lessons learned on the Audi A8 resulted in a significant improvement in the body-in-white of the A2 (Table 3.7).[33]

Table 3.7 Comparison of Audi spaceframe production of the first and second generation ASF[33]. © Light Metal Age, reproduced with permission

1st generation ASF® (Audi A8)	2nd generation ASF® (Audi A2)
• 80 cars/day • Low automation BiW (20%) • Overlap MIG joints • Cast nodes requiring tolerance compensation • Self-piercing rivets • Complex machined extrusions • Small vacuum high pressure die castings • High share of bent extrusions • 334 parts joined by 1100 rivets, 70 m MIG seams, 500 spot welds and 178 clinches • Total weight = 249 kg (549 lb)	• 300 cars/day • High automation BiW (80%) • T-joint MIG fillet welds • High precision parts requiring no tolerance compensation • Self piercing rivets and laser welding • Hydroformed/pierced extrusions • Large vacuum high pressure die castings • Only six bent extrusion • 238 parts joined by 1800 rivets, 20 m MIG seams and 30 m laser welds • Total weight = 156 kg (344 lb)

In North America, the most visible aluminum alloy spaceframe body-in-white was that of the limited production (<5000 units) 2004 Ford GT sports car, which also was a hybrid of castings, extrusions, and panels.[34] Except for the limited production of the Ford GT, the aluminum spaceframe body-in-white was never seriously exploited in automotive production in North America, although much effort was expended in studying its features. For example, the Precept demonstration automobile developed by GM in the late 1990s had a spaceframe consisting of aluminum alloy stampings, extrusions, and castings joined together by a variety of methods.[35] It weighed 152 kg (335 lb), which was a 45% reduction in weight relative to a comparable steel spaceframe, and contained 64 kg (44%) 6111-T4 and 5754-O aluminum sheet stampings, 49 kg (34%) 6061-T6 and 6063-T6 extrusions, and 32 kg (22%) A356-T6 castings. Precept body structure metrics included 192 parts, 600 self-piercing and blind rivets, 1600 resistance spot welds, and some MIG welds at selected locations. Like other well designed aluminum spaceframes, the Precept body-in-white had exceptional static torsion and bending stiffness.

In Japan, among the first major applications of aluminum alloy sheet to body panels was the 1985 Mazda RX7, followed by the aluminum intensive 1990 Honda Acura NSX (Fig. 3.9). However, aluminum alloy panels did not immediately become popular in Japanese automobiles due to their high cost. Over time, aluminum alloy panel applications have increased in Japan at each automotive OEM, primarily to help meet rising vehicle weight reduction needs.[36] Among the innovations used in joining aluminum alloy body panels and structural components in Japanese automobiles is friction stir welding (FSW) and friction stir spot welding (FSSW), the latter shown in Fig. 3.10.[37]

3.9 The aluminum intensive 1990 Honda Acura NSX body-in-white. © Light Metal Age, reproduced with permission.

3.10 Mazda 2003 RX-8 aluminum rear door assembled by fixed pin FSSW. © Light Metal Age, reproduced with permission.

As mentioned above, much of the joining of aluminum alloy sheet in automotive vehicles involves the use of adhesives, riveting, and MIG/TIG and laser welding. This is in sharp contrast to the robotic resistance spot welding commonly used almost exclusively in joining steel sheet in automotive component assembly. Although much work has been done by aluminum companies in promoting the resistance spot welding of aluminum alloy sheet, this process is not as commonly used in joining aluminum alloy panels in the production of automotive vehicles; however, it has been used in the aluminum alloy lift gates for all the large GM SUVs, welding through three thicknesses with multiple alloys, i.e. 5182 and 6111. Weld bonding, combining resistance spot welding with adhesive bonding to provide continuous joints that provide enhanced fatigue, torsional stiffness, and crash impact energy absorption, has been widely used in the GM Electric Vehicle and in many of the Ford aluminum intensive vehicles.[38]

However, both the FSW and FSSW processes involve solid state joining of similar or dissimilar aluminum alloys and are uniquely suited to aluminum alloy sheet and plate. These solid state processes for joining aluminum alloys have attracted wide attention in the automotive industry in recent years due to the high integrity of the FSW and FSSW joints.[39]

Substantial progress has been made in forming aluminum alloy sheet into automotive components. For example, the BMW 7 series rear axle subframe contains four seam welded aluminum alloy 5454-O tubes that are hydroformed (Fig. 3.11). These tubes are made from 2 and 3.5 mm thick 5454-O sheet, seam welded into tubes of 140 mm diameter that are hydroformed and joined by MIG welding to extruded or cast sections that comprise the subframe, resulting in a 35% weight saving over the steel alternative.[40]

Other major aluminum alloy sheet developments that have been actively pursued by global automotive manufacturers are tailored aluminum alloy blanks that incorporate not only fusion welded blanks but also new concepts

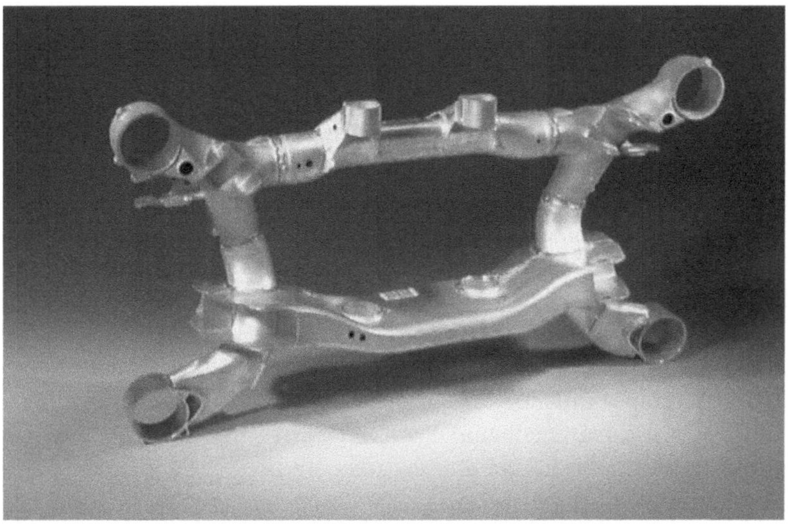

3.11 BMW 7 series rear axle subframe incorporating hydroformed seam welded aluminum alloy 5454-O tubes. © Light Metal Age, reproduced with permission.

such as FSW of dissimilar aluminum alloys and sheet thicknesses, tailor rolled blanks, and adhesively bonded and formable patch blanks.[41]

Superplastic forming (SPF) of integral automotive body panels made of 5083-H18 or-H19 aluminum alloy sheet can produce deep contoured shapes from a single sheet with no spring back and excellent repeatability with only about 10% of matched die stamping tool costs.[42] Superplastic forming of aluminum alloys is done at elevated temperatures at low strain rates, and for many years was used in aerospace applications. It has, within the past decade, been used for manufacturing body panels on such niche vehicles as the Panoz Roadster and Esperante, Morgan Aero 8, Bentley Arnage, Ford Racing Puma, Aston Martin Vanquish and others, as well as to fabricate the inner and outer trunk lids for the 2002 GM Aurora. Although a typical SPF cycle takes about 20–30 min, the automotive industry has been actively reducing cycle times to a few minutes in what is now called quick plastic forming (QPF).[43]

Although the industry has progressed on many fronts in advancing the state of the art in forming and joining aluminum alloy sheet, one of the roadblocks to expanded usage of such sheet in penetrating body panels and body-in-white applications is cost relative to competing materials, such as steel and polymer matrix composites.[44] Twin roll casting not only has the immediate advantage of lower cost relative to conventional rolling of DC cast aluminum alloy ingot, but it also has the long-term benefit of being capable of utilizing greater amounts of recycled automotive aluminum scrap due to the fact that the rapid casting rate promotes a finer metallurgical structure.[45]

3.4.3 Aluminum alloy extrusions in lightweight automotive vehicles

Of all the aluminum alloy classes, extruded forms of 6xxx, (Al–Mg–Si) and 7xxx (Al–Zn) series alloys rule the day in present vehicle applications for engine cradles, platform frame rails and cross members, space frames, seat frames, radiator frames, bumper beams, and rod and bar used in machined ABS valve bodies, sleeves and connectors, and impact extrusions. Aluminum alloy impact extrusions use slugs either sliced from extruded bars or alternatively punched from continuously cast and hot rolled plate. The 6xxx and 7xxx alloys are heat treatable by solution annealing and either natural or artificial aging (T temper in Table 3.3) or in rare cases by full annealing (O temper in Table 3.3). Typical mechanical properties of some 6xxx and 7xxx extrusion alloys used in automotive applications are presented in Table 3.8.

Aluminum alloy extrusion use in automotive vehicles, bus, and heavy truck industries amounts to some one million tonnes worldwide. The automotive industry utilizes aluminum alloy extrusions in applications such as drive shafts, transmission parts, air intake manifolds, sun roofs, air bag modules,

Table 3.8 Typical mechanical properties of aluminum extrusion alloys.[47,48] © Light Metal Age, reproduced with permission

Alloy and temper	Ultimate tensile strength (MPa)	0.2% Yield strength (MPa)	Elongation in 50 mm or 2 in. (%)	Ultimate shear strength (MPa)	Modulus of elasticity tension/ compression (GPa)
6005-T5	305	270	12	200	69
6061-T1	150	90	20	100	69
-T4	240	145	22	165	69
-T6	310	275	12	205	69
6063-T4	170	90	22	105	69
-T5	185	145	12	115	69
-T6	240	215	12	150	69
7004-T5	400	340	15	220	72
7005-T53	395	350	15	225	72
7116-T5	360	315	14	200	70
7029-T5	430	380	15	270	70
7129-T5	430	380	14	270	70

window frames and posts, bumpers and bumper reinforcement beams, seat frames and rails, steering components, door frames, motor mounts, heat exchanger tubing, antilock brake systems, luggage racks, trim, shock absorber housings and, to be sure, frame and body structures. (The truck trailer and railroad industries have utilized aluminum alloy extrusions in cost-effective structural applications for decades, primarily to take advantage of their high strength-to-weight ratio, corrosion resistance, attractive appearance, wide range of finishes, close tolerances, assured and uniform quality, design flexibility, and recyclability.)

Aluminum extrusions open up advantageous design possibilities for the automotive designer in the development of structural shapes that meet both functional and aesthetic requirements for vehicle (Fig. 3.12). Since extrusion dies are much less expensive than stamping dies and casting molds, the designer can try multiple solutions and prototypes for any given design problem at a small fraction of the cost of stampings or castings.

Although aluminum alloy extrusions can be used to manufacture fully functional components by joining together extrusions of different shapes, either cut to size or formed, they are easily joined to aluminum alloy castings and sheet stampings in structural assemblies exemplified by the aluminum intensive vehicles described previously in this chapter. The many applications of aluminum alloy extrusions in automotive vehicles offer an extensive background for designers to follow or expand upon. Properties of aluminum alloy extrusions, general design principles, and methods of forming and joining of aluminum alloy extrusions are readily available.[47–49]

The 6xxx or Al–Mg–Si alloys are most commonly extruded and have

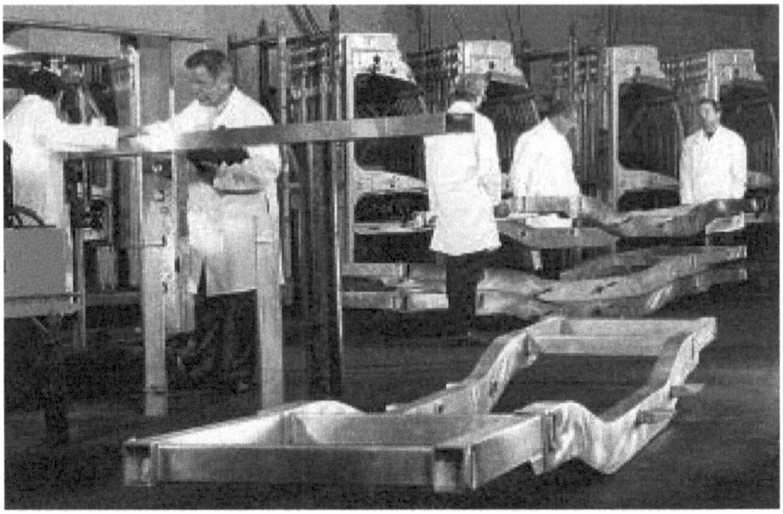

3.12 Advanced concept Ford Explorer prototype frames made in 1994 completely from extruded profiles.[46] © Light Metal Age, reproduced with permission.

excellent corrosion resistance; they are easily welded by MIG, TIG, laser, electron beam, or any fusion welding process as well as by the solid state friction stir welding processes. The most common 6xxx extrusion alloys used in automotive structural applications are 6005, 6005A, 6061, 6063, 6082, and 6351, heat treated by quenching at the press with forced air or water and subsequently oven aged for 4–8 h to achieve T5 or T6 tempers. Alloy 6063 has the lowest strength of this series of aluminum alloys, but it is the easiest to extrude and therefore is used for very complex and thin wall shapes. The 6061 alloy is the most popular structural alloy extruded in the US, although 6082 alloy (6351 alloy is a close US equivalent) is the most popular structural alloy extruded in Europe. Alloys 6005 and 6005A are intermediate in strength and applications to 6063 and 6061 or 6082 alloys.

The 7xxx or Al–Zn–Mg alloys 7004, 7005, 7029, 7116, and 7129 have higher strengths relative to the 6xxx alloys but are not as easily extruded, particularly into complex hollow shapes, are less resistant to corrosion, and only 7004 and 7005 are considered as weldable. Both 6xxx and 7xxx alloys are most formable in the annealed (O temper) condition or immediately after extrusion (F temper) or solution heat treating (W temper), and most forming operations are done under these conditions at extrusion plants prior to natural aging (T4) or artificial aging to high strength tempers (T5, T53, or T6). However, forming of age hardenable aluminum alloy extrusions by bending, flanging, hydroforming, swaging, etc. can be challenging if the formed shape requires substantial plastic deformation and the material is not in the F or W temper or rarely an O temper, that is, in an age hardened or possibly an age hardened and cold worked temper.

A special heat treatment (Retrogression Heat Treatment or RHT) has been developed to facilitate forming of age hardened aluminum alloy extrusions and has been widely used by automotive OEMs and first and second tier automotive component manufacturers in difficult forming operations.[50–55] By returning (retrogressing) the age hardened alloy shape through localized induction heating to its freshly quenched and most ductile condition, RHT allows age hardened material, even at maximum strength, to form extremely well, even exceeding that achieved by full annealing (O temper).

The RHT process has been used to make the advanced concept prototype Ford Explorer aluminum extruded frames shown in Fig. 3.12.[50] In this case, the RHT process was used on 6061-T6 hollow extrusions to bend and join cross members to rails, as it was used for the Panoz Roadster frame (Fig. 3.13).[51] It was also used successfully in the flattening and bending of 6005-T6 thin wall hollow extrusions for seat backs for an aluminum intensive PNGV automobile (Fig. 3.14) and for cold forming splines in the successfully mass produced 6061-T6 Visteon propshaft (Fig. 3.15).[51]

As demonstrated in Figs. 3.12 and 3.13, the localized RHT process makes it possible to fabricate and assemble aluminum extrusions into 3-D frame structures. Assembly can be done by any number of joining methods but, in particular, RHT can be applied to the ends of hollow extrusions to form joints by compression, essentially achieving a compression fit (CF) joint. The

3.13 Panoz Roadster frame fabricated and assembled by RHT process from 6061-T6 extruded hollow shapes. © Light Metal Age, reproduced with permission.

3.14 PNGV seat backs formed from 6005-T6 thin wall hollow extrusions by RHT process. © Aluminum Extruders Council reproduced with permission.

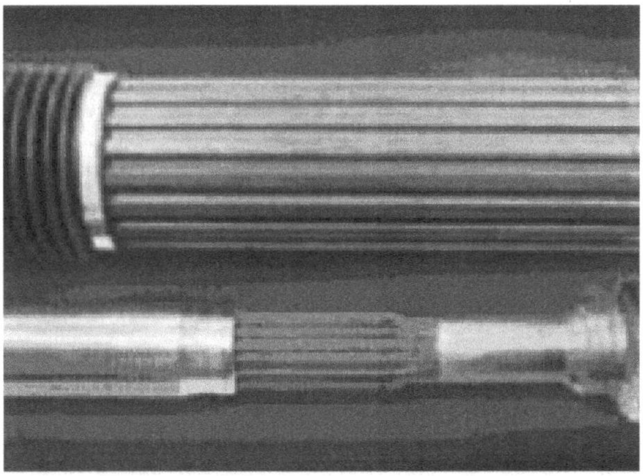

3.15 Visteon propshaft made of seamless extruded 6061-T6 tube with splines cold formed by RHT process. © Aluminum Extruders Council reproduced with permission.

CF joints on the Panoz Roadster were tested in a 100 000 mile simulation trial without loosening or loss in torsional stiffness of the frame. Also, the localized RHT process was successfully used in forming crumple zones into 6061-T6 hollow extrusions (Fig. 3.16) that were introduced fore and aft of the engine in the Panoz Roadster, thus obviating the need to cut and add special crumple inducing components into the side rails.[56]

3.16 Compression crumple test conducted on one particular RHT formed crumple zone configuration in a hollow 6061-T6 automotive frame extrusion. © Thermal Processing Technology Center, Illinois Institute of Technology, reproduced with permission.

Another important technology used by the aluminum extrusion industry for aluminum alloy bumper reinforcements in particular, and other curved profiles, is direct rounding of profiles during extrusion. Direct rounding of extrusions is done at the press, thus saving a post-extrusion bending operation, by either modifying extrusion dies to gently curve extrusions *in-situ*[57] or by using a guide tool located at the mouth of the press to apply a lateral force to the profile as it exits the die[58,59] to achieve rounded shapes (Fig. 3.17). The latter process, called curved profile extrusion (CPE), was used successfully on various rounded extrusions for the Audi A2.

By combining an in-line solutionizing treatment with cold forming and subsequent aging (called ExtruForm® Raufoss Technology AS, Norway), thermomechanically processed 6082 aluminum alloy extruded profiles have been shown to exceed mechanical properties of 6082-T6 forgings.[60] Such ExtruForm components have been used as lower control arms in automotive vehicles.

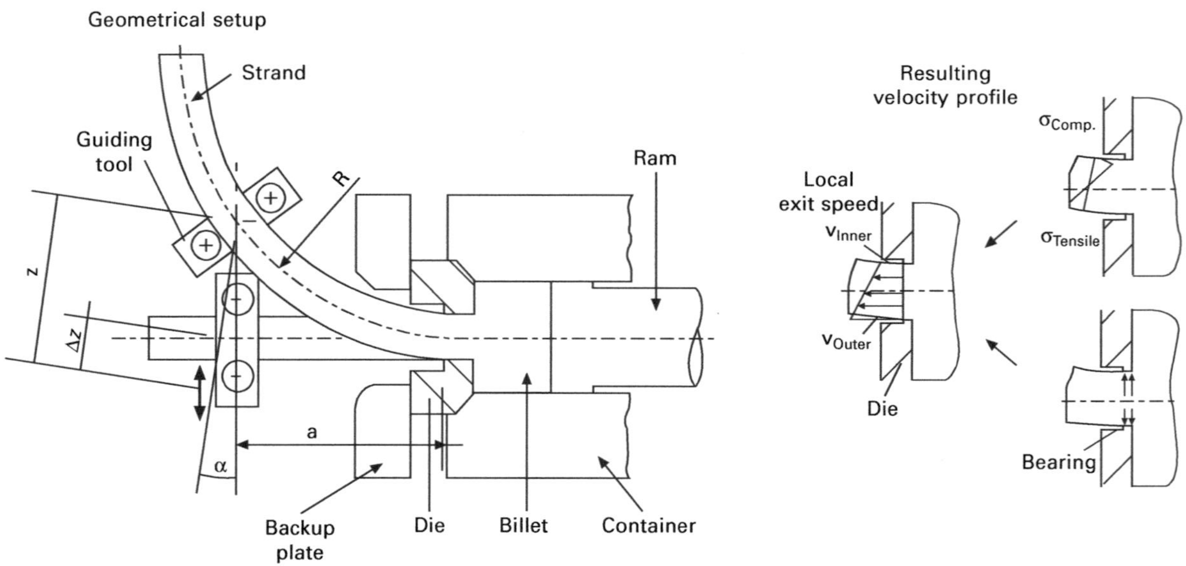

3.17 Basic principle of the CPE process in development at the University of Dortmund, Germany. © Aluminum Extruders Council, reproduced with permission.

3.5 Substituting aluminum alloys for competitive materials

From an automotive engineering perspective, substituting aluminum for cast iron and steel is not necessarily straightforward in terms of design, processing, and assembly. Direct substitution does not work, since the properties of aluminum alloys and their manufacturing and joining processes differ. Factors to be considered by engineers and purchasing managers in selecting aluminum over cast iron and steel are design flexibility, cost, lightweighting potential, performance, manufacturability, maintenance and repair requirements, crashworthiness, aesthetics, and recycling potential. Systematic techniques for evaluating competitive materials on the basis of comparative properties and costs utilize ratio analysis to good effect.[61]

To many automotive engineers and purchasing agents, aluminum alloys and their availability in so many types of alloys and forms appear to be somewhat challenging relative to the decades-long familiarity with conventional deep drawing quality sheet steel and its processing into stamped and spot welded components. At the same time, fiber reinforced composite materials are competing with aluminum alloys as a lightweighting option for automotive vehicles. Familiarity with aluminum alloy types, heat treatments, casting and metalworking processes, and special considerations with regard to joining and assembly, are necessary before a targeted substitution can be considered, and many of the references provided here can serve in this regard. Information is readily available from aluminum trade associations as well as from aluminum companies themselves. The following sections give some general considerations that automotive engineers must heed in substituting aluminum alloys for steels and polymer based composite materials.

3.5.1 General comparison of aluminum alloys with steels in automotive vehicles

In lightweighting automotive vehicles, the most common substitution for steel is made with aluminum alloys. Aluminum alloys are available to the transportation industry as flat sheet and plate or in a variety of shapes (lineal extrusions, roll formed sheet, castings, and forgings), while steel sheet and plate are available mainly as mill products (flat rolled sheet and plate, shape rolled, or roll formed sheet). When substituting aluminum and aluminum alloys for steel, the following comparative properties need to be considered:

- Density of aluminum and aluminum alloys is one-third that of steel
- Elastic modulus of aluminum and aluminum alloys is one-third that of steel
- Hardness of aluminum alloys is lower (however, hardness of the anodic coating on aluminum alloys is higher than the oxide coating on steel)

- Specific fatigue strength of aluminum alloys is about one-half that of steel
- Coefficient of thermal expansion of aluminum alloys is about 1.5 times greater than that of steel for an equivalent change in temperature
- Ductility, as measured by % elongation of aluminum alloys in the annealed condition, is about two-thirds less than that of annealed low carbon steel
- Formability of aluminum alloy sheet is lower than that of annealed low carbon sheet steel
- Aluminum alloys can be used down to cryogenic temperatures without loss of ductility, while carbon steels suffer from embrittlement at low temperatures
- Steel is strain rate sensitive while aluminum alloys are not and aluminum alloy structures have been shown to absorb more energy than steel structures upon impact
- Unlike steel, aluminum and aluminum alloys are non-magnetic
- Unlike steel, aluminum and aluminum alloys are non-sparking
- Thermal and electrical conductivities of aluminum and aluminum alloys are about four times that of steel
- Damping characteristics of aluminum alloys and steel are similar
- Atmospheric corrosion resistance of aluminum and aluminum alloys is much higher than that of steel
- Aluminum alloys can be used unfinished in many applications and can accept a wide range of mechanical and chemical finishes, while steels require paint or electroplated finishes to ward off atmospheric corrosion
- Galvanic corrosion resistance of aluminum and aluminum alloys is lower than that of steel
- Recycling value of aluminum and aluminum alloys is higher than that of steel

In reviewing the above properties of aluminum and aluminum alloys in comparison to steel, it becomes apparent that, when taking into consideration the differences in many of the properties such as thermal expansion and galvanic corrosion, vehicle designers should be careful in joining these particular materials.

3.5.2 General comparison of aluminum alloys with fiberglass reinforced plastics in automotive vehicles

Fiberglass reinforced plastics (FRPs), made primarily with a polyester or vinyl ester resin matrix and chopped E-glass reinforcing fibers, have provided vehicle manufacturers with another lightweighting material option, especially for automotive body panels and SUV flat beds. The introduction of FRPs has required automotive vehicle manufacturers, who have been fabricating and assembling their vehicles with aluminum alloys, to completely change their manufacturing and assembly operations. The physical and mechanical properties as well as costs of chopped fiber FRPs vary significantly depending on volume percentage and distribution of the glass fibers and the FRP manufacturing process. Flat sheet molding compound (SMC) sheet competes directly with aluminum alloy sheet, pultruded FRPs compete with aluminum alloy extrusions, and shapes produced from bulk molding compound (BMC) compete with aluminum alloy castings.

As with steel, the comparative properties of aluminum alloys and chopped fiber FRPs, such as SMC, must be considered by automotive designers. Some of the more critical comparative properties of aluminum alloys and FRPs are as follows:

- The density of FRPs varies but can be comparable or even lower than that of aluminum alloys depending on fiberglass loading
- The strength of FRPs varies but can be comparable or even higher than that of aluminum alloys depending on fiberglass orientation and loading
- The strength of aluminum alloys increases at low temperatures while FRPs decrease in strength at very low temperature and become brittle
- Aluminum alloys retain strength better than FRPs at high temperatures
- The strength-to-weight ratio of aluminum alloys and FRPs used in vehicle structures is rather comparable
- The modulus of elasticity of aluminum alloys is about 80–100% higher than that of FRPs
- The ductility of aluminum far exceeds that of FRPs
- Formability of aluminum alloys is high, while FRPs cannot be formed but must be molded into shapes
- Machinability of aluminum alloys is excellent compared with FRPs
- Weldability of aluminum alloys by any of the conventional welding methods is excellent while FRPs cannot be welded
- Joining of FRPs is limited to adhesive bonding and mechanical fasteners while all mechanical methods of joining can be used on aluminum alloys including interlocking extruded shapes and hemmed sheet
- Coefficient of thermal expansion of FRPs is as much as three times higher than that of aluminum alloys
- Under atmospheric exposure the corrosion resistance of aluminum alloys is excellent and can be further improved along with appearance by anodizing or other coatings, while FRPs do not corrode like metals but possess poor weatherability

- Aluminum alloys and FRPs are non-sparking and non-magnetic
- Aluminum alloys are non-combustible, while FRPs are combustible and emit toxic fumes during combustion
- Aluminum alloys are highly recyclable and have high scrap value, while FRPs are not and entail disposal and landfilling costs

3.6 References

1. *The Aluminum Association Auto & Light Truck Group (ALTG)*: www.autoaluminum.org.
2. Stempel, B. (Novelis), representing ALTG, 'The Road Ahead Looks Brighter and Lighter,' presentation made at the *American Metal Market's North American Aluminum Conference*, June 26–27, 2007, Chicago, IL.
3. Schultz, R. (Ducker Worldwide North America), *Ducker Worldwide Materials Growth Report*, 2008, richards@ducker.com.
4. Scheps, R. (Alcoa, representing ALTG), 'The Battle for Market Share,' presentation made at the *American Metal Market's 3rd Annual Automotive Metals Conference*, Oct. 28–30, 2008, Detroit, MI.
5. Bull, M. (Novelis, representing ALTG), 'Going Green: The Aluminum Perspective,' presentation made at the *American Metal Market's 3rd Annual Automotive Metals Conference*, Oct. 28–30, 2008, Detroit, MI.
6. Martchek, K. (Alcoa), 'Keys to the Climate in the Aluminum Supply Chain,' presentation made at the *American Metal Market's 3rd Annual Automotive Metals Conference*, Oct. 28–30, 2008, Detroit, MI.
7. Alu*Drive,* Summer 2008, European Aluminium Association (www.aluminium.org).
8. *Aluminum Standards and Data 2008 (standard and metric)*, The Aluminum Association, 2008.
9. Hesse, W. (Ed.), *Aluminium Schlüssel/Key to Aluminium Alloys*, Aluminium-Verlag Marketing & Kommunikation GmbH, Düsseldorf, Germany, 8th edition, 2008.
10. Hesse, W. (ed.), *Aluminium Material Data Sheets*, Aluminium-Verlag Marketing & Kommunikation GmbH, Düsseldorf, Germany, 5th edition, 2007.
11. Davis, J.R. (ed.), *Aluminum and Aluminum Alloys*, ASM Specialty Handbook, ASM International, 1993.
12. Sequeira, W., Pikhovich, V., and Weiss, D., 'Finding New Strength in Aluminum,' *Modern Casting*, Feb., 2006, 36–41.
13. Vinarcik, E.J., *High Integrity Die Casting Processes*, John Wiley & Sons, Inc., 2003.
14. Benedyk, J.C., 'Squeeze Casting,' Paper No. 86, *6th SDCE International Die Casting Congress*, Nov. 16–19, 1970, 10 pp.
15. Benedyk, J.C., 'Squeeze Casting: Combining Forging Properties in a Large Casting,' ASME Publication 72-DE-7, presented at the *Design Engineering Conference & Show*, Chicago, IL, May 8–11, 1972.
16. Benedyk, J.C., 'Manufacturing Possibilities with Squeeze Casting,' *Creative Manufacturing Engineering Program*, Paper No. CM71-840, Soc. of Manufacturing Engrs., 1971.
17. Benedyk, J.C., 'Automotive Aluminum Casting Trends and Developments,' *Light Metal* Age, Vol. 58, October, 2000, 36–41.

18. Benedyk, J.C., '21st Auto Aluminum Design & Fabrication Seminar,' *Light Metal Age*, Vol. 59, February, 2001, 72–80.
19. *Secured Lender Order of Sale of AEMP Aluminum Casting Operation*, Hilco Industrial, LLC., February, 2003.
20. Kopper, A., Donahue, R., and Midson, S., *Case Studies of Large Components Produced by High-pressure Die Casting and Slurry-on-Demand Casting*, 6th SDCE International Die Casting Congress SAE Technical Paper 2005–01-1691.
21. von Zengen, K.-H., 'Aluminum in the European Auto Industry,' presentation made at the *22nd Annual Automotive Aluminum Design and Fabrications Seminar*, October, 2001.
22. *Aluminum for Automotive Body Sheet Panels*, Publication AT3, December 1998, The Aluminum Association.
23. *Automotive Aluminum Crash Energy Management Manual*, Publication AT5, December 1998, The Aluminum Association.
24. *Repair of Aluminum Automotive Sheet by Welding*, Publication AT1, March 1994, The Aluminum Association.
25. *Aluminum Industry Roadmap for the Automotive Market: Enabling Technologies and Challenges for Body Structures and Closures*, May, 1999, The Aluminum Association.
26. Schultz, R.A., 'Aluminum for Light Vehicle Bodies in North America: an Objective Look at the Next Ten Years,' Presentation at the *Management Briefing Seminars World Class Manufacturing Session*, Traverse City, MI, Aug. 8, 2000.
27. Stempel, B. (Novelis), representing ALTG, 'Aluminum Flat Rolled Products,' presentation made at the *23rd Metal Bulletin International Aluminum Conference*, September, 2008, Montreal, Canada.
28. Benedyk, J.C., 'Novelis Fusion™ Process Breakthrough in the Simultaneous DC Casting of Multiple Al Alloy Layers for Rolling Ingot,' *Light Metal Age*, Vol. 64, August, 2006, 48–50.
29. White, M., 'The New Jaguar XJ Sedan,' presentation made at the *22nd Annual Automotive Aluminum Design and Fabrication Seminar*, October, 2001, Livonia, Michigan.
30. Young, C.S., 'The P2000 Body Structure,' presentation made at the *21st Annual Automotive Aluminum Design and Fabrication Seminar*, October, 2000, Livonia, Michigan.
31. *Aluminum Vehicle Structure – Manufacturing and Lifecycle Cost Analysis*, IBIS Associates, Waltham, MA, October, 2005.
32. *DOE Automotive Lightweighting Materials: Peer Review Panel 2007 – Final Report*, August, 2007, 43–46.
33. Ruch, W., 'Development of the Audi A2 Body-in-White for Volume Production,' presentation made at the *21st Annual Automotive Aluminum Design and Fabrication Seminar*, October, 2000, Livonia, Michigan.
34. Ramsden, D., 'Spaceframe Propels Ford GT,' *Modern Castings*, February, 2006, 26–29.
35. Lobkovich, T.M., 'The Application of Aluminum to the Precept Demonstration Vehicle,' presentation made at the *21st Annual Automotive Aluminum Design and Fabrication Seminar*, October, 2000, Livonia, Michigan.
36. Inaba, T., 'Automobile Aluminum Sheet,' *Automotive Engineering; Lightweight, Functional, and Novel Materials*, Series in Materials Science and Engineering, Eds. B. Cantor, P. Grant, and C. Johnston, Taylor & Francis, 2008, 19–27.

37. *Mazda News Release*, February 27, 2003.
38. Clark, J.A., 'Welding Automotive Aluminum,' presentation made at the *22nd Annual Automotive Aluminum Design and Fabrication Seminar*, October, 2001, Livonia, Michigan.
39. Benedyk, J.C., 'SAE Celebrates 100th Anniversary at SAE 2005: Recent Advancements in Automotive Light Metals Part IV: Friction Stir Welding Developments,' *Light Metal Age*, Vol. 64, April, 2006, 34–43.
40. Rempe, W. and Benedyk, J.C., 'Hydroforming Aluminum Tube,' presentation made at the *22nd Annual Automotive Aluminum Design and Fabrication Seminar*, October, 2001, Livonia, Michigan.
41. Benedyk, J.C., 'Tailored Aluminum Blanks for Automotive Applications: Continuous Improvement and New Concepts,' *Light Metal Age*, Vol. 58, December, 2000, 73–78.
42. *Superform Metals: Aluminum Superplastic Forming*, brochure of Superform Metals (www.superform-aluminium.co.uk).
43. Benedyk, J.C., 'Superplastic Forming of Automotive Parts from Aluminum Sheet at Reduced Cycle Times,' *Light Metal Age*, Vol. 60, June, 2002, 28–31.
44. Sherman, A., 'Perspectives on Automotive Applications of Aluminum,' presentation made at the *22nd Annual Automotive Aluminum Design and Fabrication Seminar*, October, 2001, Livonia, Michigan.
45. Romano, E. and Romanowski, C., 'Twin Roll Casting, Possibly the Most Economic Solution to Manufacturing Aluminum Automotive Sheet,' *Light Metal Age*, Vol. 67, February, 2009, 32–42.
46. Haddad, C., *21st Century Car Project*, Ford Motor Company, 1994.
47. *The Aluminum Extrusion Manual*, The Aluminum Association & AEC (Aluminum Extruders Council).
48. *Aluminum Automotive Extrusion Manual*, Publication AT6, December 1998, The Aluminum Association.
49. Barbareschi, G., 'Aluminum Extrusion Fabrication Solutions for the Automotive and Transportation Markets,' *Proc. 9th International Aluminum Extrusion Technology Seminar and Exposition*, 2008, Aluminum Extruders Council.
50. Benedyk, J.C., 'Retrogression Heat Treatment Applied to Aluminum Extrusions for Difficult Forming Applications: Part I, Process Description and Advantages,' *Light Metal Age*, Vol. 54, October, 1996, 8–11.
51. Benedyk, J.C., 'Retrogression Heat Treatment as a Means of Improving Formability of Aluminum Extrusions,' *Proc. 9th International Aluminum Extrusion Technology Seminar and Exposition*, 2008, Aluminum Extruders Council.
52. Benedyk, J.C., US Patent 4 766 664.
53. Benedyk, J.C, US Patent 5 458 393.
54. Benedyk, J.C., US Patent 5 720 511.
55. Benedyk, J.C., US Patent 5 911 844.
56. Jiménez, M., Takahashi, K., and Thach, T., '*Extruded Aluminum for Automotive Crash Absorbing Structures*,' MMAE 402 Final Report, Illinois Institute of Technology, May 14, 1997.
57. Späth, W.E., 'Designing with Curved Extrusions,' *Proc. 5th International Aluminum Extrusion Technology Seminar and Exposition*, 1992, Aluminum Extruders Council.
58. Klaus, A. *et al.*, 'Direct Rounding of Profiles during Extrusion,' *Proc. 7th International Aluminum Extrusion Technology Seminar and Exposition*, 2000, Aluminum Extruders Council.

59. Jensrud, O. and Høiland, T., 'A New Advanced Profile-shaping Technique for Design and Manufacturing of Automotive Components,' *Proc. 9th International Aluminum Extrusion Technology Seminar and Exposition*, 2008, Aluminum Extruders Council.
60 Becker, D., Schikorra, M., and Tekkaya, A.E., 'Flexible Extrusion of 3-D Curved Profiles for Structural Components,' *Proc. 9th International Aluminum Extrusion Technology Seminar and Exposition*, 2008, Aluminum Extruders Council.
61. F. Ashby, *Materials Selection in Mechanical Design*, Elsevier (2005).

4

Magnesium alloys for lightweight powertrains and automotive structures

B. R. POWELL, P. E. KRAJEWSKI, and A. A. LUO,
General Motors, USA

Abstract: This chapter introduces magnesium, the lightest of the structural automotive metals. It provides an overview of alloy nomenclature and properties, and the major casting, sheet forming, and extrusion processes. Descriptions of automotive magnesium applications produced by each process are provided and there is a summary that describes the challenges in alloy and process development that need to be overcome if the magnesium content in automotive sub-system applications is to be increased.

Key words: magnesium, casting, sheet forming, extrusion, hydroforming, automotive.

4.1 Introduction

This chapter reviews material characteristics, specific alloys, and applications for magnesium (Mg), the lightest structural metal. The elemental density of magnesium is 1.74 g/cc, which is one-third that of aluminum (Al) and less than one-fourth that of iron and steel, the other major structural metals in automotive use today. Magnesium is less dense than most glass fiber-reinforced automotive polymers and similar in density to that of carbon fiber composites, although magnesium alloys can cost considerably less.

4.1.1 Magnesium extraction and consumption

Magnesium is a greyish-white metal that makes up 2.7% of the earth's crust (Okamoto, 1988, pp 1–3). Due to its high chemical activity, it is never found as a pure metal in nature. Instead, it occurs commonly as magnesite ($MgCO_3$), dolomite ($MgCO_3.CaCO_3$), brucite ($Mg(OH)_2$), and as the silicates serpentine($(Mg,Fe)_3Si_2O_5(OH)_4$) and olivine ($(Mg, Fe)_2SiO_4$) (Pidgeon *et al.*, 1946, pp 4–22). It is also found in sea water, this being a major commercial source of magnesium. The first reported isolation of magnesium metal was accomplished in 1828 by Antoine-Alexander Bussy, who reduced magnesium chloride with potassium (Bussy, 1831). Today, there are four common commercial processes for extracting magnesium: the Dow Process, which extracts magnesium from sea water by electrolysis of magnesium chloride

(Emley, 1966, pp 41–44); the Magnola Process which also uses electrolysis of magnesium chloride, but obtains the chloride through a conversion of magnesium silicates in asbestos tailings (Habashi, 2006, p 37); and two thermal processes, the Pidgeon Process and the Magnetherm Process (Habashi, 2006, pp 34–36), that use magnesite and/or dolomite ore, which are reduced with ferro-silicon in a retort furnace. For further information about these processes, see Polmear (1999, pp 3–5).

Magnesium production capacity in 2008 was slightly greater than 700 000 metric tons; about two-thirds of it produced in China, nearly all of that using the Pidgeon Process. This tonnage represents a significant increase in production compared with ten years ago when only 350 000 tons were produced. The history of magnesium production is shown in Fig. 4.1. Magnesium has seen a clear upward trend in production over the past 70 years with most of the increase coming in the past ten years.

4.1.2 Magnesium alloys, properties, and processes overview

For all structural applications, magnesium is alloyed with other metals to provide the proper strength, corrosion resistance, formability, etc. A brief summary (Polmear, 1999, pp 14–15) of some of the typical alloying and impurity elements and their respective effects on magnesium is presented below (letters in parenthesis indicate alloy designation in commercial practice):

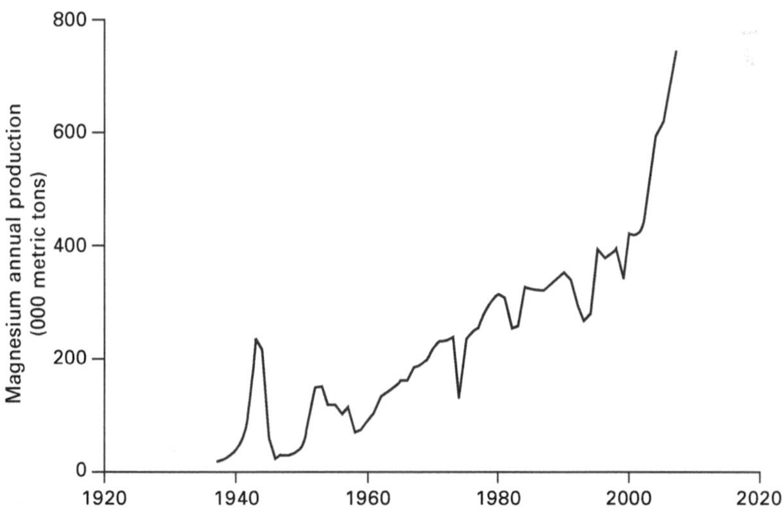

4.1 Magnesium annual production data 1938–2008 (adapted from DiFrancesco, 2008).

- Aluminum (**A**) is the most common alloying element. It improves strength, hardness and corrosion resistance, but reduces ductility. An aluminum content of about 5–6% yields the optimum combination of strength and ductility for structural applications. Increasing aluminum content widens the freezing range and makes the alloy easier to cast, but more difficult to extrude due to increased hardness.
- Zinc (**Z**) is used to increase strength. However, it reduces ductility, increases hot-shortness of Mg–Al based alloys, can lead to microporosity, and can accelerate corrosion.
- Manganese (**M**) slightly increases yield strength but has no effect on tensile strength. Its most important function is to improve the corrosion resistance of Mg–Al based alloys by removing iron and other heavy-metal elements into relatively harmless intermetallic compounds, some of which separate out during melting.
- Rare earth additions (**E**) provide precipitation strengthening and can improve both strength and ductility through alteration of crystallographic texture. Rare earth additions can also improve creep resistance.
- Silicon (**S**) increases fluidity and slightly improves creep strength.
- Zirconium (K) is an important grain refiner for sand and metal mold, gravity and low pressure casting of magnesium alloys.
- Tin (**T**) increases ductility.
- Lithium (**L**) reduces density and can improve ductility by creating a cubic structure at over 11% additions. Adding lithium significantly increases cost and reduces corrosion resistance.
- Thorium (**H**) and Yttrium (**W**) increase creep resistance. The alkaline earths calcium (**X**) and strontium (**J**) have also been shown to increase creep resistance.
- Iron (no designation) reduces the corrosion resistance of magnesium alloys and is kept below 50 ppm in all alloys used today (ASTM International, 2008). Similarly, copper (**C**) and nickel (no designation) reduce the corrosion resistance of Mg alloys. Copper has a letter designation because of its use in ZC63, an early sand casting alloy.

Magnesium alloys are identified using a combination of letters (above) and numbers which describe the major alloying elements and the percentages of those elements. ASTM Standard B 951-08 defines the protocol for use of letters and numbers for naming magnesium alloys (ASTM International, 2008). Per this standard, magnesium alloy families are represented by two letters representing the major alloying elements. (Note: some developmental alloys are represented by three letters or use other nomenclatures, but those included in the ASTM standard are limited to two letters). The letters are shown in Table 4.1. The letters of the alloy represent the two highest concentration alloying elements and are arranged in order of decreasing

Table 4.1 Letters representing alloying elements for magnesium (adapted from ASTM International, 2008)

A – aluminum	J – strontium	R – chromium
B – bismuth	K – zirconium	S – silicon
C – copper	L – lithium	T – tin
D – cadmium	M – manganese	V – gadolinium
E – rare earths	N – nickel	W – yttrium
F – iron	P – lead	X – calcium
H – thorium	Q – silver	y – antimony
	Z – zinc	

concentration. Following these letters are numbers which show the amount of each of the two alloying elements. The most well-known magnesium alloy is AZ91. This magnesium alloy contains nine percent (by weight) aluminum and one percent zinc. It also contains manganese, but this is not indicated by the name. Other examples include AZ31 which is three percent aluminum and one percent zinc, and ZK30 which is three percent zinc and less than one percent zirconium. Following the numbers, a letter may appear which indicates the order of the alloy within the chronology of that particular alloy development. Hence, AZ91B is a more recent version of AZ91A, and so on. These versions may vary in the range and amounts of secondary alloying elements, to satisfy cost targets or to provide some subset of properties or processing benefits. For example, the atmospheric corrosion behavior of magnesium alloys is determined to a great extent by their 'purity' (Hillis, 1983). Thus, the main difference between AZ91C and AZ91D is not in the addition of alloying elements, but instead it is the imposition of maximum values for the impurities copper, iron, and nickel, see Table 4.2.

AZ91E is also shown in the table. It is the current composition specification for sand cast AZ91, whereas AZ91D is the current composition specification for high pressure die cast AZ91. The lower allowable amount of copper in AZ91E is specified because at the lower rate of solidification during sand casting; the larger size of copper intermetallics increases the susceptibility of this alloy to micro-galvanic corrosion. Similarly, the higher range of manganese is to react with iron in the melt and remove it from the alloy. Actually the specification for this alloy states that, if either the maximum iron or minimum manganese specification is not met, then there is an alternative specification, setting an upper limit to the iron to manganese ratio.

Magnesium has a hexagonal close packed (HCP) crystal structure, which is different from most structural metals such as aluminum or iron, which are cubic. The hexagonal nature of the magnesium structure makes deformation at room temperature difficult because there are fewer slip systems for deformation compared with aluminum or iron. Other well-known elemental materials which have limited ductility due to their hexagonal structure are titanium and zirconium. Ice also has this property. The mechanical properties

Table 4.2 Chemical compositions of AZ91C and AZ91D alloys (adapted from ASTM International, 2008)

Element	AZ91C	AZ91D	AZ91E
Aluminum	8.3–9.2	8.5–9.5	8.3–9.2
Copper	0.08	0.025	0.015
Iron	not specified	0.004	0.005
Manganese	0.15–0.35	0.17–0.40	0.17–0.5
Nickel	0.010	0.001	0.0010
Silicon	0.20	0.08	0.20
Zinc	0.45–0.9	0.45–0.9	0.45–0.9
Other elements	0.30 total	0.01 each	0.01 each; 0.30 total

of a variety of magnesium alloys are shown in Table 4.3. The individual alloys will be discussed in more detail later in this chapter.

4.1.3 Automotive applications of magnesium

Magnesium has been used in a wide range of non-automotive commercial products. The first commercial uses were pyrotechnics, both civilian and military. Due to its high chemical reactivity, magnesium has also had a role in organic chemistry and pharmaceuticals, as well as in the electrochemical industry. Like zinc, magnesium has been used as a sacrificial anode to protect other metals in corrosive environments. Finally, in the metallurgy industry, the major non-component uses of magnesium are as an alloying element for aluminium, to which it imparts strength and corrosion resistance; in steel melt processing to accomplish desulfurization; and in iron melt processing to produce nodular iron. The worldwide demand (in metric tons) for these products in 2007 is shown in Fig. 4.2. Some of the uses of magnesium as a structural material in non-automotive applications are presented in Table 4.4.

The advantages of magnesium as a structural material are summarized in Table 4.5. Some of the key advantages include specific strength, specific stiffness, fluidity, hot formability, machining, and damping. Each of these advantages is briefly addressed in the table, but it is these advantages that drive consideration for magnesium in automotive applications.

Magnesium has a long history of automotive use dating back to the mid 1930s when it was introduced as an engine block in the Volkswagen Beetle. Since that time, there has been a steady stream of applications covering a wide range of powertrain, chassis, and body structure applications. A pictorial summary of many of these historic and current applications is provided in Fig. 4.3. Most major vehicle components have been made with magnesium either as prototype or production applications. The majority of these applications have been cast components which take advantage of the excellent fluidity of magnesium and the ability to cast very complex and thin walled shapes.

Table 4.3 Nominal compositions and typical room-temperature mechanical properties of magnesium alloys (adapted from ASM International, 1992)

Alloy	Nominal composition, weight percent							Yield strength					
							UTS	Tensile	Comp-ressive	Bearing	Elon-gation	Shear strength	Hardness
	Al	Mn	Th	Zn	Zr	Other	MPa	MPa	MPa	MPa	%	MPa	HRB
Sand casting and permanent mold casting alloys													
AZ81A-T4	7.6	0.13	–	0.7	–	–	275	83	83	305	15	125	55
AZ91C-T4	8.7	0.13	–	0.7	–	–	275	145	145	360	6	145	66
EQ21A-T6	–	–	–	–	0.7	1.5 Ag 2.1 Dy	235	195	195	–	2	–	65–85
EZ33A-T5	–	–	–	2.7	0.6	3.3 RE	160	110	110	275	2	145	50
HK31A-T6	–	–	3.0	–	0.7	–	220	105	105	275	8	145	55
WE54A-T6	–	–	–	–	0.7	5.2 Y 3.0 RE	250	172	172	–	2	–	75–95
ZC63A-T6	–	0.50	–	6.0	–	2.7 Cu	210	125	–	–	4	–	55–65
ZK61A-T6	–	–	–	6.0	0.7	–	310	195	195	–	10	180	70
High pressure die casting alloys													
AM60	6.0	0.13	–	–	–	–	205	115	115	–	6	–	–
AS21	1.7	0.4	–	–	–	1,1 Si	240	130	130	–	9	–	–
AZ91D	9.0	0.13	–	0.7	–	–	230	150	165	–	3	140	63
Extruded bars and shapes													
AZ31B-F	3.0	–	–	1.0	–	–	260	200	97	230	15	130	49
AZ61A-F	6.5	–	–	1.0	–	–	310	230	130	285	16	140	60
ZK60A-T5	–	–	–	5.5	–	–	365	305	250	405	11	180	88
Sheet and plate													
AZ31B-H24	3.0	–	–	1.0	–	–	290	220	180	325	15	160	73
HK3A-H24	–	–	3.0	–	0.6	–	255	200	160	285	9	140	68
HM21A-T8	–	0.6	2.0	–	–	–	235	170	130	270	11	125	–

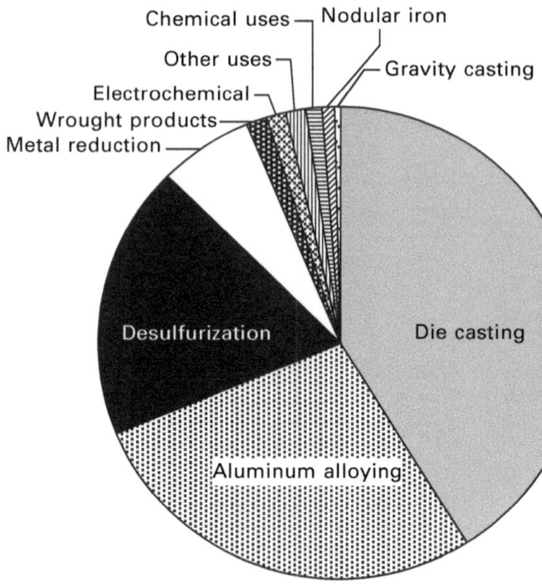

4.2 Magnesium consumption by end use in 2007 (Brown, 2008) (used with permission from TMS).

Table 4.4 Uses of magnesium as a structural material in non-automotive applications

Military and Aerospace	Consumer products
• Aircraft air frames	• Power tools
• Engines	• Cameras
• Transmission cases	• Hand luggage
• Missile skins and frames	• Appliance parts
• Electronic housings	• Cell phones
	• Portable computers

Despite the large number of applications which have been explored or used with magnesium, it remains a relatively small percentage of the materials in a typical vehicle as shown in Fig. 4.4. There are many reasons for the lack of use including the high cost of magnesium, low formability limiting its use in sheet metal components, low corrosion resistance, and limited overall design and manufacturing experience with the material compared with steel. Despite these limitations, magnesium has shown significant recent growth in applications as evidenced by the number of applications on the bottom half of Fig. 4.3. The growth and potential for further growth will be discussed in the remaining sections of this chapter.

Table 4.5 Key advantages for improved properties, design, and manufacturing with magnesium (adapted from ASM Internation, 1999)

Property	Advantage
• Specific strength	• Magnesium has a specific strength that is similar to cast iron, and similar or greater than many traditional automotive aluminum alloys and thus can provide more mass reduction relative to aluminum
• Specific stiffness	• Magnesium has a higher specific stiffness than many polymeric materials and composites, thus allowing improved mass reduction.
• Fluidity	• The relatively high fluidity of magnesium allows extremely thin walled castings (1.5 mm), which enhances mass reduction opportunities.
• Hot formability	• Wrought magnesium can be formed into very complex shapes using elevated temperature forming processes.
• Machining	• Machining tools last longer with magnesium than aluminum, reducing costs. The only issue is added care required with machining chips.
• Damping	Magnesium alloys have excellent damping capability compared with other materials making them attractive
• Low temperature properties	• Magnesium does not exhibit a brittle to ductile transition so it can be used at very low service temperatures

4.2 Cast magnesium

Of the 370 MT of magnesium consumed in 2002, 132 MT were as castings; and of these, die castings accounted for 130 MT (Webb, 2003). In the automotive industry, high pressure die casting has been the preferred high volume manufacturing method for magnesium. Since 2000, however, other casting processes have been introduced for automotive applications. The growth in these areas is small but increasing. Accordingly, this section will introduce the reader to magnesium casting alloys and casting processes, review the historical and current automotive applications of cast magnesium, and will conclude with a discussion of the challenges and opportunities for automotive magnesium castings.

4.2.1 Cast magnesium alloy nomenclature and alloy families

The protocol for naming magnesium alloys was introduced in the beginning of this chapter. The AZ alloy family (magnesium–aluminum–zinc) is the most well known of the magnesium alloy families because of its wide use in both gravity and low pressure casting, and in high pressure die casting. Some of the alloys in this family have also been used for automotive applications. The AZ alloys have a long and rich history of use in transportation. AZ and the other major families of magnesium casting alloys are shown

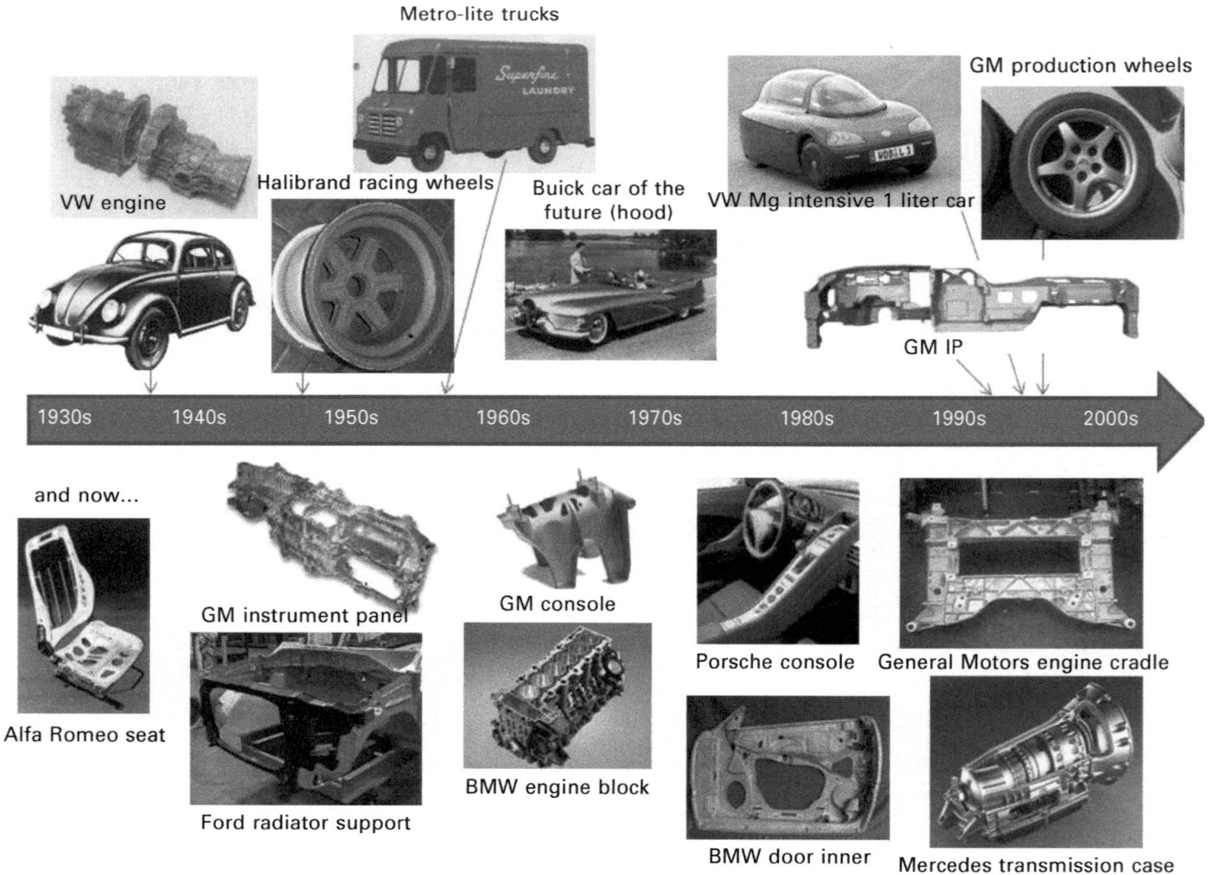

4.3 Pictorial summary of past and current magnesium automotive applications.

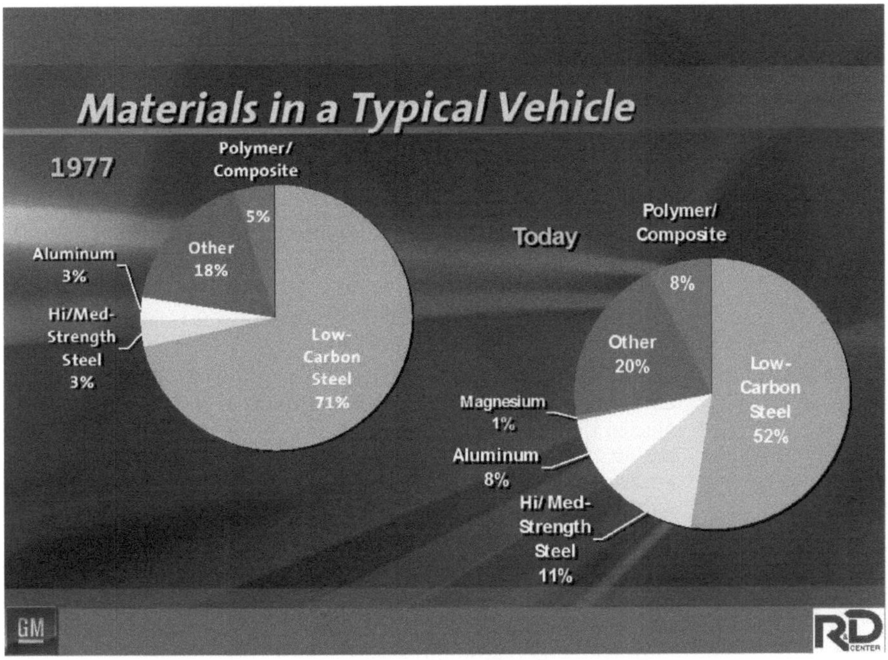

4.4 Breakdown of materials in a typical automobile (after Taub, 2007) (used with permission of General Motors).

in Table 4.6. Their application, strengths, and examples of automotive applications are included in the table. The typical casting process is also shown. Specific alloys and nominal compositions for many of the casting alloys are also provided in Table 4.3.

Both AZ and AM alloy families are versatile. They can be cast using a wide range of processes, from gravity and low pressure sand casting to high pressure die casting. As will be seen, they also form the basis of important wrought magnesium alloy families. Among the casting alloys, aluminum is added to magnesium to increase fluidity for castability and to also provide strength to the cast product. Increasing the aluminum content increases tensile strength, but at the expense of ductility. Accordingly, AZ91 alloys are used for applications where strength is the design criterion. At lower aluminum levels, five to six percent, the ductility of the alloys (AM family) makes them especially attractive for applications requiring crashworthiness; e.g. steering wheels, instrument panels, and seat frames. However, these alloys do not perform well at elevated temperatures (>125 °C) due to their poor creep resistance. While aluminum improves the castability of the melt, it forms a very low melting point phase with magnesium, ($Mg_{17}Al_{12}$) upon solidification. This phase melts at 450 °C and forms a eutectic with magnesium at 437 °C. Its presence in the microstructure lowers creep resistance of the

Table 4.6 Magnesium casting alloy families (adapted from ASM International, 1999)

Alloy family	Application	Casting process
AZ (Mg–Al–Zn–Mn)	• High strength, low ductility components in thermal environments below 125 °C • Alloys – AZ91D and AZ91E • Brake brackets, clutch brackets, covers, transfer cases and gearbox housings, intake manifolds, valve covers, manual transmission cases, wheels, CVT transmission cases, trim, air bag housings	• Gravity or low pressure • Sand or metal mold • High pressure die casting • Squeeze casting • Thixomolding
AM Mg–Al–Mn)	Lower strength, but higher ductility component; e.g., crash and impact risk; also in thermal environments below 125 °C • Alloys – AM50 and AM60 • Steering wheels, seat frames, instrument panels, cross-car beams, trim, roof frame	• Gravity or low pressure • Sand or metal mold • High pressure die casting • Squeeze casting • Thixomolding
AS (Mg–Al–Si–Mn)	Replaces AZ alloys for thermal environments about 125 °C due to improved creep strength • Alloys – AS21, AS31, and AS41 • Volkswagen air-cooled engine of 1970s (AS21 and AS41) and currently the Mercedes automatic transmission case (AS31)	• High pressure die casting
AE (Mg–Al–E–Mn)	Higher creep-strength in powertrain operating environment to 150 °C, but expensive due to E content • Alloys – AE 42, AE44 • Corvette engine cradle (AE44)	• High pressure die casting
AX and AJ (Mg–Al–Sr/Ca–Mn)	High creep-strength for powertrain components, but potentially at lower cost than E-containing alloys • Alloys – AJ52, AJ62, AXJ530, MRI 153M and MRI? 230D (Note: The AXJ and MRI alloys are developmental and do not conform to ASTM standard nomenclature.) • BMW composite engine (AJ62)	• High pressure die casting
Zr refined (Mg–Zn–Zr)	These alloys are all generally high strength and creep resistant, but higher cost, and thus used in aerospace and military applications. The major alloy systems are listed below. For specific examples see Table 4.3 • ZK – (Mg–Zn–Zr) • ZE – (Mg–Zn–E–Zr) • WE – (Mg–Y–E–Zr) • QE – (Mg–Ag–E–Zr)	• Gravity and low pressure casting in sand or metal molds

AZ and AM alloys. While lower Al contents reduce the amount of $Mg_{17}Al_{12}$ formed upon solidification, lower Al is not sufficient to achieve good creep resistance in magnesium alloys. An example is the air-cooled Volkswagen magnesium engine, which used AZ91. It was introduced in the 1970s when engine temperatures were low enough that creep was not a problem. However, as engine performance (and temperature) increased, a new family of alloys was developed with lower aluminum content but containing silicon instead of zinc. This AS family of alloys comprised first AS41 and later AS21 (Mg with two per cent Al and one per cent Si), which had better high-temperature properties than AS41. Both alloys relied on the reduced Al content (reduced presence of low-melting $Mg_{17}Al_{12}$) and the formation of Mg_2Si to impart creep strength to the alloy (Hollrigl-Rosta, 1980a). Although these alloys achieved greater creep resistance, they were not resistant enough for future engine applications.

The creep-resistant alloy AE42 alloy was developed in the 1970s. Rare earth additions had been shown to impart creep resistance in non-Al containing magnesium alloys because the rare earth Mg_9E precipitates at the grain boundaries (Nelson, 1970; Wei and Dunlop, 1992, p 335) In aluminum-containing alloys, it was discovered that the aluminum reacted with the rare earth and formed $Al_{11}E_3$ precipitates under high pressure die casting conditions and greatly improved creep resistance. However, the rare earth additions increased the cost of the alloy and made it susceptible to hot-cracking (Mercer, 1990). Furthermore, the compressive creep resistance of AE42 decreased abruptly above 150 °C (Sieracki *et al.*, 1996).

While the exact mechanism for the abrupt decrease in creep strength is still debated, Powell and co-workers (2002) reported that this breakdown was accompanied by a decomposition of $Al_{11}E_3$ and the formation of Al_2E and the undesirable $Mg_{17}Al_{12}$ phase. As stated earlier, $Mg_{17}Al_{12}$ is the low-melting-temperature phase that is present in AZ91D, AM60, and AM50 and to which is attributed the poor creep behavior of these alloys. New alloys were subsequently developed that substituted the lower cost alkaline earth elements Ca, Sr, and Ba for rare earth addition, since these elements also formed the $Al_{11}E_3$-type phase, maintained stability above 150 °C, and thus retained creep strength at elevated temperatures suitable for powertrain and underhood applications (Buschow, 1967). This development was subsequently extended to the development of the AXJ series of alloys for creep resistance, specifically AXJ530 (5% Al, 3% Ca, and 0.2% Sr), which demonstrated creep resistance approaching that of aluminum 380 (Powell *et al.*, 2001; Luo *et al.*, 2001). AXJ530 is a developing alloy. When it is entered into the ASTM tables, its designation will conform to the standard nomenclature. The same can be said for the MRI alloys (see below). Microstructural analysis demonstrated that the Ca addition favored the formation of a grain boundary eutectic structure comprising $(Mg,Al)_2Ca$ instead of $Mg_{17}Al_{12}$ between the primary magnesium grains, and which was more resistant to decomposition during elevated temperature exposure, Fig. 4.5.

Other calcium-containing alloys, MRI153M and MRI230D were also developed and commercialized (Aghion *et al.*, 2007). A strontium-containing alloy, AJ62, was also developed (Kunst *et al.*, 2006) and is being used in the BMW composite engine (see Table 4.6).

The remaining alloy families in Table 4.6 contain zirconium. They do not contain aluminum. If present, aluminum will react with the zirconium and precipitate it from the melt (Emley, 1966, pp 183–185). The zirconium is added during casting of these alloys for grain refinement. None of these alloys are used for high pressure die casting since grain refinement is necessary only at the lower solidification rates of gravity and low pressure casting. The zirconium-refined alloy families have found extensive use in aerospace and military applications, but not in the automotive industry due to their high cost (due to the cost of the alloying elements yttrium, silver, and rare earths, and the cost of the zirconium grain refiner). The use of rare earths in high pressure die casting alloys also increases cost, but not as much because these alloys use rare earths in the ratio that they occur in nature, in

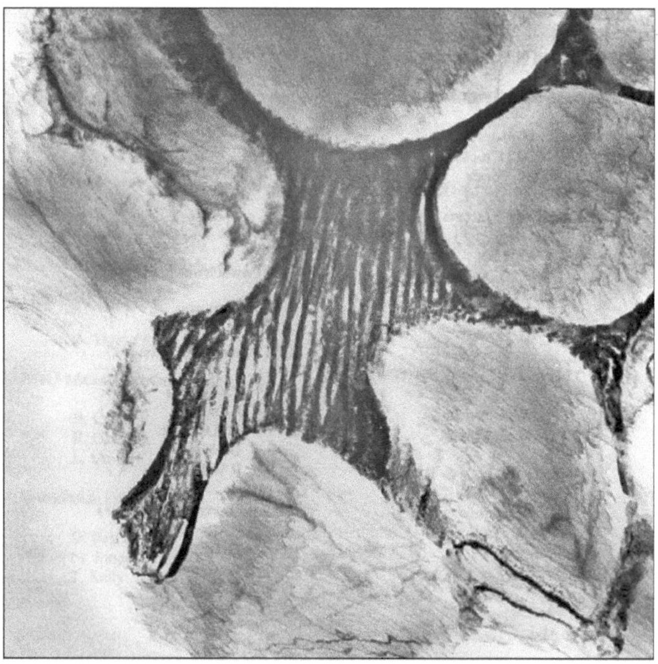

4.5 Analytical electron micrograph of AXJ530 alloy showing primary Mg grains and a grain boundary eutectic structure comprising $(Mg,Al)_2Ca$ and Mg (Powell, 2001) (used with permission of SAE International).

mischmetal. A typical mischmetal contains 50% cerium, 30% lanthanum, 15% neodymium, and 5% praesodymium. Refinement of mischmetal into individual rare earth elements substantially increases the cost of the alloy. Because the zirconium-refined alloys are not used for automotive applications they will not be discussed further in this section. However, the interested reader is directed to the literature (e.g. ASM International, 1999).

4.2.2 Magnesium casting principles

The steps in casting molten metal consist of preparing the mold or die, melting the metal, melt processing (to remove sources of defects and to modify, refine, or otherwise adjust the chemistry of the alloy), pouring or forcing the melt into the mold or die, solidification and removal of the casting. Casting is a very complex process and even the briefest introduction to its theory and practice is beyond the scope of this book. For further information about casting processes, see *The Metals Handbook Ninth Edition, Volume 15, Casting* (ASM International, 1988). The principles and practice of magnesium casting are generally the same as those for aluminum, but with a few significant differences which will become apparent in this section. The casting characteristics of magnesium are:

- The melting point of pure magnesium is 650 °C. When alloyed, magnesium forms eutectic-type systems, see Fig. 4.6. The melting ranges for die cast magnesium alloys are from 450 °C to 600 °C; those of the zirconium sand casting alloys are typically 550 °C to 650 °C.
- Most magnesium casting alloy families are quite fluid at the casting temperature and are susceptible to oxidation. Due to the small density difference between the metals and their oxides (or dross), any oxide particles or films formed on the melt surface are of neutral buoyancy and thus are likely to become entrapped in the casting. The fluidity of the metal increases the ease of entrapment. These oxides can have a deleterious effect on the mechanical properties of the casting and should be avoided. Magnesium melts are often fluxed and/or passed through filters before entering the casting cavity in the mold.
- An additional method for reducing the occurrence of oxides in the casting involves minimizing the turbulence of the molten metal front during mold filling. This is accomplished in gravity casting (sand or metal mold) by means of unpressurized gating systems. The design of an unpressurized gating system is such that the metal front decelerates as it flows through the gating system and into the mold cavity. The deceleration is achieved by continuously enlarging the area of the metal front as it moves though the runners (increasing the runner cross-sectional area) until the runners open into the casting cavity, these openings being termed 'gates'.

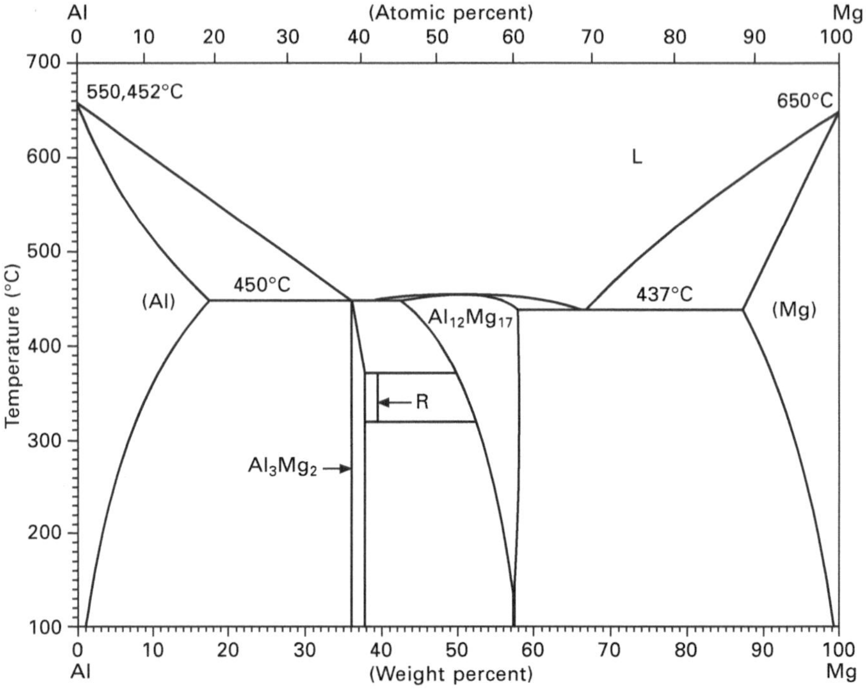

4.6 The magnesium–aluminum phase diagram (after Okamoto, 1998, p.598, with acknowledgement).

- Consistent with controlling metal flow to reduce turbulence is the principle of moving the metal always uphill, at least as much as possible. The principle is difficult to implement in gravity casting and so leads naturally to the idea of low pressure casting, in which the metal is pressure-driven upwards into the casting cavity.

The casting practices described above also apply to aluminum. Magnesium casting methods differ from those for aluminum in two ways. The first is due to the fact that the oxide film that forms on molten magnesium is neither as strong, nor as impermeable, as the film that forms on molten aluminum. This has two effects. The first is an advantage for magnesium because the weakness of the oxide film reduces its detrimental effect on the fluidity of the metal. For this and other reasons, magnesium has a greater fluidity than aluminum, with the result that it is possible to cast more complicated shapes and thinner walls with magnesium than is possible for aluminum; less than 1 mm thick versus greater than 3 mm, respectively, for high pressure die casting. The thin walls that can be cast with magnesium enable greater design flexibility in final components.

The disadvantage of the weak, permeable oxide film is that it creates a safety concern that must be addressed during all steps of the casting process.

The film is not an effective barrier to reaction between molten magnesium and other materials in the casting environment, such as the sand mold or the foundry tools and crucibles. Nor is the melt protected from reaction with the atmosphere. These reactions must be prevented since, unlike molten aluminum, which forms an impermeable oxide film and has a low vapour pressure at casting temperatures, e.g. 10E-6 Pa at 660 °C. (Magnesium has a very high vapor pressure at its melting point, 360 Pa (Margrave, 1967), and this vapor burns.)

To prevent molten magnesium fires, it is necessary to reduce the effective magnesium vapor pressure over the molten metal. One approach is to add beryllium (usually this is already present in the ingot, but it can be added directly to the melt), which, when present even at the 5 ppm level segregates to the melt surface and reacts with the oxide film to strengthen it, thereby reducing its permeability. However, this practice applies only to high pressure die casting because beryllium can interfere with grain refinement, which is necessary for gravity and low pressure casting (Kaye and Street, 1982, p 146).

Second, fluxes or a reactive cover gas are used during melting and casting. Originally the flux was added as a mixture of sulphur and salts of chloride and fluoride and boron as boric acid. The composition of these flux mixtures was developed for optimum melting, spreading, and viscosity, so that the molten flux would be effective and would remain on the melt surface. Unfortunately, the chloride residues contribute to more corrosion of the casting. Since the 1970s, most fluxes have been replaced with 'cover gas' mixtures containing sulphur hexafluoride (SF_6) at about 0.5% in a carrier gas of argon, air, carbon dioxide, etc. This method provides very good melt protection and became the standard protection method. Unfortunately, SF_6 is now recognized as a powerful greenhouse gas. It is gradually being replaced with sulfur dioxide and various hydrocarbon formulations.

Salt-based flux mixtures are still used to protect the sand mold from reaction with the melt. Magnesium will reduce silicon dioxide and any moisture that is present. Safe practice in the foundry requires the use of these 'inhibitors' in the sand and purging of the sand mold cavity with the cover gas mixture before pouring. Salt fluxes are also kept nearby to spread over magnesium fires that can still occasionally occur.

All foundry tools are kept dry and clean because molten magnesium will react with either the moisture or metal oxides. In the presence of iron oxide, for example, magnesium will reduce the oxide with substantial heat evolution, e.g. thermite. A final word about magnesium fires is in order. Because of magnesium's ability to reduce water and liberate hydrogen, magnesium fires are never fought with water. Instead, flux mixtures are used.

With respect to melt treatment, both magnesium and aluminum reduce and absorb hydrogen from water vapor. When aluminum solidifies, the solubility of hydrogen in it decreases by about 70% (Jorstad, 1993) and results in gas porosity, which usually renders the casting as scrap. The decrease in the hydrogen solubility in magnesium is not so much; less than 20% (Japanese Casting Society, 1996). Furthermore, the absolute solubility of hydrogen in solid magnesium is much greater than that in aluminum, see Fig. 4.7. The result is that gas porosity in cast magnesium is less of an issue than it is in cast aluminum.

4.2.3 Magnesium casting processes

This section surveys the processes in use today to cast magnesium. Each of the processes has been in industrial use for many years but all of the processes are becoming truly advanced due to several factors: (i) improved process monitoring and control, which is due to advances in sensors and computer software; (ii) more rapid development of the optimum mold/die design resulting from the use of fill and simulation software; and (iii) a greater understanding of the interrelationship between properties, microstructure, and processing, leading to better properties of the castings.

Gravity vs. low pressure casting

The distinction between gravity and low pressure casting is that the force of gravity is used to fill the mold in gravity casting. A schematic of a gravity casting mold is shown in Fig. 4.8. Molten metal is poured into a vertical

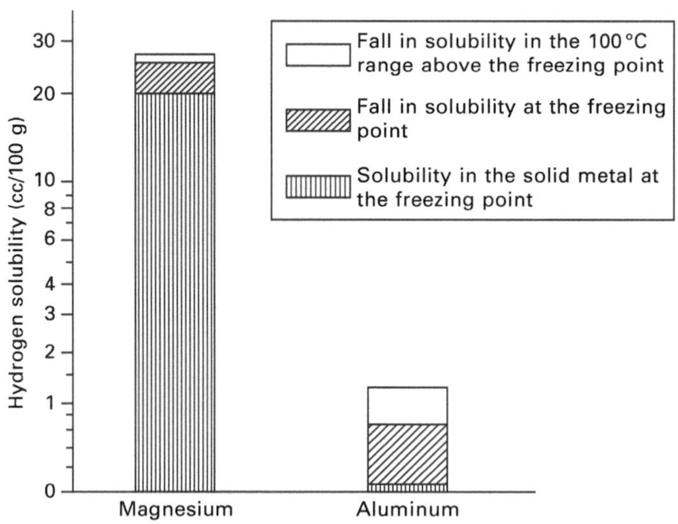

4.7 Solubility of hydrogen in magnesium and aluminum (Kaye, 1982) (with permission from Elsevier).

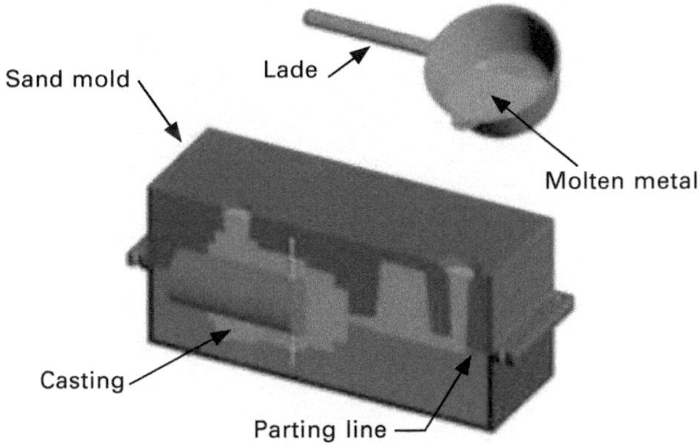

4.8 Schematic of a gravity casting mold showing sprue for receiving poured metal and the runner for horizontal flow of metal from sprue into casting cavity. (with kind permission from Custom Part Inc., www.custompart.net).

opening (sprue) and flows through the bottom of the sprue and horizontally along a single runner into the casting cavity. The large vertical cavity above the runner in this figure is a riser which provides a reservoir of molten metal to 'feed' the casting during solidification. The volume of the metal in the cavity will decrease by several percent due to thermal contraction of the metal and solidification shrinkage. If material is not continuously fed while the casting is solidifying, porosity will occur in the casting that will degrade the quality of the cast part. As can be seen, molten metal typically flows both up and down in gravity casting despite the need to minimize turbulence.

In low pressure casting, the mold is located above the sprue and metal flows 'up' the sprue and into the runner system and the casting cavity(s), Fig. 4.9. The metal flow for the arrangement shown in Fig. 4.9 is accomplished by pressurizing the furnace, which is located below the mold. The rate of metal flow is controlled by the rate of pressurization of the furnace. Metal flow can also be directed by electromagnetic pumping, but otherwise the principle of low pressure casting is the same.

High pressure die casting

Metal flow, die filling, and solidification during high pressure die casting are entirely different from gravity or low pressure casting. Molten metal is injected into the metal mold cavity at very high speed; the linear velocity can be tens of meters per second. Injection is accomplished by pouring the metal into a cylindrical tube (shot sleeve in Fig. 4.10) and using a high speed piston at several m/s to drive the molten metal into the runner system. The result is that the fill time (the time to fill the casting cavity) is measured in tens or

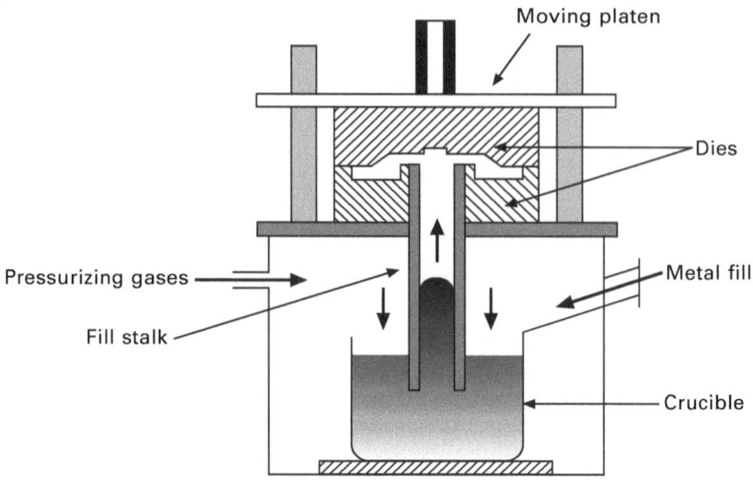

4.9 Schematic of furnace and sprue for low pressure casting. The mold will be located above and in line with the sprue (courtesy Kurtz, Inc.).

hundreds of ms, instead of tens of seconds as in gravity and low pressure casting. These conditions result in the metal being 'virtually' sprayed into the casting cavity where the atomized droplets of metal hit, condense, and solidify rapidly on the internal walls of the casting cavity.

Feeding the casting during high pressure die casting is not accomplished with risers. Rather, extra metal in the shot tube/runners is forced into the full mold under high impact loading; on the order of ~70 MPa. Metal cooling rates are correspondingly much higher; ~100 °C/s during high pressure die casting versus less than 10 °C/s during gravity or low pressure casting. The poor thermal conductivity of sand further lowers the cooling rate to about 1 °C/s in sand casting. The differences in cooling rates drastically affect the solidification rates of the metal and the subsequent tensile properties of the casting. Most of the structural integrity of the high pressure die cast magnesium part is in the skin of the casting, the skin being the thin external surface of the casting which is characterized by very low porosity and very fine grain size. Extensive porosity can occur beneath the skin (core) due to the filling and solidification conditions. As long as the core is not exposed by machining away the skin or other means, the mechanical and sealing integrity of the casting can be maintained. Often however, the design of the part will require some machining; thus controlling the amount and location of porosity in the cast product is critical.

Advantages of high pressure die casting include higher productivity because the filling and solidification rates, as well as the time to eject (remove) the part from the mold, are much faster than for gravity or low pressure casting. The chief disadvantage is gas entrapment in the casting, which occurs when molten metal is injected into the cavity and results in casting defects

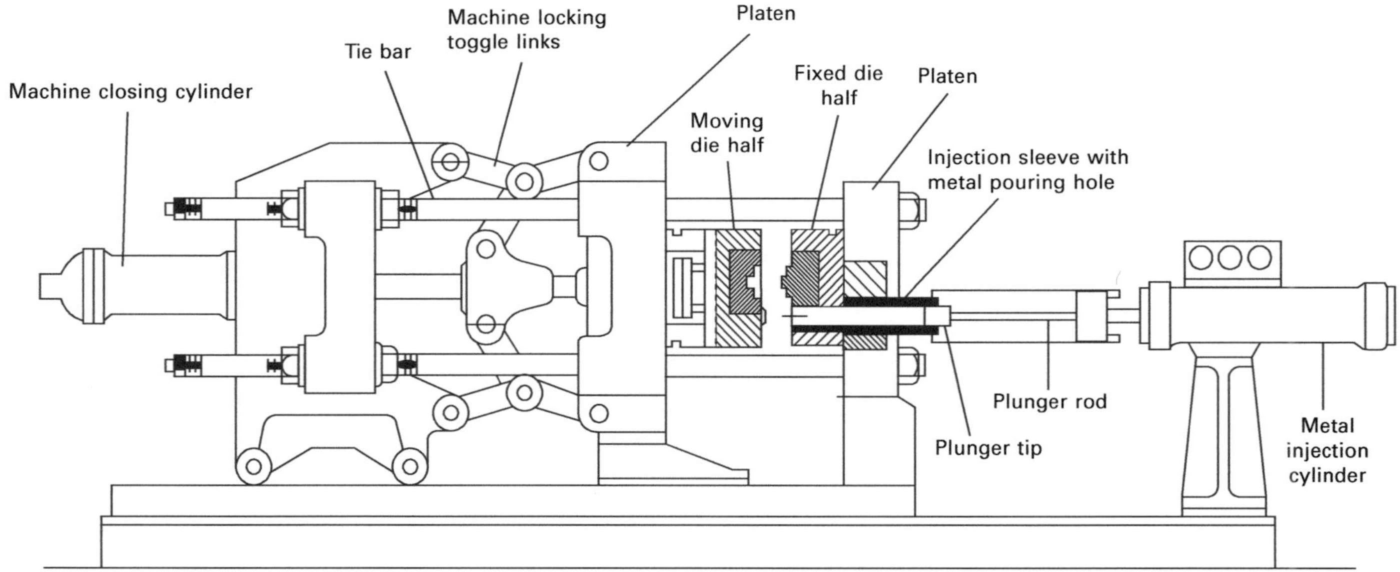

4.10 Schematic of a cold chamber, high pressure die casting system (Kaye, 1982, p 6) (with permission from Elsevier).

which lower the strength and ductility of the part. In addition to the casting defects, the presence of entrapped gas also precludes heat treating to improve properties. Vacuum die casting, where the die cavity is evacuated prior to and during metal fill has been developed (Vinarcik, 2003, pp 29–49; Luo and Sachdev, 2008), but the level of vacuum achieved is still not enough to eliminate resultant porosity. The development of 'supervacuum die casting,' may further reduce porosity and improve properties (Brown *et al.*, 2009). At the time of this writing, essentially all of the magnesium parts cast for use in the automotive sector are high pressure die cast. If vacuum and supervacuum die casting can be commercialized, the range of automotive applications for magnesium will increase.

Squeeze casting

Squeeze casting is a hybrid of low pressure casting and high pressure casting, and it has the potential to completely eliminate the gas defects associated with high pressure die casting, and to enable heat treatment of the castings. In squeeze casting, the die is filled slowly with metal to maintain laminar flow. Once the cavity is full, the pressure on the melt is increased to over 100 MPa and maintained to feed the casting to compensate for shrinkage until the casting has solidified. Die design for squeeze casting is different from that for die casting, and includes thick gates and a large shot end biscuit to ensure that the gates do not freeze before the casting in the cavity has solidified and to ensure feeding the shrink during solidification. Other advantages of squeeze casting are contained in Table 4.7 (Kainer and Benzler, 2003).

The limitations of squeeze casting include less flexibility in part geometry, lower productivity, high machining requirements, and greater cost. However, as a potential casting process for safety-critical parts in automotive systems such as space frame joints, squeeze casting may find its niche.

Table 4.7 Advantages of squeeze casting for magnesium (Kainer and Benzler, 2003)

- Reduced porosity
- Hot cracking prevention for alloys with wide freezing ranges
- Increased strength and ductility due to:
  - Fine-grained microstructure
  - Defect-free microstructure
- Possibility of heat treatment to improve properties
- Offers a wider range of alloy choices, including those that may be difficult to cast by other means
  - Creep-resistant alloys
  - Thixotropic melts
- Metal matrix composites can be produced

Semi-solid casting

Semi-solid casting is a casting process that involves filling a mold with the metal in a partially molten state in which globules of solid are homogeneously dispersed in the liquid. The benefits of semi-solid casting are: (i) reduced shrinkage due to the lower casting temperature of the melt, (ii) low gas porosity enabling heat treating of the castings, (iii) high mechanical properties due to the uniquely fine microstructure of the resulting castings, and (iv) extremely fine surface finish. In some cases semi-solid castings can compete with forgings. Semi-solid aluminum casting was developed in the 1970s by Flemings and co-workers, it was referred to as rheocasting. In that process, an alloy containing two phases, the alpha phase with a higher melting point than the eutectic, is stirred at a preselected temperature between the liquidus (all liquid above this temperature) and solidus (all solid below this temperature) of the starting alloy to provide about 35% liquid, to obtain the homogeneous distribution of the solid phase dispersed in the liquid, Fig. 4.11. The particular distribution and volume fractions of the two phases render the material thixotropic, where its viscosity decreases with shear rate, to enable the mixture to be injected at high rates into a die cavity while maintaining laminar flow. The key to doing this successfully is thermal management of the runner system and the die cavity. The process later evolved into Thixocasting, where carefully prepared electromagnetically stirred billets that contain globular alpha aluminum are dispersed within a lower melting point eutectic phase. These billets are reheated to a temperature to melt the eutectic and inject the mostly solid mass at high velocity into the die cavity.

A yet third process was developed under the umbrella of semi-solid forming, called thixomolding, and has gained greater use for magnesium compared with the other two processes, which are used primarily for aluminum. In thixomolding (see Fig. 4.12), magnesium chips are fed into a modified plastic injection molding machine and are melted in the barrel while an internal screw advances the mass towards the shot end (LeBeau *et al.*, 1998). By the time the material mass reaches the shot end, it contains about 5–10% solid, which is injected in to the die cavity. The much lower metal temperature compared to die casting allows the casting of extremely thin wall magnesium castings.

Because of the demands of die thermal management, thixomolding has generally been limited to small parts such as cell phones and other consumer electronics such as computer cases, which have the common characteristic of being very thin wall castings. Also, the molding temperature being lower than other casting processes, the mold life can be extended and the cast parts have fine and clean surfaces which are cosmetically appealing. Finally, the lower fill rate of the semi-solid process leads to less entrapped gas and shrinkage, giving better properties and enabling post-casting heat treating.

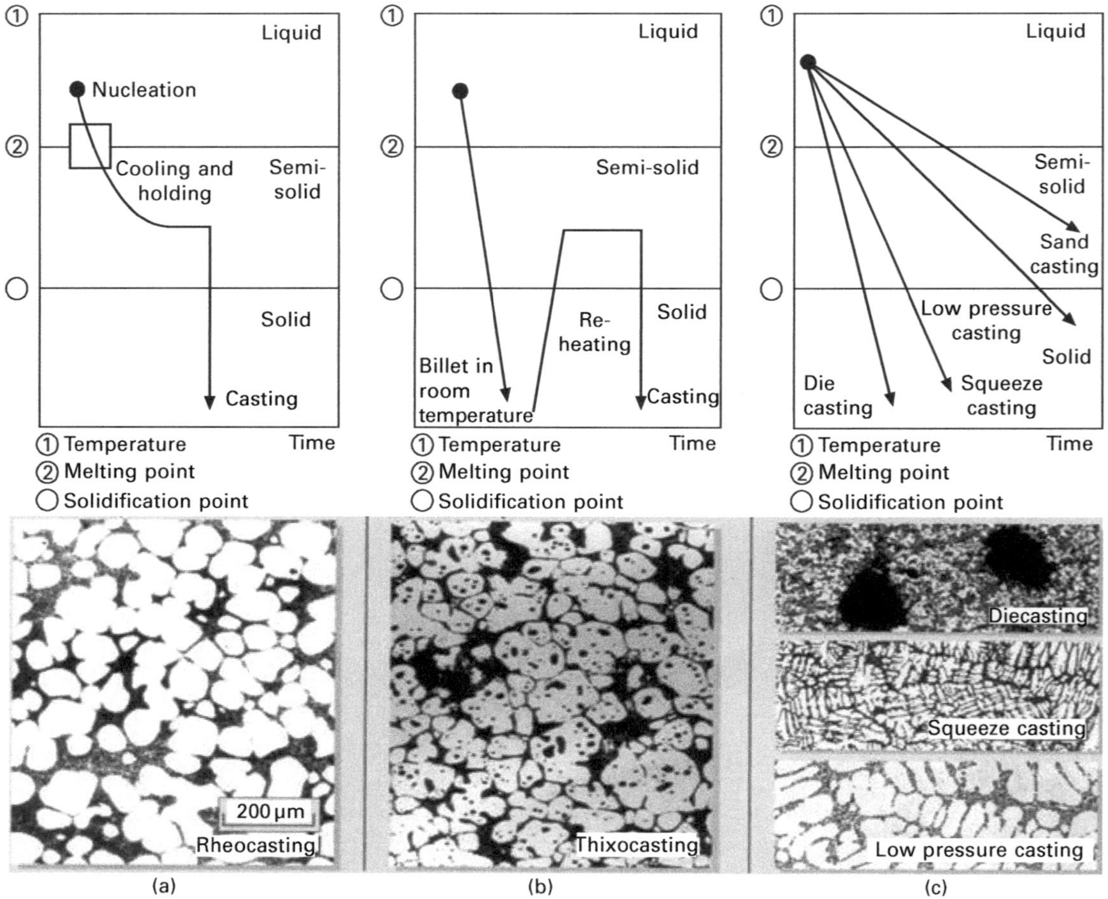

4.11 Comparison of microstructures resulting from various casting processes (courtesy Ube Machinery, Inc.).

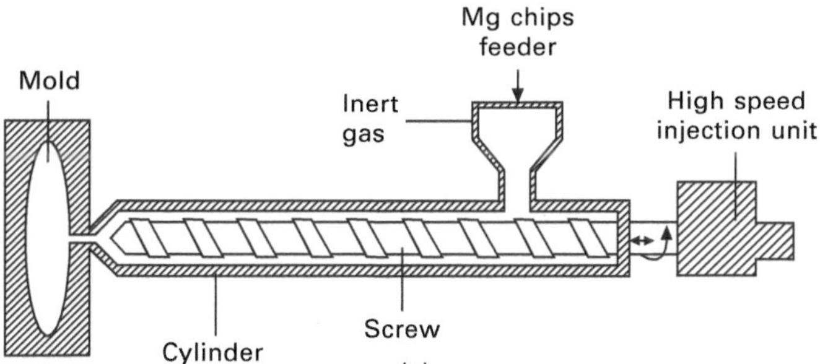

(a)

(b)

4.12 Schematic of (a) the thixomolding equipment with (b) a representative image of the metal chips/feed used in the process (used with permission from R. Decker, Thixomat).

Over time, as die thermal management techniques evolve, it is expected that the advantages of thixomolding will extend to automotive components.

4.2.4 Automotive applications of cast magnesium

One of the earliest applications of magnesium as an automotive material was a sand cast crankcase on the 1931 Chevrolair. Commercially viable applications appeared in London in the 1930s (Emley, 1966, p 828). These applications included lower crankcases for city buses, and transmission housings for tractors, see Fig. 4.13. The crankcase shown in the figure weighed 26 kg. Over 500 000 transmission housings were cast. Both components were sand cast. Other magnesium components included gear boxes, clutch housings, sumps, chassis parts, wheels, and truck body components, usually sand cast with Commercial C alloy, AZ81. Crankcases and housings were also produced in Germany, but by high pressure die casting (HPDC). The alloy used for these

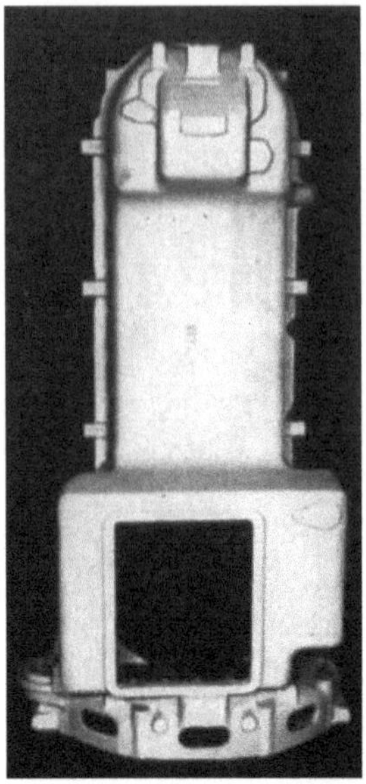

(a)

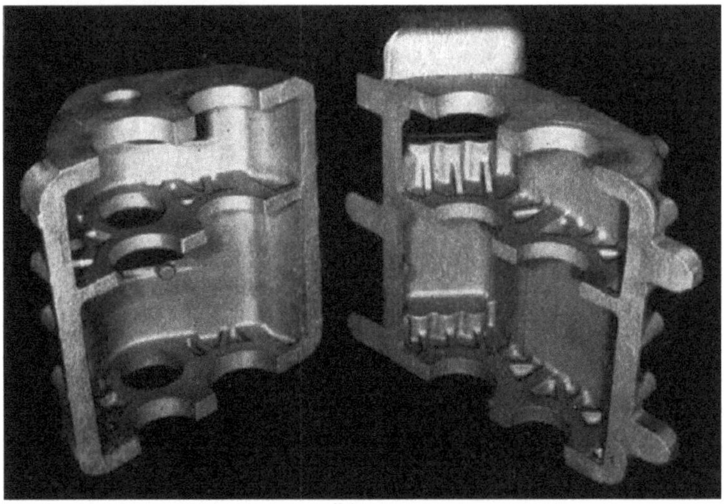

(b)

4.13 Fully commercialized cast magnesium powertrain components used in England in the 1930s, (a) a lower crankcase and (b) a transmission housing (Emley, 1966). (Used with permission from Elsevier.)

castings was AZ81. Together, these components weighed 17 kg, which was 50 kg less than the iron components they replaced (Schumann and Friedrich, 1998). An example of a large HPDC casting is shown in Fig. 4.14.

Magnesium usage grew throughout the 1930s and then grew exponentially to nearly 250 000 MT production during World War II as mass reduction, desulfurization, and alloying became critical in both land and air weaponry,

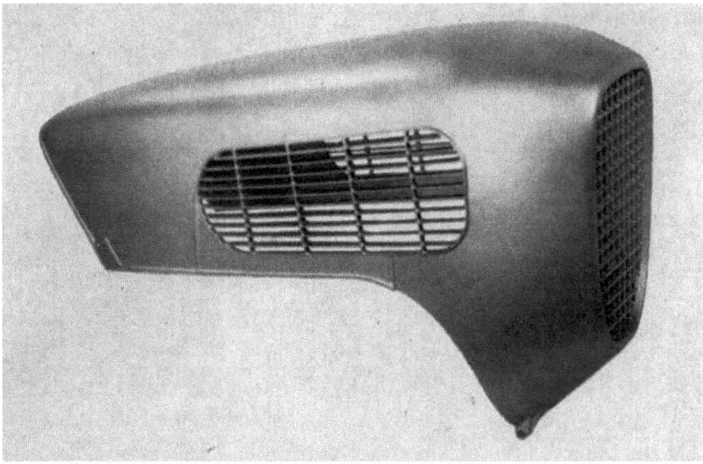

4.14 A high-pressure die cast tractor hood. The nominal wall thickness for this casting was 2.2–2.5 mm (Emley, 1966). (Used with permission from Elsevier.)

see Fig. 4.1. After the War, magnesium demand collapsed and remained low until the mid-1950s. With the introduction of the Volkswagen Beetle, automotive magnesium consumption again accelerated. Worldwide magnesium for all applications reached 240 000 MT in 1971, of which 42 000 MT were for the Beetle air-cooled engine and gearbox, which together weighed about 20 kg (Friedrich and Schumann, 2001). However, then several factors emerged and combined to cause the reduction and eventual elimination of magnesium as a structural powertrain component. These factors included greater power requirements for the engine, which increased both its operating temperature and load and which ultimately resulted in the conversion of the engine from air cooling to water cooling; the AZ81 alloy, and later the AS41 or the AS21 alloys, lacked sufficient creep resistance in the required operating environment (Hollrigl-Rosta, 1980). The use of water cooling put magnesium at a disadvantage compared with other engine materials because of its poor corrosion resistance. At that time, the effect of iron, copper, and nickel impurities (in ppm amounts) on promoting the corrosion of magnesium had not been recognized (Hillis, 1983). By the time the 'high purity' alloys AZ91D and AM60B, which replaced AZ91C and AM60A, respectively, were developed in the 1980s, the cost of magnesium alloys had begun to increase, and the use of magnesium in automotive applications decreased although some applications remained. These included cam covers, brackets, and manual transmission cases, which used the new, greater corrosion-resistant magnesium alloys that resulted from the work of Hillis and others. With time, high pressure die cast automotive applications have increased, as shown in Section 4.1. A lot of the increase is attributed to the many advances made in the development of casting fill and solidification software, in addition to the high purity alloys that became available. The simulation tools have enabled a much broader range of cast products, both from the perspective of complexity and properties. These applications, which accounted for a better than ten percent annual growth in magnesium usage per year throughout the 1990s, included instrument panel structures and cross-car beams, seat frames, steering wheels, intake manifolds, and numerous brackets and covers. General Motors usage of magnesium in 2005 is shown in Fig. 4.15.

Other significant applications include the Mercedes 7-speed Tiptronic automatic transmission case (Greiner *et al.*, 2004), see Fig. 4.16, the BMW composite engine (Hoeschl *et al.*, 2006), Fig. 4.17, and the Corvette engine cradle, Fig. 4.18 (Li *et al.*, 2005). The BMW composite engine is an inline block, and uses an aluminum inner casting to carry the head bolts, bulkheads, water cooling passages, and cylinder bores. The BMW engine block is 57 percent lighter than the iron block it replaced. The Corvette engine cradle was the result of a project which was jointly sponsored by the US Department of Energy (DOE) and the US Council for Automotive Research (USCAR), and led by the OEMs in the US. Two other USCAR projects that are ongoing are also pushing the envelope for automotive uses of magnesium. The first is the Magnesium Powertrain Cast Components (MPCC) Project which has the objective of demonstrating the readiness of magnesium alloys for completely replacing the major aluminum components of a V block engine (Powell *et al.*, 2005). The MPCC cylinder block achieved a mass reduction of 25% (29% for all of the cast aluminum components which were replaced by magnesium). A prototype engine made with a low pressure sand cast cylinder block, a Thixo-molded rear seal carrier, and high pressure die cast oil pan and front cover, with all other parts carried over from the baseline aluminum engine, has been completed (Powell, 2009), see

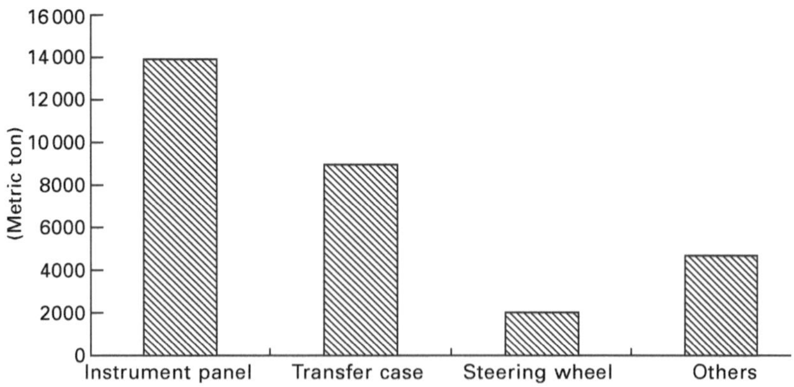

4.15 Usage of magnesium in General Motors North American vehicles in 2005 (courtesy World Wide Purchasing, General Motors Corporation).

4.16 The Mercedes 7-speed automatic transmission case (after Greiner, 2004). (Used with permission of Daimler.)

4.17 The Mercedes 7-speed automatic transmission case (after Greiner, 2004). (Used with permission of Daimler.)

Fig. 4.19. The prototype engine is being dynamometer tested for durability. The other USCAR project is the Front End Project (Luo and McCune, which, 2008a) is described later. Briefly, that project seeks to create a completely magnesium front end structure, thereby reducing mass significantly (see the schematic of the front end in Fig. 4.20). Projects such as these are good examples of collaborative projects that are advancing the field. For more information about the US DOE Lightweighting Program, see reference Carpenter *et al.*, 2008.

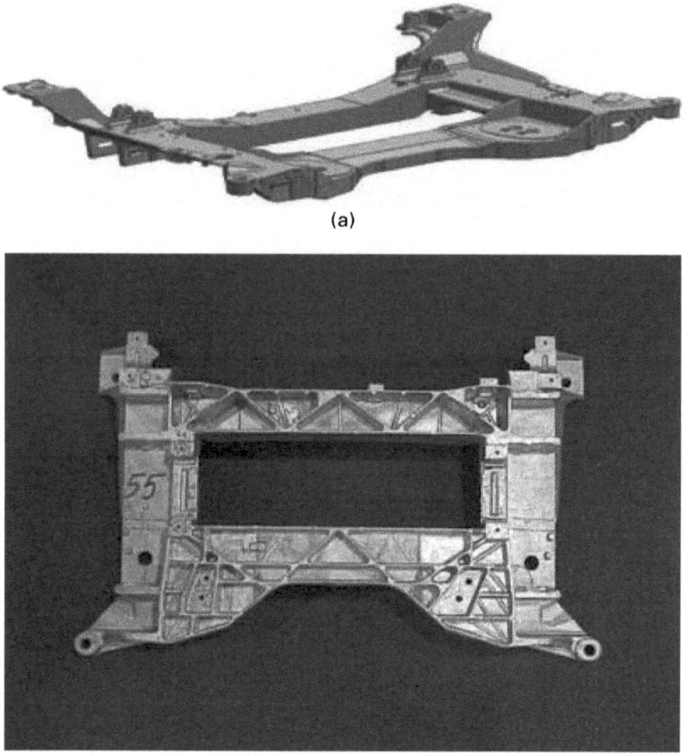

(a)

(b)

4.18 The Corvette engine cradle (a) in schematic and (b) actual casting (courtesy General Motors).

4.3 Sheet magnesium

4.3.1 Alloy families, nomenclature and properties

The number of commercially available magnesium sheet alloys is very limited. The most commonly used alloy is the aluminum- and zinc-containing AZ31B. Other alloys which are currently commercially available are AZ61, HM21, HK31 and ZM21. The materials are typically available in either an annealed, O temper, or in a partially-hard, H-24 temper. Typical microstructures of AZ31 in these two temper conditions are shown in Fig. 4.21. The annealed O temper material has a homogeneous distribution of grains with clearly visible boundaries, while the, as rolled, H24 temper material is highly worked and it is difficult to discern individual grains. In the O temper, the AZ31 material provides similar yield strength to work hardenable aluminum sheet alloys such as AA5754 or AA5182, but lower strength than age-hardenable alloys such as AA6111. One issue with the mechanical properties of sheet magnesium is

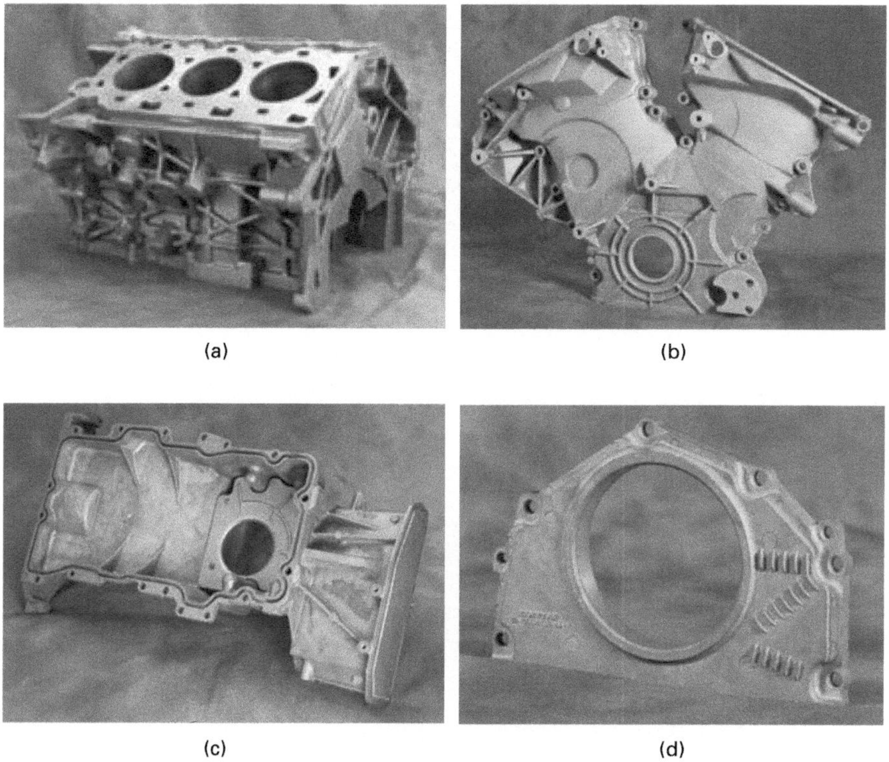

4.19 Magnesium powertrain components from the USCAR Magnesium Powertrain Cast Components Project; (a) cylinder block, (b) front engine cover, (c) oil pan and (d) rear seal carrier (Powell, 2009). (Used with permission from Ford Motor Co.)

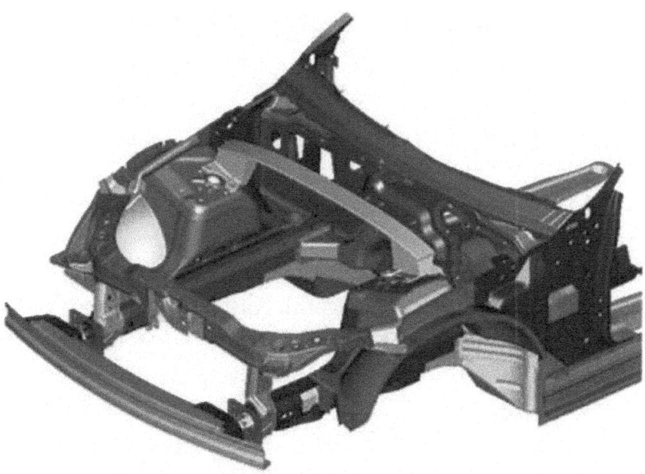

4.20 Schematic of the front end of a production sedan. (Used with permission from General Motors Inc.)

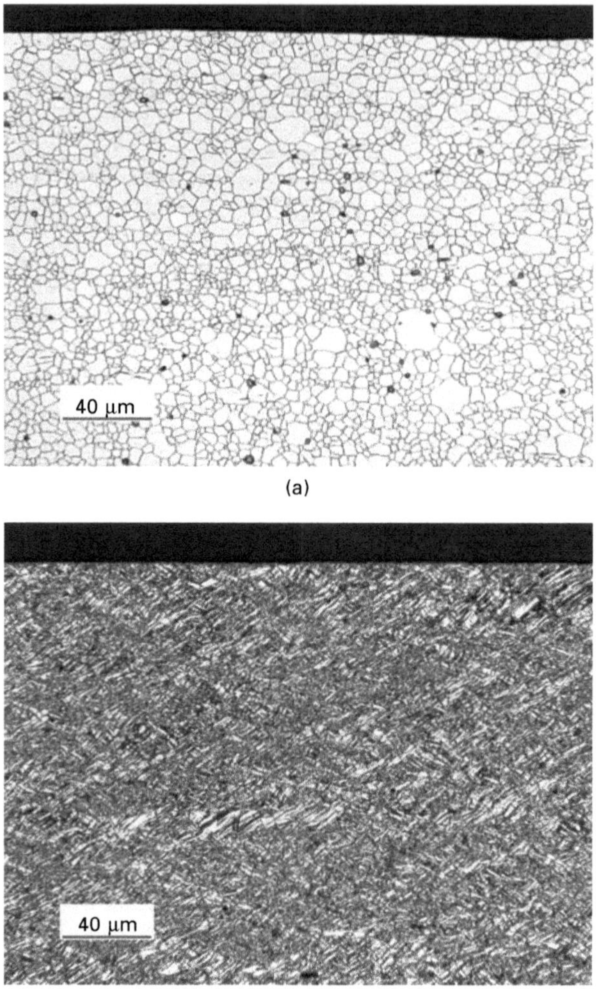

4.21 Typical microstructures in AZ31 sheet showing (a) annealed, O temper and (b) as-rolled, H-24 temper.

that the properties are typically anisotropic, meaning the strength or ductility varies with direction on the sheet. For example, yield strength is typically approximately 10% lower transverse to the rolling direction compared with parallel to the rolling direction. The anisotropy is the result of two main features, (i) irregular or non homogeneous grain structure and (ii) texture differences, between the rolling and transverse to rolling directions. An example of grain inhomogeneity is shown in Fig. 4.22, which shows regions of very fine grains oriented in bands through the material.

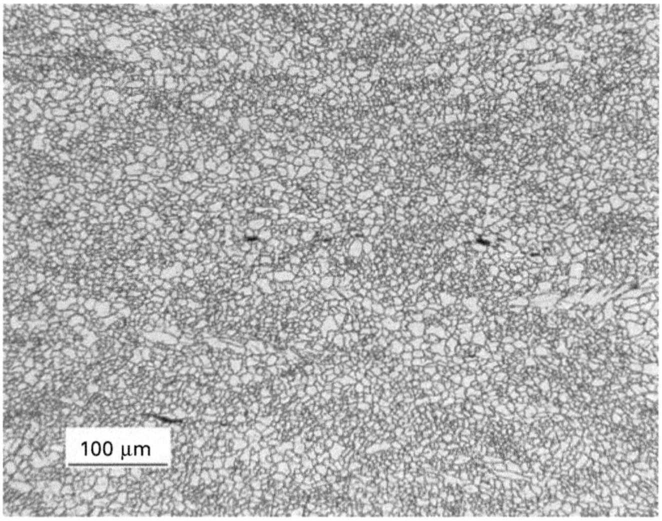

4.22 Micrograph of rolled and annealed AZ31 material showing inhomogeneous grain structure with bands of fine grains.

4.3.2 Magnesium sheet forming processes

Traditionally, sheet magnesium has been made by ingot metallurgy and subsequent hot rolling. This can be a very expensive process, as the limited formability of magnesium at room temperature necessitates that the reduction of the ingot to the final sheet be done at elevated temperatures. As a result, the conversion cost (cost to convert raw magnesium material into sheet product) is higher than for converting steel or aluminum into sheet. Recently, a significant amount of work has been ongoing globally to develop a lower cost, strip casting process whereby magnesium sheet can be cast to almost net shape, and then require minimum rolling to the final gage. More information on this process can be found elsewhere (Krajewski *et al.*, 2007a). Because of the high cooling rates in strip casting, these materials can have a finer grain size than ingot cast materials which can be an advantage to both strength and formability. An example of a fine grain structure produced by strip castings is shown in Fig. 4.23.

4.3.3 Forming processes using magnesium sheet

The majority of the processes used to convert sheet metal into automobile components occur at room temperature including, stamping, flanging, bending, hemming and trimming. The processes are very robust with high formability materials such as mild steel, and have also been used for less formable materials such as aluminum and high strength steel. Unfortunately, the limited

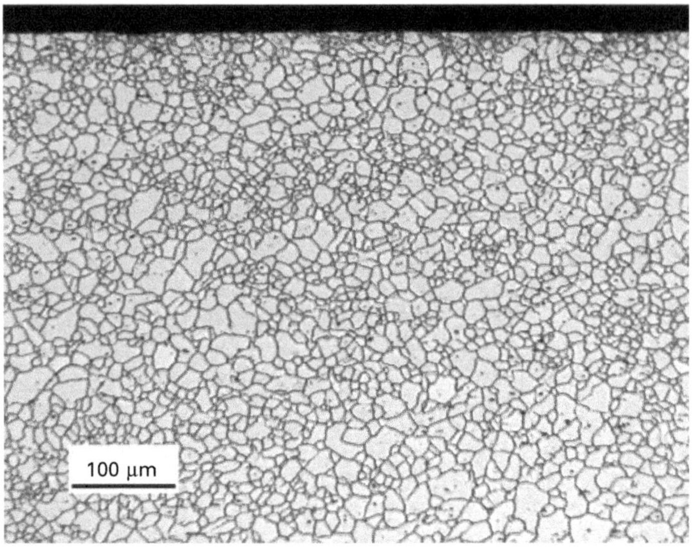

4.23 Micrograph of strip cast AZ31 material with very fine grain structure.

formability of magnesium due to its HCP structure (discussed earlier) makes the use of these processes very difficult. An example of the problem is shown in Fig. 4.24 where the results of a forming trial using a simple rectangular pan for a mild steel and AZ31B magnesium sheet are compared. The pan could be formed to a 125 mm depth with the steel, but split after only about 12 mm with the magnesium (Krajewski *et al.* 2006).

Another example of the limited formability of magnesium sheet can be seen by comparing the minimum bend radius for magnesium with those of other materials. *The Metals Handbook* suggests that, for a 90 degree bend, the minimum bend radius for magnesium sheet is between 5*t* and 13*t*, where *t* is the metal thickness. This depends on the alloy and temper chosen, with AZ31B-O being the best. For comparison, aluminum and steel can achieve bending radii approaching 1*t* for a 180 degree bend, which is essentially bending the material flat over itself.

The limited formability at room temperature with magnesium sheet necessitates the use of elevated temperature forming processes. Increasing the temperature significantly increases bendability, thereby reducing the achievable bend radius. This is summarized in Table 4.8, taken from *The Metals Handbook* (ASM International, 1998). All of the applications described previously were created using some variant of an elevated temperature forming process. A brief summary of the relevant elevated temperature forming processes now follows.

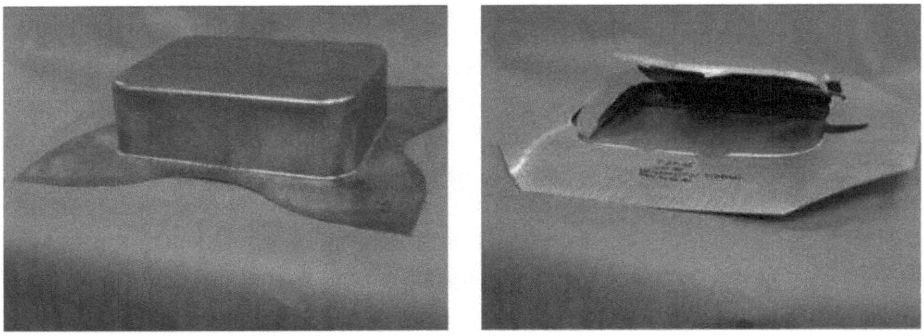

(a) (b)

4.24 Comparison of a forming trial on a 125 mm deep pan with (a) mild steel and (b) AZ31B magnesium.

Table 4.8 Recommended minimum radii for 90° bends in magnesium sheet (adapted from ASM International, 1998)

Alloy and temper	Forming temperature (bend radii as multiples of sheet thickness)							
	20 °C (70 °F)	95 °C (200 °F)	150 °C (300 °F)	205 °C (400 °F)	260 °C (500 °F)	315 °C (600 °F)	370 °C (700 °F)	425 °C (800 °F)
AZ13B-O	5.5*t*	5.5*t*	4*t*	3*t*	2*t*	...	...	...
AZ13B-H24	8*t*	8*t*	6*t*	3*t*	2*t*	...	...	...
HK31A-O	6*t*	6*t*	6*t*	5*t*	4*t*	3*t*	2*t*	1*t*
HK31A-H24	13*t*	13*t*	13*t*	9*t*	8*t*	5*t*	3*t*	...
HK31A-T8	9*t*	9*t*	9*t*	9*t*	9*t*	8*t*	6*t*	4*t*

Superplastic forming

Superplastic forming (SPF) involves slowly forming a sheet of material in a single sided tool or die using gas pressure at a temperature of about 500 °C. A typical forming cycle for SPF is approximately 30 minutes or longer. The process requires material which has been specially processed to have a fine grained or dual phase microstructure (ASM International, 1998). Since the 1970s, SPF has also been used by the automotive industry to produce complex, lightweight *aluminum* panels for niche products with annual volumes of less than 1000 units (Barnes, 2002). The technology has been optimized around niche vehicles where tooling costs are extremely low, cycle times are long, labor content and material price are high, and parts can be reworked after production because each vehicle is handcrafted. Superplastic formability trials on automobile components have been carried out on a number of magnesium alloys including AZ31B, ZK10, and ZE10 (Barnes, 2007, pp 440–454). The excellent high temperature ductility of magnesium sheet enables this process to be used, often without special processing of the magnesium sheet. The technology should be able to make viable low to niche volume automotive panels with magnesium sheet.

Quick plastic forming

Quick plastic forming (QPF) was developed as a gas forming technology that could be used in the mainstream automotive industry (Rashid *et al.*, 2001; Schroth, 2004, p 9; Krajewski and Schroth, 2007, pp 3–12). As in SPF, forming is done at elevated temperatures using gas pressure. The temperatures used in QPF (~450°C) are typically lower than those used for SPF (~500°C). However, much faster forming cycles are achieved (~1–5 minutes). The faster cycle time is achieved through process automation, thermal control of the sheet metal and dies, and a number of other improvements detailed elsewhere (Krajewski and Schroth, 2007, pp 3–12). These improvements enable the production of components at automotive-type volumes, up to 75 000/year.

As with SPF, QPF has also been implemented to produce *aluminum* components for the Chevrolet Malibu Maxx and aluminum decklids for the Cadillac STS. The QPF process has been used to produce many prototype magnesium inner closure panels using the same bill of process (BOP) which was used for the production of aluminum QPF panels (Carter *et al.*, 2008). As with SPF, this technology is ready for use with magnesium sheet but at much higher volumes. Examples of QPF prototype products are shown in a later section.

Warm forming

Warm forming refers to elevated temperature stamping that gives higher forming rates but lower formability than processes such as QPF or SPF. It is typically carried out at temperatures between 200 °C and 350 °C using heated, matched die sets, and has been the standard process for stamping complex shapes from sheet and plate magnesium in non-automotive industries including aerospace components, luggage, and computer cases. Warm forming guidelines for draw depths, stretching limits, and bending radii, have been summarized by Emley (1966, pp 584–603). Spinning, hammer forming, and rubber pad forming have also been used at 'warm' temperatures. Warm stamping of magnesium can enable a wide range of moderately complex automobile components to be formed. An example of a part formed with this process is the door inner panel shown in Fig. 4.25, formed using AZ31B (Krajewski *et al.*, 2006). The issue currently preventing warm forming from being used is the lack of an integrated warm forming process. This includes the need to rapidly heat blanks, control die temperature and dimensions, as well as developing a lubricant.

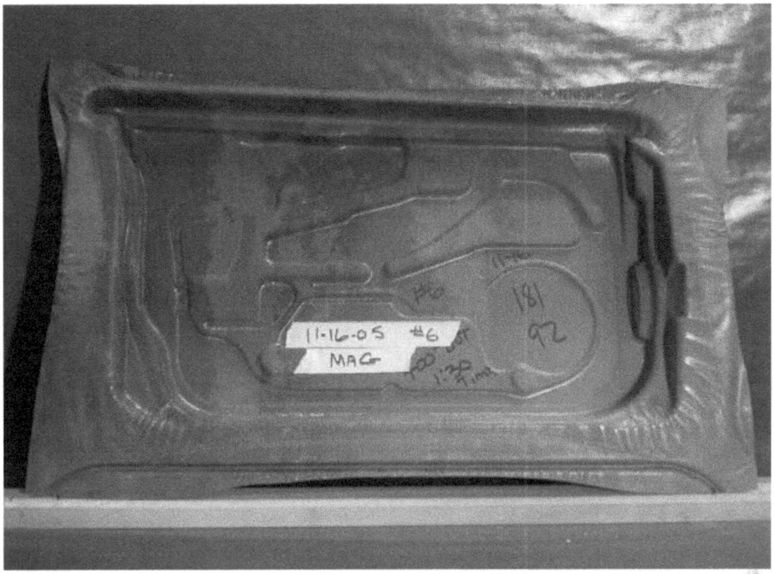

4.25 Magnesium door inner panel made by warm forming (Krajewski, 2006).

Thermohydroforming

Another process for elevated temperature forming of magnesium sheet is thermohydroforming. Thermohydroforming is a combination of warm forming and traditional sheet hydroforming (Groche *et al.*, 2002).

Using heated synthetic oil as a forming medium, the magnesium sheet blank is formed into the desired shape using the fluid pressure. There are many varieties and applications of conventional room temperature hydroforming, but thermohydroforming remains only a research topic due to limitations in developing suitable oil which is stable at the warm forming temperatures, and handling the oil in a safe manner in a production environment.

Warm hemming

A critical process for assembling automotive closure panels is hemming, in which an outer panel (exterior of the vehicle) is bent to 180° over an inner panel to create a closure assembly. For a uniform 180° flat hem, the flange die radius should be approximately equal to or less than the thickness of the outer panel, t, which would allow for an outside metal radius in the order of $2t$ or less. In steel and aluminum, this process occurs at room temperature. Bending wrought magnesium alloys with such a sharp radius, however, would cause unacceptable cracking on the outer surface unless the bending was performed at an elevated temperature. A recent study (Carsley and Kim,

2007, pp 331–338) showed that successful hemming of AZ31B-O, required temperatures of 270–280 °C to achieve an acceptable condition with good surface appearance. Bending at these temperatures ensures that the material is annealed, that the grain structure is recrystallized, and that additional slip systems are activated to accommodate the uniform bending strain without plastic localization (necking) or cracking failure at the outer surface of the panel. While this laboratory study provided encouragement that warm hemming can be performed, a viable production method is needed.

Warm clinching

In addition to hemming, clinching is a critical technology to assemble automobile components. For example, mechanical clinching has been used extensively to attach reinforcements to aluminum inner closure panels. The extremely high localized deformation in mechanical clinching operations makes them very difficult to execute at room temperature with magnesium. Warm clinching of magnesium sheet to either aluminum sheet or magnesium sheet has been demonstrated on a laboratory scale (Behrens and Hubner, 2005, pp 59–62). In this work, clinching equipment was modified by the addition of electrical resistance heaters to heat the punch and die to 390 °C. Water cooling and insulation were added to keep the rest of the equipment cool. When clinching magnesium (punch side) to aluminum (die side), a minimum temperature of 275 °C was required to form good joints. For clinching magnesium to magnesium, a minimum of 200 °C was required. As with hemming, robust, production-viable methods of warm clinching need to be developed.

4.3.4 Automotive applications of magnesium sheet

Magnesium sheet has been used as a structural material in the transportation industry since World War II. The most famous application was the B36 bomber, which contained 9000 lbs of magnesium sheet (Barnes, 1992, pp 29–43). The first commercial ground transportation applications were developed in the early 1950s. Metro-lite trucks were manufactured between 1955 and 1965 and featured magnesium sheet panels as well as structures made from magnesium plate and extrusions (Barnes, 1992, pp 29–43). These trucks had increased payload capacity and were excellent applications for magnesium because they did not require extreme formability. However, sheet magnesium has not been used in high volume production in the mainstream automobile industry. The only current production application of magnesium sheet is the center console in the low volume Porsche Carerra as shown in Fig. 4.26.

While production applications are limited, numerous prototype components have been made using sheet magnesium. General Motors made prototype

4.26 Sheet magnesium center console cover in Porsche Carrera GT automobile. (Copyrighted by Dr. Ing. H.c. F. Porsche AG. Used with permission.)

hoods for the Buick LeSabre in 1951, various body panels for the Chevrolet Corvette SS Race Car in 1957, and hoods for the Chevrolet Corvette in 1961 (see Fig. 4.27).

More recently, Volkswagen has made a prototype hood for the Lupo, shown in Fig. 4.28 (Moll *et al.*, 2004). GM has made numerous panels including a hood, door inner panel, decklid inner, liftgate, and various reinforcements, some of which are shown in Fig. 4.29 (Krajewski, 2001; Verma and Carter, 2006; Carter *et al.*, 2008). Chrysler LLC has performed a number of studies using magnesium sheet including the inner panel shown in Fig. 4.30 and a magnesium intensive body structure (see also Logan *et al.*, 2006).

The majority of the applications just discussed were 'inner' panels which create the structure of the vehicle closures but are not visible on the outside of the vehicle. This is due to two factors. First, the surface quality of the currently available magnesium sheet requires significant finishing compared to aluminum or steel, and second, the limited formability at room temperature makes assembly processes for outer panels, such as hemming, difficult. This means that the first commercial applications for magnesium sheet are likely to be inner panels.

(a)

(b)

4.27 (a) 1951 Buick LeSabre concept car with magnesium and aluminum body panels, (b) 1961 Chevrolet Corvette with prototype hood made from magnesium sheet, and (c) 1957 Chevrolet Corvette SS Race Car with 'featherweight magnesium body'. (Copyright 2007 GM Corp. Used with permission, GM Media Archive.)

(c)

4.27 Continued

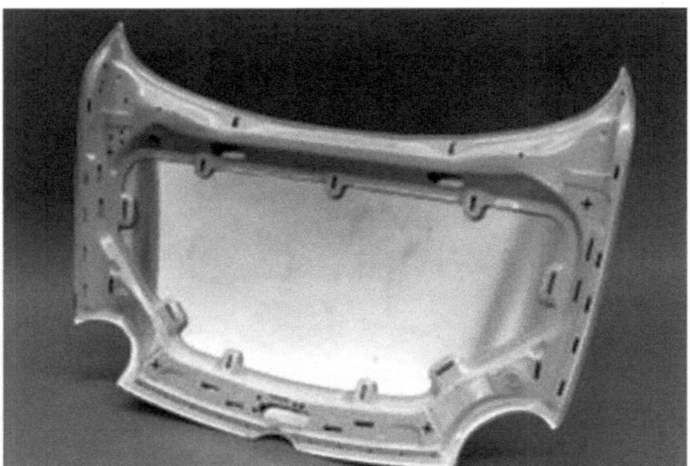

4.28 VW Lupo magnesium hood (Moll F *et al.*, 2004). (Used with permission from John Wiley and Sons.)

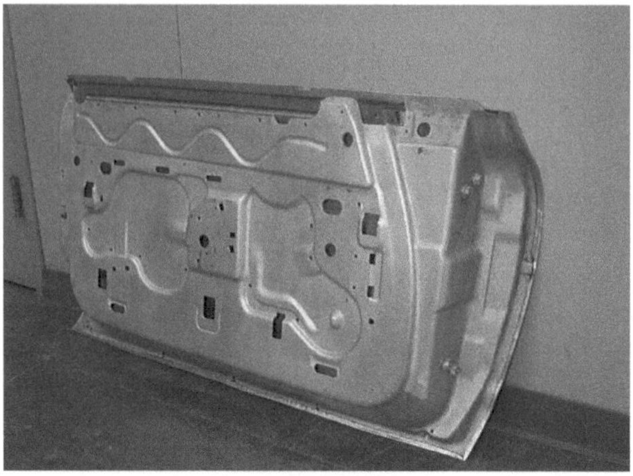

(a)

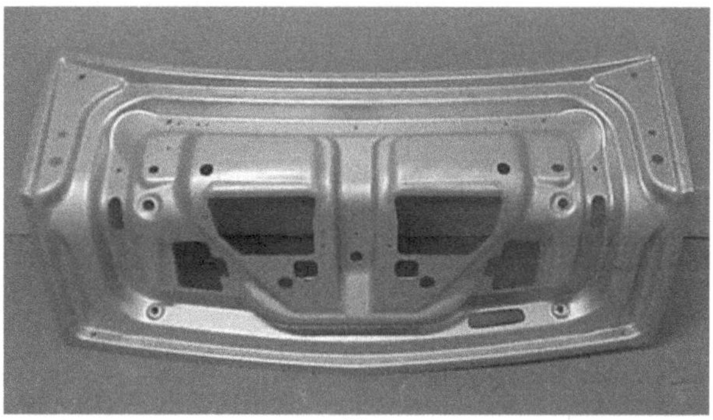

(b)

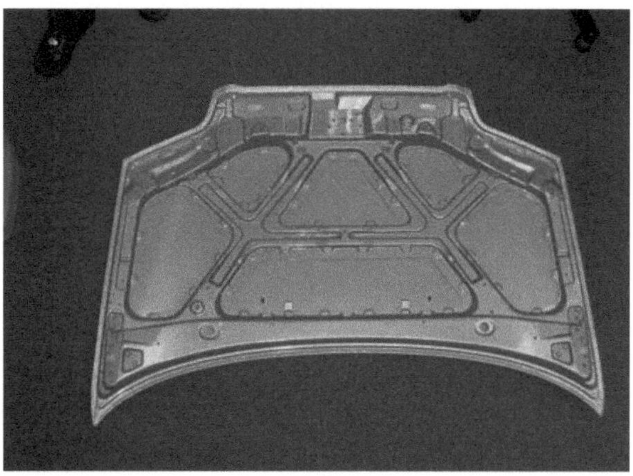

(c)

4.29 Magnesium sheet panels formed recently by General Motors: (a) door inner panel, (b) decklid inner panel, and (c) hood. (Krajewski 2001; Verma, 2006; Carter, 2008.)

4.30 Inner panel drawn by Daimler–Chrysler using magnesium sheet. (Used with permission from Chrysler Corp.)

4.4 Extruded magnesium

4.4.1 Alloys and properties

Table 4.9 lists the nominal composition and typical room-temperature tensile properties of common extruded magnesium alloy tubes (ASM International, 1999; Timminco Corporation, 1998; Luo and Sachdev, 2007, pp 321–326). Of the commercial extrusion alloys, AZ31 is most widely used in non-automotive applications. The higher aluminum content alloys, AZ61 and AZ80, offer greater strength than the AZ31 alloy, but have much lower extrudability. The high-strength Zr-containing ZK60 was designed for applications in racing cars and bicycles, such as wheels and stems (Timminco Corporation, 1998). The extrusion speed of ZK60A tubes is extremely low, rendering it uneconomical for automotive applications.

AM30 (3 per cent Al, 0.4 per cent Mn), a new extrusion magnesium alloy developed by General Motors (Luo and Sachdev, 2007, pp 321–326), is aimed to provide a good balance of strength, ductility, extrudability and corrosion resistance. Depending on the extrusion temperature used, the AM30 alloy was shown to extrude 20–30% faster without visible defects compared to the conventional AZ31 alloy (Luo and Sachdev, 2007, pp 321–326). Figure 4.31 summarizes the relative extrudability of magnesium alloys compared to aluminum alloy 6063, based on information from the literature (ASM International, 1999; Timminco Corporation, 1998; Busk, 1987); and the authors' experience (Luo and Sachdev, 2007, pp 321–326).

Table 4.9 Nominal compositions and typical room-temperature tensile properties of extruded magnesium alloy tubes (adapted from ASM International 1999; Luo and Sachdev, 2007 Timminco, 1998)

Alloy	Temper	Composition (wt. per cent)				Tensile properties		
		Al	Zn	Mn	Zr	Yield strength, (MPa)	Tensile strength, (MPa)	Elongation (per cent)
AZ31	F	3.0	1.0	0.20	–	165	245	12
AZ61	F	6.5	1.0	0.15	–	165	280	14
AZ80	T5	8.0	0.6	0.30	–	275	380	7
ZK60	F	–	5.5	–	0.45	240	325	13
ZK60	T5	–	5.5	–	0.45	268	330	12
AM30	F	3.0	–	0.40	–	171	232	12

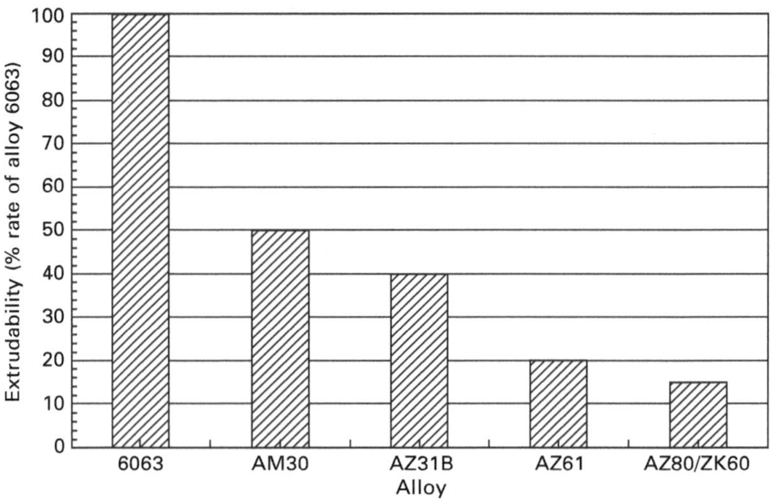

4.31 Relative extrudability of magnesium alloys compared to alloy 6063 (Luo, 2008).

4.4.2 Extrusion processes

Special equipment is not required to extrude magnesium alloys. The alloys can be warm or hot extruded in hydraulic presses to form bars, tubes, and a wide variety of profiles (ASM International, 1999). While hollow magnesium extrusions can be made with a mandrel and a drilled or pierced billet, it is generally preferable to use a bridge die where the billet is broken into three to four streams that get their internal shape as the streams flow over the supporting mandrel whose end face sits essentially flush with the die exit face that contains the external profile of the extrusion.

The hydrostatic extrusion process, typically used for copper tubing fabrication, is a much faster extrusion process compared with conventional direct extrusion. It was recently reported that seamless magnesium tubes were extruded using the hydrostatic process at speeds up to 100 m/min, due to the absence of friction between the billet and container since the billet is suspended in hydraulic oil (Savage and King, 2000, pp 609–614). Although the process is capable of extrusion ratios up to 700, the outer diameter of the tubes produced by this process is limited to about 45 mm, even with a large 4000-tonne press (Savage and King, 2000, pp 609–614).

4.4.3 Magnesium tube bending

Bending at room temperature

The rotary draw bending process, as shown in Fig. 4.32, is generally used for bending aluminum and magnesium tubes. In this process, the tube is clamped against the bend die insert with the clamp die, and a multi-ball steel mandrel is positioned inside the tube. The pressure die holds the collet end of the tube against the wiper die as the bend die rotates and the tube is drawn forward. Bending of conventionally extruded magnesium AZ31 and AM30 tubes at room temperature is generally unsuccessful; a 2D/90° bend could not be consistently achieved (Luo and Sachdev, 2005a, pp 477–482; Luo *et al.*, 2005b, pp 145–148). Fig. 4.33 compares the longitudinal cross

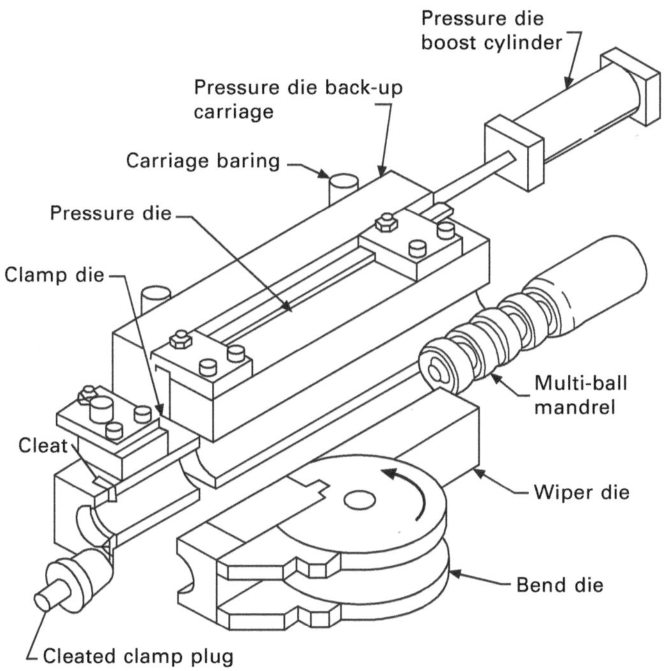

4.32 Schematic of the tooling used for bending tubes (Luo and Sachdev, 2008).

section micrographs of an AZ31 tube before and after a 30° bend at room temperature. The micrograph in Fig. 4.33(b) clearly shows the onset of strain localization initiating a crack prior to fracture. There is no evidence of any grain elongation, even near the outermost surface where the strain is the largest, suggesting that twinning is the predominant deformation mode at room temperature.

Bending at elevated temperatures

A moderate temperature (100–200°C) bending process was developed at GM for magnesium tubes, using a Pines rotary draw bending machine

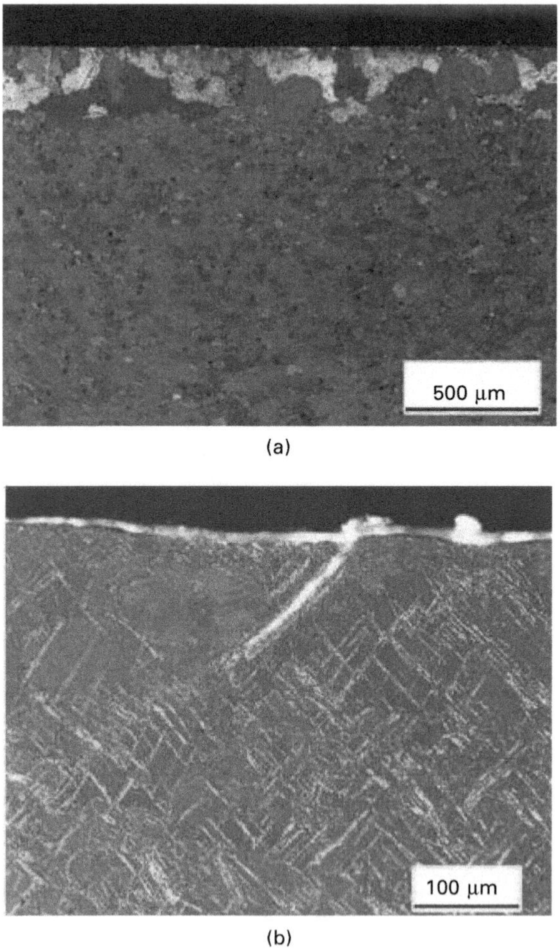

4.33 Optical micrographs showing the microstructure of an AZ31 tube: (a) before bending; and (b) after bending at room temperature (fractured at 30°) (Luo, 2008).

with heated tooling, as shown in Fig. 4.34 (Luo, 2005 and Sachdev, pp 477–482). In this bending process, three parts of the bend tooling, the bend die, the pressure die, and the mandrel, were heated, and the temperature for each heating zone was controlled separately. Using the optimal parameters developed for the moderate temperature bending process (Luo and Sachdev, 2005a, pp 477–482), a 2D/90° bend, as shown in Fig. 4.35, was consistently achieved with AZ31 and AM30 tubes at about 150 °C. No surface cracks were detected in these bent tubes.

In the cold rotary draw bending process, the tension (outer) side of the tube is thinned while the compression (inner) side is thickened. Figure 4.36 shows the degree of thinning along the tension side of magnesium alloy tubes bent

4.34 A rotary draw bending machine with heated tooling for magnesium tube bending.

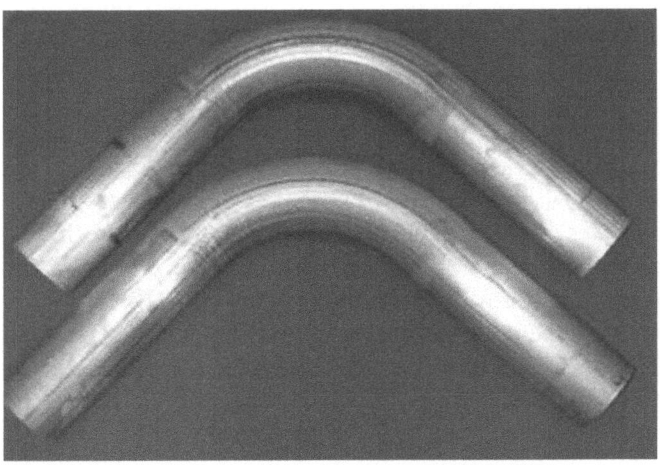

4.35 AM30 and AZ31 tubes bent (2D/90°) at 150 °C (Luo and Sachdev, 2008).

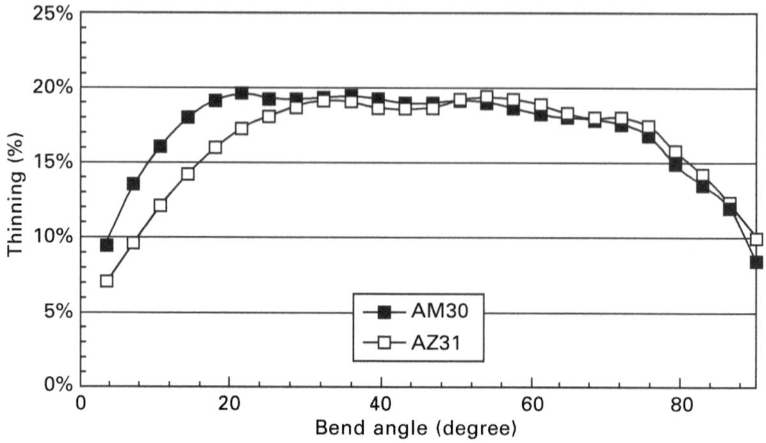

4.36 Thinning distribution in magnesium tubes bent at 150 °C (Luo and Sachdev, 2008).

to the 2D/90° condition at 150 °C. The smooth thinning distribution in Fig. 4.36 suggests uniform deformation during bending at elevated temperatures. The improved bendability of the magnesium alloy tubes is due to the higher ductility and formability reported in the literature (Luo and Sachdev, 2004, pp 79–85; Krajewski, 2001, 175–179).

4.4.4 Forming of magnesium extrusions

Elevated temperatures increase the formability of magnesium alloys. More complex geometries and greater circumferential expansion can be achieved at elevated temperatures using pressurized fluids such as oil or gas. Recently, a warm gas forming process at a much lower temperature range (150–350 °C) was developed using an Interlaken 5000-KN press (Fig. 4.37) (Martin *et al.*, 2005, pp 51–56). This press is equipped with two separate controllers and pressure intensifiers to perform the tests with either gas at elevated temperature or water at room temperature. For warm forming, nitrogen is pre-charged into the system with a pre-charge control valve, and forming under gas pressure or volume control is achieved with a pressure intensifier. Fast forming cycles of about 10 seconds can be obtained with either operating mode. Two end-feed actuators are used to seal and provide axial feeding during forming. The forming-die holder is heated with electric resistance elements and a temperature controller is used to measure and control the temperatures independently in the six zones of the die. The die cavity has two rectangular zones, a large and a small zone, and die inserts are used to create different geometries required for tube forming. Complete expansion involves an average tube expansion of 25 per cent and 50 per cent in the smaller and larger cavity regions, respectively. Figure 4.38 compares an

4.37 Warm gas forming press at CANMET, Ottawa, Canada (Martin, 2005).

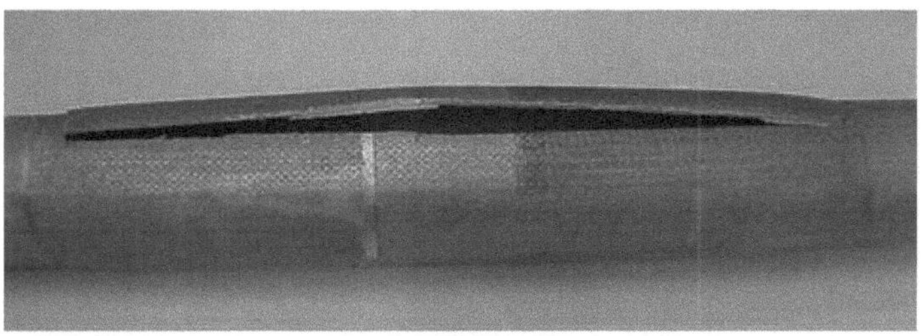

(a)

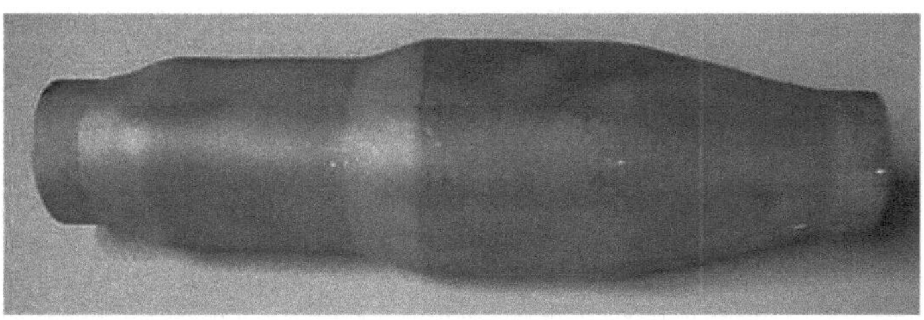

(b)

4.38 (a) Room temperature hydroforming and (b) warm gas forming of magnesium alloy AZ31 tubes (Luo, 2008).

AZ31 magnesium alloy tube hydroformed at room temperature with a similar tube gas-formed at 250 °C. While both tubes failed at an extrusion seam, gas forming could provide a circumferential expansion of 28% compared to only 8% with hydroforming at room temperature. Further optimization of extrusion and warm forming process would improve the formability of the magnesium tubes, especially if controlled end feeding could be accomplished. Alternatively, seamless extrusions can be considered.

Seamless extrusions of AZ31 magnesium alloy were the subject of another study of warm hydroforming (gas formed using nitrogen) (Ben-Artzy *et al.*, 2006, pp 253–258). Figure 4.39(a) shows the schematic of the warm forming process. The tooling consists of two units, the clamping and the forming dies. The clamping unit ensures the sealing of the tube and the supply of the pressure medium. The tube is put into the forming unit, which consists of a main body and a free expansion guided zone equipped with cooling and heating devices. The guided zone makes it possible that tubes of different outer diameters can be formed under plane strain condition. Band heaters are used to heat the forming unit and the sealing punches. Figure 4.39(b) shows an AZ31 tube formed at 350 °C with 80 per cent circumferential expansion.

4.4.5 Automotive applications

Magnesium extrusions are used in aerospace, nuclear, luggage, hand-tools, bicycle and motorcycle applications, but there have been no reported applications of magnesium extrusions in the automotive industry. Table 4.10 summarizes the potential applications of magnesium extrusions in automotive interior, and body and chassis areas, some of which may involve hydroforming processes.

Magnesium extrusions have been used to make prototype parts, such as bumper beams and most parts of a spaceframe for the VW 1-Liter Car (Schumann and Friedrich, 2003). However, the production of magnesium tubes/extrusions in automotive structures would require more development to meet all performance and cost targets as well as a supply base for high volume automotive production. Recently, a US–Canada–China collaborative effort has been focused on the development of enabling technologies for magnesium body applications using front end body structure as a test bed. A magnesium front end would weigh about 35 kg and provide up to 40 kg of mass saving, compared with the equivalent steel construction (Luo and McCune, 2008a). Bending and gas forming processes of magnesium tubes will be further developed and validated in this project for automotive applications.

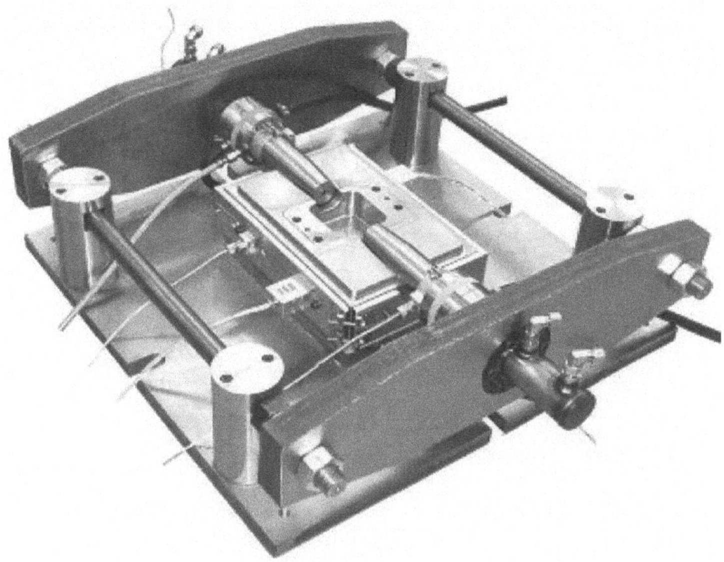

(a)

(b)

4.39 Warm gas forming of magnesium alloy tubes (a) lower die and (b) AZ31 tube formed at 350°C (80 per cent circumference expansion) (Ben-Artzy, 2006). (Used with permission from TMS.)

Table 4.10 Potential automotive applications of magnesium extrusions

System	Component
Interior	Instrument panel Seat components HVAC (heating, ventilation and air-conditioning) components
Body	Roof frame Bumper beam Radiator support Shotgun Frame rail
Chassis	Engine cradle Subframe

4.5 Future trends

While magnesium is the lightest structural metal and the third most commonly used metallic material in automobiles following steel and aluminum, many challenges remain in various aspects of alloy development and manufacturing processes to exploit its high strength-to-mass ratio for widespread lightweight applications in the automotive industry.

4.5.1 Material challenges

Compared with the numerous aluminum alloys and steel grades, there are only a limited number of low-cost magnesium alloys available for automotive applications. The conventional Mg–Al based cast alloys offer moderate mechanical properties due to the limited age-hardening response of this alloy system. Since the development of vacuum die casting and other high-integrity casting processes, magnesium castings can be heat-treated with no blisters. Alloy systems with significant precipitation hardening such as Mg–Sn (Mendis *et al.*, 2006, pp 163–171; Luo and Sachdev, 2009, pp. 437–443) and Mg–rare earths (Fu *et al.*, 2008, pp 182–192) should be developed with improved mechanical properties. New alloys with improved ductility, fatigue strength, creep resistance and corrosion resistance should also be explored.

For wrought magnesium alloys, the strong plastic anisotropy and tension–compression asymmetry due to texture remain obstacles for many structural applications. Microalloying with elements such as Ce (Mishra *et al.*, 2008, pp. 562–565), to improve the plasticity of magnesium alloys, has proven to be an effective approach in wrought magnesium alloy development. Other alloys systems such as Mg–Zn–rare earths have shown more 'isotropic' mechanical properties. Although additions of more than 11% Li can transform the HCP α-Mg solid solution into highly workable and more isotropic body-centered cubic alloys, automotive applications of this alloy will remain difficult.

The properties of magnesium alloys can be significantly enhanced if micro- and nano-particles are introduced to form metal matrix composites (MMC). Micro- and nano-sized particles offer strengthening mechanisms in different length scales and provide a tremendous opportunity for a new class of engineering materials with tailored properties and functionalities for automotive applications. Computational thermodynamics and kinetics will be used in the design and manufacture of these MMC materials

4.5.2 Process challenges

The current success of magnesium is primarily attributed to its superior die-castability compared with aluminum alloys. However, more challenges exist in gravity, low-pressure and squeeze casting of magnesium alloys due to the need to compensate for larger shrinkage when compared with aluminum alloys. Melt handling, molten metal transfer, grain-refinement, and kind and amount of die lubrication coating as well as casting parameters need to be developed specifically for magnesium alloys to fully utilize their intrinsic properties.

Various forming processes need to be optimized for magnesium alloys. Elevated temperature forming is needed for most extrusion and sheet components. Research efforts have been directed to lowering the forming temperatures and reducing the cycle times. New forming processes should be developed to utilize the dramatically improved formability of magnesium alloys at certain ranges of temperature and strain rate. Room temperature (RT) or near-RT forming techniques are also being explored for new magnesium alloys such as Mg–Zn–Ce alloy (Bohlen *et al.*, 2007, pp 2101–2112).

Fabrication of Mg-based MMCs includes liquid-mixing and preform infiltration. Both processes rely on the ability to achieve controlled dispersion (distribution) of the reinforcement phase in the metal matrix. Dispersion can be enhanced by new techniques such as ultrasonic, friction-stir and vacuum die casting for MMC fabrication.

Welding and other joining methods comprise the very important final step in assembling magnesium components and sub-systems. There is a wide range of joining techniques that can be potentially used for Mg-to-Mg, Mg-to-Al and Mg-to-steel joining. However, to date there have not been welded magnesium structures in automotive applications. Some of the techniques being developed for magnesium include arc-welding, laser welding, resistance spot welding (RSW), friction stir welding (FSW), adhesive bonding, self-piercing rivets (SPR) and mechanical fastening (http://www.aist.go.jp/aist_e/latest_research/2008/20081105/20081105.html).

4.5.3 Performance challenges

There are several performance-related challenges that need significant research efforts. Some of them are highlighted in two current USAMP projects, the Magnesium Powertrain Cast Components Project (Powell, 2009; Beals *et al.*, 2007) and the Magnesium Front End Research and Development Project (Luo *et al.*, 2008b, pp 3–10).

Crashworthiness

Magnesium castings have been used in many automotive components such as the instrument panel beams and radiator support structures. High-ductility AM50 or AM60 alloys are used in these applications and performed well in crash simulation and tests; and many vehicles, with these magnesium components achieved five-star crash rating. However, there is limited material performance data available for component design and crash simulation. A recent study shows that magnesium alloys can absorb significantly more energy than either aluminum or steel on an equivalent mass basis (Easton *et al.*, 2008, pp 57–62). While steel and aluminum tubes fail by progressive folding in crash loading (more desirable situation), magnesium alloy tubes (AZ31 and AM30) tend to fail by sharding or segment fracture (Easton *et al.*, 2008, pp 57–62; Wagner *et al.*, 2009). However, the precise fracture mechanisms for magnesium under crash loading are still not clear, and material models for magnesium fracture are needed for crash simulation involving magnesium components. Additionally, new magnesium alloys need to be developed to have progressive folding deformation in crash loading.

Noise, vibration and harshness (NVH)

It is well known that magnesium has high damping capability, but this can be translated into better NHV performance only for mid range sound frequency; 100–1000 Hz (Logan *et al.*, 2006). Low-frequency (<100 Hz) structure-borne noise can be controlled by component stiffness between the source and receiver of the sound. The lower modulus of magnesium, compared to steel, is often compensated by thicker gauges and/or ribbing designs. For high-frequency (>1000 Hz) airborne noise, a lightweight panel, regardless of material, would transmit significantly more road and engine noise into the occupant compartment unless the acoustic frequencies could be broken up and damped. Magnesium, with its low density, is disadvantaged for this type of applications unless new materials with laminated structures are developed for sound insulation; greater damping.

Fatigue and durability

Fatigue and durability are critical in magnesium structural applications and there are limited data in the literature, especially on wrought alloys. The effect of alloy chemistry, processing and microstructure on the fatigue characteristics of magnesium alloys needs to be studied. Casting, extrusion and sheet products need to be characterized sufficiently to establish links between microstructural features and fatigue behavior. Multi-scale simulation tools can be used to predict the fatigue life of magnesium components and sub-systems, which can be validated for automotive applications.

Corrosion

Pure magnesium has the highest standard reduction potential of the structural automotive metals (see Table 4.11). As noted earlier, while pure magnesium (at least with very low levels of iron, nickel, and copper) has atmospheric corrosion rates that are similar to those of aluminum, magnesium's high reduction potential makes it very susceptible to galvanic corrosion; that is, when magnesium is in electrical contact with other metals below it in the reduction potential table. The impact of this susceptibility to galvanic corrosion on the application of magnesium in exposed environments is severe in both the macro-environment and the micro-environment. In the macro-environment, magnesium alloys must be electrically isolated from other metals to prevent the creation of galvanic couples; e.g. the use of steel bolts in direct contact with magnesium. Isolation can be achieved by replacing the bolt with a less reactive metal, as has been done in the Mercedes automotive transmission case, where steel bolts have been replaced with aluminum bolts (Greiner *et al.*, 2004). Isolation can also be achieved by coating the 'other' metal. Finally, isolation can be achieved by the use

Table 4.11 Standard reduction potential of common metals (Hawke *et al.*, 1999)

Electrode	Reaction	Potential (V)
Li, Li^+	$Li^+ + e^- \Rightarrow Li$	–3.02
K, K^+	$K^+ + e^- \Rightarrow K$	–2.92
Na, Na^+	$Na^+ + e^- \Rightarrow Na$	–2.71
Mg, Mg^{2+}	$Mg^{2+} + 2e^- \Rightarrow Mg$	–2.37
Al, Al^{3+}	$Al^{3+} + 3e^- \Rightarrow Al$	–1.71
Zn, Zn^{2+}	$Zn^{2+} + 2e^- \Rightarrow Zn$	–0.76
Fe, Fe^{2+}	$Fe^{2+} + 2e^- \Rightarrow Fe$	–0.44
Cd, Cd^{2+}	$Cd^{2+} + 2e^- \Rightarrow Cd$	–0.40
Ni, Ni^{2+}	$Ni^{2+} + 2e^- \Rightarrow Ni$	–0.24
Sn, Sn^{2+}	$Sn^{2+} + 2e^- \Rightarrow Sn$	–0.14
Cu, Cu^{2+}	$Cu^{2+} + 2e^- \Rightarrow Cu$	0.34
Ag, Ag^+	$Ag^+ + e^- \Rightarrow Ag$	0.80

of shims or spacers of compatible materials of sufficient geometry and size to prevent electrical contact in the presence of salt water, as shown, for example, for the Corvette cradle (Fig. 4.16). While the casting cost can be competitive with aluminum, the isolation strategies required can often make the application more expensive and thus restrictive in its use. Much work is being conducted to address corrosion issues with magnesium, but these are beyond the scope of this chapter (Song, 2005).

A major challenge in magnesium automotive applications is to establish the surface finishing and corrosion protection processes. The challenge is two-fold since surface treatments to magnesium play roles in both manufacturing processes (e.g. adhesive bonding) as well as in the product life cycle that demands corrosion resistance. Furthermore, the current manufacturing paradigm for steel-intensive body structures employs chemistries in the paint shop that are corrosive to magnesium. Conditions are further aggravated by galvanic couples containing primarily steel fasteners. Future research will explore novel coating and surface treatment technologies including pretreatments such as micro-arc anodizing, non-chromated conversion coatings, and 'cold' metal spraying of aluminum onto magnesium surfaces. Since most studies of corrosion protection and pre-treatment of magnesium have focused on die castings, the behavior of sheet, extrusion and high-integrity castings will be explored for process compatibility.

In summary, the future success of magnesium as a major automotive material will depend on how the technical challenges are addressed. These challenges are huge and global, and would require significant collaboration among industries, governments and academia from many countries. It is very encouraging that many of these international and interdisciplinary collaborations are being nurtured for magnesium and it is expected that future developments will enable us further utilize the benefits of magnesium, the lightest structural metal.

4.6 Acknowledgments

The authors gratefully acknowledge the careful read and insightful comments by Dr. Anil K. Sachdev, General Motors Research & Development Center.

4.7 References

Aghion E, Moscovitch N, Arnon A, and Bronfin B (2007), 'The Capability of MRI Magnesium Alloys to Address High Temperature Application Requirements', in Beals R, Luo A, Neelameggham N, and Pekguleryuz M *Magnesium Technology 2007,* TMS, Warrendale, Pennsylvania, 253–255.

ASM International (1988), *Metals Handbook,* **15**, ASM International, Materials Park, Ohio.

ASM International (1992), *Metals Handbook,* **2**, ASM International, Materials Park, Ohio.

ASM International (1998), David J R, *Metals Handbook, Desk Edition*, 2nd Edition, ASM International, Materials Park, Ohio.

ASM International (1999), in Avedesian M and Baker H *Magnesium and Magnesium Alloys,* ASM International, Materials Park, Ohio.

ASTM International, Committee B07 on Light Metals and Alloys (2008), *Standard Practice for Codification of Unalloyed Magnesium and Magnesium – Alloys, Cast and* Wrought', ASTM International Designation B951-08, ASM International, West Conshohocken, Pennsylvania.

Barnes A J (2002), 'Industrial Applications of Superplastic Forming', *23rd Annual Aluminum Design and Fabrication Seminar*, October 23, 2002, Livonia, Michigan.

Barnes L T (1992), 'Rolled Magnesium Products: What Goes Around, Comes Around', *Proceedings of the International Magnesium Association*, Chicago, IL, 29–43.

Barnes A JA (2007), 'Superplastic Forming 40 Years and Still Growing,' *JMEPEG*, **16**, 440–454.

Beals R S, Liu Z-K, Jones J W, Mallick P K, Emadi D, Schwam D, and Powell B R (2007), 'US Automotive Materials Partnership, Magnesium Powertrain Cast Components Project: Fundamental Research Summary', Journal of Minerals, Metals, and Materials, **59** (8) 43–8.

Behrens B-A and Hubner S (2005), 'Heated Clinching of Magnesium Sheet Metal', *Annals of the German Academic Society for Production Engineering*, Production Engineering Research Development, **XII** (1), 59–62.

Ben-Artzy A, Spinat E, Dahan O, Siegert K, Jager S, Mueller K, Altan T (2006), 'High Internal Pressure Forming of Magnesium Tubes', in *Magnesium Technology 2006*, eds. Luo A A, Neelameggham N R and Beals R S, TMS, Warrendale, PA, 253–258.

Bohlen J, Nuernberg M, Senn J W, Letzig D, and Agnew S R (2007), 'The Texture and Anisotropy of Magnesium-Zinc-Rare Earth Alloy Sheets', *Acta Materialia*, **55** (6), 2101–2112.

Brown, R E (2008), 'Future of Magnesium Developments in 21st Century', presentation at *Materials Science & Technology Conference*, Pittsburgh, PA, USA, October 5–9.

Brown Z, *et al.* (2009) 'Development of Super-vacuum Die Casting Process for Magnesium Alloys', to be published in *Cast Expo '09 Proceedings*, North American Die Casting Association, Wheeling, Illinois.

Buschow K H J and van Vucht J H N (1967), 'Systematic Arrangement of the Binary Rare Earth–Aluminum Systems', *Philips Res. Reports*, **22,** 233–45.

Busk R S (1987), *Magnesium Product Design*, Marcel Dekker, Inc., New York and Basel.

Bussy A, *Mémoire sur le Radical métallique de la Magnésie*, 1831.

Carpenter J A, Jackman J, Osborne R J, Powell B R, Li N, and Sklad P (2008), 'Automotive Research and Development in North America,' *Die Casting Engineer*, (3), 54–9.

Carsley J E and Kim S (2007), 'Warm Hemming of Magnesium Sheet,' *Journal of Materials Engineering and Performance*, **16,** 331–338.

Carter J, Krajewski P E, and Verma R (2008), 'The Hot Blow Forming of AZ31 Mg Sheet: Formability Assessment and Application Development', *Journal of Minerals, Metals, and Materials*, **60** (11), 77–81.

DiFrancesco C A and Kramer D A (2008), 'Magnesium Metal Statistics', U.S. Geological Survey. Available from http://minerals.usgs.gov/minerals/pubs/commodity/magnesium/

Easton M, Beer A, Barnett M, Davies C, Dunlop G, Durandet Y, Blacket S, Hilditch T, and Beggs P (2008), 'Magnesium Alloy Applications in Automotive Structures', *Journal of Minerals, Metals, and Materials*, **60** (11), 57–62.

Emley E (1966), *Principles of Magnesium Technology,* 1st Ed., Pergamon Press, Ltd., Oxford, England, 828.

Friedrich H and Schumann S (2001), 'Research for a "new age of magnesium" in the automotive industry', *Journal of Mat. Proc. Tech.,***117**, 276–281.

Fu P, Peng L, Jiang H, Chang J, and Zhai C (2008), 'Effects of Heat Treatments on the Microstructures and Mechanical Properties of Mg–3Nd–0.2Zn–0.4Zr (wt.%) Alloy', Materials Science and Engineering A, **486,** 183–192.

Greiner J, Doerr C, Nauerz H, and Graeve M (2004), 'The New "7G-TRONIC" of Mercedes-Benz: Innovative Transmission Technology for Better Driving Performance, Comfort, and Fuel Economy', *SAE Technical Paper No. 2004-01-0649,* SAE International, Warrendale, Pennsylvania.

Groche R, Huber R, Dorr J, and Schmoeckel D (2002),'Hydromechanical Deep Drawing of Aluminium Alloys at Elevated Temperatures', *Ann. CIRP*, **51/1**, 215–218.

Habashi F (2006), 'A History of Magnesium', in Pekguleryuz M and Mackenzie L *Magnesium Technology in the Golden Age*, Montreal, METSOC, 31–42.

Hawke D L, Hillis J E, Pekguleryuz M, and Nakatsugawa I (1999), 'Corrosion Behavior' in Avedesian M M and Baker H, ed., *Magnesium and Magnesium Alloys,* ASM International, Materials Park, Ohio, 194.

Hillis J (1983), 'The Effects of Heavy Metal Contamination on Magnesium Corrosion Performance', *SAE Technical Paper No. 830523*, SAE International, Warrendale, Pennsylvania.

Hoeschl M, Wagener W, and Wolfe J (2006), 'BMW's Magnesium–Aluminum Composite Crankcase: State-of-the-Art Light Metal Casting and Manufacturing', *SAE Technical Paper No. 2006-01-0069,* SAE International, Warrendale, Pennsylvania.

Hollrigl-Rosta F (1980), 'Magnesium in Volkswagen', *Light Metal Age*, 22–29.

Hollrigl-Rosta F, Just R, Kohler J, and Melzer, H-J (1980a), 'Magnesium in Volkswagen', *Metall,* **12,** 12.

Japanese Casting Society (1996), *Japan Casting Handbook,* Japanese Casting Society, **4**, 8.

Jorstad J (1993), in Zalensas D *Aluminum Casting Technology,* 2nd Ed., The American Foundrymen's Society, Des Plaines, Illinois, 21.

Kainer K and Benzler T (2003), 'Squeeze-casting and Thixo-casting of Magnesium Alloys', in Kainer K *Magnesium – Alloys and Technologies,* Wiley-VCH Verlag GmbH & Co. KGaA, Weinheim, Germany, 58.

Kaye A and Street A (1982), *Die Casting Metallurgy,* Butterworth Scientific, London.

Krajewski P E (2001), 'Elevated Temperature Forming of Sheet Magnesium Alloys', *SAE International Technical Paper No. 2001-01-3104,* Warrendale, Pennsylvania, 175–179.

Krajewski P E, *et al.* (2006), 'Warm Forming of Aluminum: Summary of USAMP Project AMD307', presentation at *Materials Science & Technology Conference*, Oct. 15, Cincinnati, Ohio.

Krajewski P E and Schroth J G (2007), 'Overview of Quick Plastic Forming Technology', *Materials Science Forum*, **551/552**, 3–12.

Krajewski P E, Kim S, Carter J T, and Verma R (2007a),'Magnesium Sheet: Automotive Applications and Future Opportunities', *Trends in Metals & Materials Engineering*, **20** (5), 60–68.

Kunst M *et al.* (2006), 'Creep Deformation and Mechanisms of AJ (Mg–Al–Sr) Alloys', in Pekguleryuz M and Mackenzie L *Magnesium Technology in the Golden Age,* Montreal, METSOC, 647–661.

LeBeau S E, Yamamoto Y, and Sakamoto K (1998), 'Thixomolding of Magnesium Automotive Components', *SAE International Technical Paper No. 980087*, SAE International, Warrendale, Pennsylvania.

Li N, Osborne R, Cox B, and Penrod D (2005), 'Magnesium Engine Cradle – The USCAR Structural Cast Magnesium Development Project', *SAE International Technical Paper No. 2005–01–0337*, SAE International, Warrendale, Pennsylvania.

Logan S, Kizyma A, Patterson C, and Rama S (2006), 'Lightweight Magnesium-Intensive Body, Structure', *SAE International Technical Paper No. 2006-01-0523*, SAE International, Warrendale, Pennsylvania.

Luo *et al.* (2001), 'Creep and Microstructure of Magnesium–Aluminum-Calcium-based Alloys, *Met. and Mat. Trans.*, 33A, 567–574.

Luo A A and Sachdev A K (2004), 'Mechanical Properties and Microstructure of AZ31 Magnesium Alloy Tubes', in *Magnesium Technology 2004*, ed. Luo A A, The Minerals, Metals and Materials Society (TMS), Warrendale, PA, 79–85.

Luo A A and Sachdev A K (2005a), 'Development of a Moderate Temperature Bending Process for Magnesium Alloy Extrusions', presented at *The International Conference on Magnesium*, Beijing, China, Sept. 20–24, 2004, and published in *Materials Science Forum*, **488/489**, 477–482.

Luo A A, Sachdev A K, Mishra R K and Kubic R C (2005b), 'Bendability and Microstructure of Magnesium Alloy Tubes at Room and Elevated Temperatures', in *Magnesium Technology 2005*, eds. Neelameggham N R, Kaplan H I, and Powell B R, TMS, Warrendale, PA, 145–148.

Luo A A and Sachdev A K (2007), 'AM30 – A New Wrought Magnesium Alloy', in Beals R S, Luo A A, Neelameggham N R, and Pekguleryuz M O, *Magnesium Technology 2007*, TMS, Warrendale, PA, 321–326.

Luo A A and Sachdev A K (2008), 'Bending and Hydroforming of Aluminum and Magnesium Alloy Tubes', in Koc M, ed., *Hydroforming for Advanced Manufacturing,* Woodhead Publishing, Cambridge, England.

Luo A A and McCune R C (2008a), 'Magnesium Front End Projects', V.S. Dept. of Energy, *Automotive Lightweighting Materials*, FY 2006 Progress Report, TMS 2008.

Luo A A, Nyberg E A, Sadayappan K, and Shi W (2008b), 'Magnesium Front End Research and Development: A Canada–China–USA Collaboration', published in *Magnesium Technology 2008*, eds. M.O. Pekguleryuz, N.R. Neelameggham, R.S. Beals and E.A. Nyberg, TMS, Warrendale, PA, 3–10.

Luo AA and Sachdev AA (2009), 'Microstructure and Mechanical Properties of Mg–Al–Mn and Mg-Al-Sn Alloys', published in *Magnesium Technology 2009*, eds. E.A. Nyberg, S.R. Agnew, N.R. Neelameggham and M.O. Pekguleryuz, TMS, Warrendale, PA, 437–443.

Margrave J (1967), *The Characterization of High Temperature Vapors,* John Wiley & Sons, New York, New York.

Martin P, Baragar D, Boyle K P, Luo A A, Jonas J J, Godet S, Neale K W (2005), 'Elevated Temperature Property Measurements for Warm Forming Aluminium Alloy Tubes', in Proceedings of *the 2nd International Light Metals Technology Conference* Ed. Kaufmann H, June 8–10, 2005, St. Wolfgang, Austria. 51–56.

Mendis C L, Bettles C J, Gibson M A, and Hutchinson C R (2006), 'An Enhanced Age Hardening Response in Mg–Sn Based Alloys Containing Zn', *Materials Science and Engineering A* **435/436** 163–171.

Mercer W E (1990), 'Magnesium Die Cast Alloys for Elevated Temperature Applications', SAE Technical Paper No. 900788, SAE International Warrendale, PA.

Mishra R K, Gupta A K, Rao P R, Sachdev A K, Kumar A M, and Luo A A (2008), 'Influence of Cerium on the Texture and Ductility of Magnesium Extrusions', *Scripta Materialia*, **59**, 562–565.

Moll F, Mekkaoui F, Schumann S, and Friedrich H (2004) 'Application of Mg sheets in Car Body Structure', in Kainer K U Ed. *Magnesium: Proceedings of the 6th International Conference Magnesium Alloys and Their Applications*, Wiley-VCH, Weinheim, Germany.

Nelson K E (1970), 'Magnesium Die Casting Alloys', *SDCE Transactions*, Paper No. 13.

Okamoto H (1988), 'Mg (Magnesium)', in *Phase Diagrams of Binary Magnesium Alloys*, A.A. Nayeb-Hashema and J.B. Clark, Ed., ASM International, 1–3.

Pidgeon L M, Mathes J C, Woldman N E, Winkler J V, and Loose W S (1999), *Magnesium*, ASM International, 4–22.

Polmear I J (1999), in *Magnesium and Magnesium Alloys*, M.M. Avedesian and H. Baker, Ed., ASM International, 3–5.

Powell B R, Luo A A, Rezhets V, Bommarito J J, and Tiwari B L (2001), 'Development of Creep-Resistant Magnesium Alloys for Powertrain Applications: Part 1 of 2', *SAE Technical Paper 2001-01-0422*, SAE International, Warrendale, PA.

Powell B R, Rezhets V, Balogh M P, and Waldo R A (2002), 'Microstructure and Creep Behavior in AE42 Magnesium Die-Casting Alloy', *TMS Journal of Minerals, Metals, and Materials*, **54** (8), 34–38.

Powell B, *et al.* (2005), 'Progress Toward a Magnesium-intensive Engine: the USAMP Magnesium Powertrain Cast Components Project', *SAE 2004 Transactions, Journal of Materials & Manufacturing Paper No. 2004-1-0654*, SAE International, Warrendale, Pennsylvania, 250–259.

Powell B R (2009), 'Magnesium Powertrain Cast Components,' published in *FY2008 Annual Progress Report for Automotive Lightweighting Materials, U.S.* Department of Energy, Washington, D.C., April.

Rashid M S *et al.* (2001), *Quick Plastic Forming of Aluminum Alloy Sheet Metal.* US Patent 6 253 588, July 3, 2001.

Savage K and King J F (2000), 'Hydrostatic Extrusion of Magnesium', in *Magnesium Alloys and Their Applications*, ed. Kainer K U, Wiley-VCH, Weinheim, Germany, 609–614.

Schroth J G (2004), in Taleff E M ed. *Advances in Superplasticity and Superplastic Forming* TMS, Warrendale, PA.

Schumann S and Friedrich H (1998), 'The Use of Magnesium in Cars – Today and in the Future,' in Mordike B and Kainer K *Magnesium Alloys and their Applications*, Werkstoff-Informationsgesellschaft, Frankfurt, Germany, 3–13.

Schumann S and Friedrich H (2003), 'Current and Future Use of Magnesium in the Automobile Industry', *Magnesium Alloys 2003,* Materials Science Forum, Trans Tech, Switzerland, **419–422**, 51–56.

Sieracki E G, Velazquez J J, and Kabiri K (1996), 'Compressive Stress Retention Characteristics of High Pressure Die Casting Magnesium Alloys', SAE Technical Publication No. 960421, SAE International, Warrendale, PA.

Song G (2005), *Advanced Engineering Materials*, **7**, 7.

Taub A I, Krajewski P E, Luo A A, and Owens J N (2007), 'The Evolution of Technology for Materials Processing over the Last 50 Years: The Automotive Example', *Journal of Minerals, Metals, and Materials*, **59** (2) 48–57.

Timminco Corporation (1998), *Magnesium Wrought Products*, Timminco Corporation Brochure, Aurora, CO.

Verma R and Carter J T (2006), 'Quick Plastic Forming of a Decklid Inner Panel with Commercial AZ31 Magnesium Sheet', *SAE International Technical Paper No. 2006-01-0525*, SAE International, Warrendale, Pennsylvania.

Vinarcik E (2003), *High Integrity Die Casting Processes,* John Wiley & Sons, Hoboken, New Jersey.

Wagner D A, Logan S D, Wang K, Skszek T, and Salisbury, C P (2009), 'Test Results and FEA Predictions from Magnesium AM30 Extruded Beams in Bending and Axial Compression', published in *Magnesium Technology 2009*, eds. E.A. Nyberg, S.R. Agnew, N.R. Neelameggham and M.O. Pekguleryuz, TMS, Warrendale, PA.

Webb L (2003), 'Magnesium Supply and Demand 2002', in *International Magnesium Association 60th Annual World Magnesium Conference,* Washington, D.C., 31–34.

Wei L Y and Dunlop G L (1992), 'Precipitation Hardening in a Cast Mg–Rare Earth Alloy', in *Magnesium Alloys and Their Applications*, DGM Informationsgesellschaft, Verlag, Germany, 335.

5 Thermoplastics and thermoplastic–matrix composites for lightweight automotive structures

P. K. MALLICK, University of Michigan-Dearborn, USA

Abstract: This chapter discusses the properties, processing methods and design issues for thermoplastics and thermoplastic–matrix composites used for automotive applications. It starts with a brief review of the important characteristics of thermoplastic polymers, differences in design considerations for metals and thermoplastics, and injection molding, the most common processing method for producing thermoplastic parts in the automotive industry. It then discusses the properties and processing of various thermoplastic–matrix composites reinforced with short fibers, long fibers, glass mat, glass fabric, unidirectional fibers, self-reinforcement, natural fibers and nano-reinforcements. It also describes various joining methods commonly used for thermoplastic–matrix composite parts.

Key words: thermoplastics; thermoplastic–matrix composites, random fiber composites, continuous fiber composites, self-reinforced composites, natural fiber composites, nanocomposites, joining.

5.1 Introduction

Thermoplastics are a class of polymers that can be softened and melted by the application of heat, and can be processed either in the heat-softened state (e.g. by thermoforming) or in the liquid state (e.g. by extrusion and injection molding). This is in contrast to the thermosets, the other class of polymers, which cannot be melted by the application of heat. Thermoplastic polymers can be processed repeatedly by the application of heat and can be recycled directly into making new products; however, it should be noted that repeated processing may cause deterioration in some of their properties. The common manufacturing processes used for making thermoplastic parts are injection molding, blow molding and thermoforming.

In addition to the advantage of recycling, thermoplastics have several other advantages over thermosets. In general, they have higher ductility and impact resistance than thermosets. They can also be joined together by a variety of welding techniques, such as resistance welding, vibration welding and ultrasonic welding. In general, the processing time for thermoplastic parts is significantly lower than that for thermoset parts. This is because processing of thermoset parts involves a chemical reaction (a curing or cross linking reaction) in the mold, which, depending on the mold temperature and part thickness, can take several minutes to several hours. Processing of thermoplastic parts does not involve any chemical reaction in the mold. If injection molding is used for making thermoplastic parts, heating required for melting the thermoplastic polymer takes place outside the mold; only cooling of the parts takes place in the mold, which is usually accomplished in less than a minute.

5.2 Thermoplastics used in automobiles

5.2.1 General characteristics

A large variety of thermoplastics is already used in automobiles (Maxwell, 1994; Lange, 2003), a list of which is given in Table 5.1. They are often combined with either mineral fillers or short fibers to increase modulus and/or strength, increase heat deflection temperature, control thermal expansion and reduce mold shrinkage. They may also contain a variety of additives, such as heat stabilizers, anti-oxidants, fire retardants, plasticizers, impact modifiers, and colorants; some to improve their processing characteristics, some to change their physical or thermal characteristics and some to improve one or more mechanical properties. Some of the thermoplastics listed in Table 5.1 are amorphous polymers, while others are semi-crystalline, containing both crystalline and amorphous phases. In general, semi-crystalline polymers have higher chemical resistance, higher coefficient of thermal expansion and higher mold shrinkage than amorphous polymers. The amount of crystallinity in semi-crystalline polymers depends strongly on the processing conditions. In general, the higher the amount of crystallinity, the higher are the modulus, yield strength, tensile strength, and resistance to solvents and chemicals, but the lower the elongation to failure and resistance to crack propagation.

Both amorphous and semi-crystalline polymers exhibit a glass transition temperature, T_g. At temperatures below T_g, many amorphous polymers behave like a glass, i.e. they are hard, brittle solids with low strain-to-failure and low impact strength. Above T_g, they become soft, ductile, rubber-like solids with relatively high strain-to-failure and high impact strength. Their modulus also reduces by several orders of magnitude as the temperature is raised above T_g (Fig. 5.1). At a much higher temperature, the amorphous polymers change from a solid to a liquid; however, unlike crystalline materials, they do not exhibit any melting point. For semi-crystalline polymers, the reduction in modulus above T_g is not as large as that for amorphous polymers. They also exhibit a crystalline melting point, T_m, above which the polymer changes from a solid material to a liquid material. At higher temperatures, both amorphous and semi-crystalline thermoplastics start to degrade irreversibly and eventually decompose.

Table 5.1 Mechanical properties of several thermoplastic polymers used in automobiles

Polymer	Acronym	Amorphous or semi-crystalline	Density (g/cc)	Modulus (GPa)	Strength (MPa)	Strain to failure strength (%)	Notched Izod impact examples* (J/m)	Automotive application
High density polyethylene	HDPE	Semi-crystalline	0.96	1	20–30	–	20–750	Windshield washer fluid bottle, ducts, fuel tank
Polypropylene	PP	Semi-crystalline	0.90	1.3	30–40	30–200	20–70	Battery case, cooling fan blades, heating ducts, bumper beam, instrument panel, splash shields, interior consoles
Polycarbonate	PC	Amorphous	1.2	2.5	60	90–130	640–960	Head lamp lens, instrument panel carrier
Polymethyl methacrylate	PMMA	Amorphous	1.17–1.20	2.8	50–70	2–6	15–20	Tail lamp lens, instrument panel clusters
Acrylonitrile butadiene styrene	ABS	Amorphous	1.05–1.07	2.2	40	90–130	320	Bumper beam, instrument panel, wheel covers, mirror housing, interior consoles
Styrene maleic anhydride	SMA	Amorphous	1.05–1.15	2.8	35–60	2–30	25–650	instrument panel carrier
Polyamide-6 (nylon-6)	PA-6	Semi-crystalline	1.14	2.5	80	45	60	Intake manifold, resonator, engine throttle body, cooling fan blades
Polyamide-6,6 (nylon-6,6)	PA-6,6	Semi-crystalline	1.14	2.8	60–80	60	100	Intake manifold, resonator, switches, gears, cooling fan blades
Polyamide-4,6 (nylon-4,6)	PA-4,6	Semi-crystalline	1.18	3	99	25	98	Electrical connectors, switch body, charge-air coolers, chain tensioners

Polyoxy-methylene (acetal)	POM	Semi-crystalline	1.41	2	61–70	25–75	65–85	Gears, window crank, seat belt components, door lock housings
Polyethylene terephthalate	PET	Amorphous	1.37	2	70	–	45	Electrical connectors, ignition coil bobbins
Polybutylene terephthalate	PBT	Semi-crystalline	1.31	1.7	55	55–60	65–70	Electrical connectors, fuse boxes, sensor housings, actuator cases, mirror housings
Polyphenylene sulfide	PPS	Semi-crystalline	1.34	3.3	69	1.6	16	Fuel rail, pump impellers, thermostat housing, throttle body, ignition coil bobbin

*In many of these applications, the polymer may be combined with fillers or short fiber reinforcements. The properties listed are without any fillers or short fiber reinforcements.

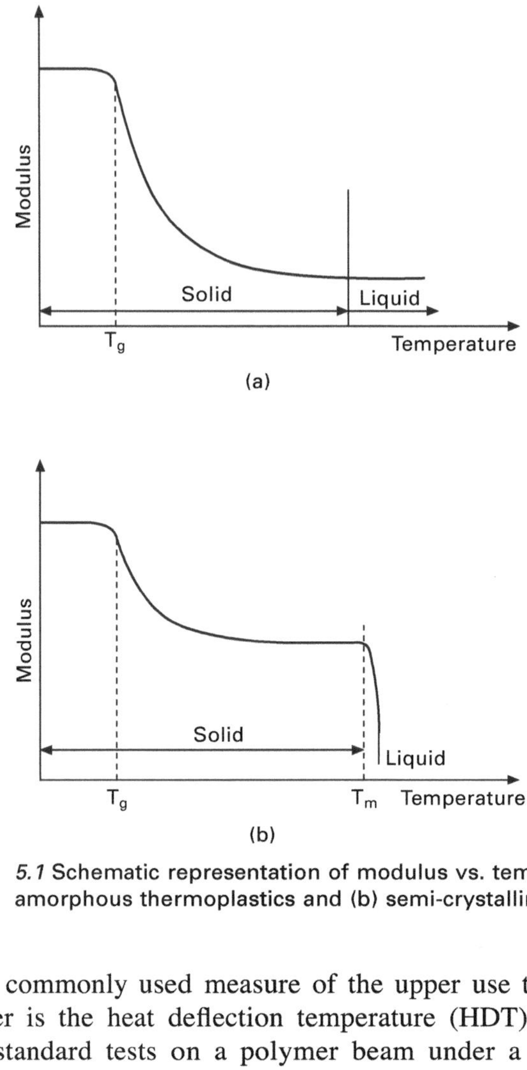

5.1 Schematic representation of modulus vs. temperature for (a) amorphous thermoplastics and (b) semi-crystalline thermoplastics.

The commonly used measure of the upper use temperature limit for a polymer is the heat deflection temperature (HDT), which is determined using standard tests on a polymer beam under a specified stress level. While HDT can help in the preliminary selection of a polymer, it does not provide any information on the polymer's behavior when the application temperature changes. The transition temperatures T_g and T_m can provide better and more fundamental information for this purpose. For example, if the application temperature of an amorphous polymer is higher than its T_g, its impact performance is significantly better, but it may undergo a large deformation due to the large reduction in modulus. For a semi-crystalline polymer, the reduction of modulus between T_g and T_m is not very large, and therefore it can still maintain its load bearing capability without exhibiting a large deformation. Table 5.2 lists T_g and T_m for a number of commonly used thermoplastics. Polyamide-4, 6 and polyphenylene sulfide (PPS) listed in Table 5.2 have high T_g and T_m and they are considered for applications requiring higher temperature resistance than the more commonly used polymers.

There is a large variety of polymer blends used in automotive applications. They are mixtures of two or more compatible thermoplastic polymers that are melt-blended to obtain properties or processing characteristics that cannot be achieved by using either of the constituent polymers alone. A few examples are given below:

- Polybutylene terephthalate (PBT) is blended with polycarbonate (PC) to improve its chemical resistance. This PC/PBT blend is used for making bumper beams.
- PC is blended with ABS to produce a combination of impact strength, modulus and heat deflection temperature that is not available in ABS alone. The ABS/PC blend is used in instrument panel structures, consoles, mirror housings and cowel panel grilles.
- Polystyrene is blended with polyphenylene oxide (PPO) to improve its processing. The PPO/PS is used in instrument panel structures, electrical connectors and wheel covers.
- PPO is blended with polyamide-6,6 (PA-6,6) to improve its creep and heat resistance. The PA-6,6/PPO blend is used in outer body panels.

5.2.2 Design considerations for thermoplastic polymers

While polymers have many advantages over metals (lower density, parts consolidation, lower energy consumption during processing, etc.), they behave differently from metals when subjected to mechanical loads or exposed to thermal or chemical environments. In this section, some of the differences between polymers and metals are briefly discussed. They apply to both thermoplastic and thermoset polymers.

(i) They have much lower modulus and strength than metals. For example, the modulus of steel is 207 GPa and the modulus of polymers is lower than 5 GPa. The modulus and strength of polymers can be enhanced by adding fibers, such as glass and carbon fibers; but even with fiber reinforcement, these two properties may not be as high as the modulus and strength of steel.

(ii) Their mechanical properties, such as modulus, yield strength, strain-to-failure and impact strength, are influenced by temperature and strain rate, and for some polymers (e.g. polyamide-6 and polyamide-6,6), also by humidity. Decreasing the temperature, for example, increases modulus and yield strength, but decreases strain-to-failure and impact strength. Similarly, increasing the strain rate increases modulus and yield strength, but decreases strain-to-failure and impact strength.

Table 5.2 Thermal properties of several thermoplastic polymers used in automobiles

Polymer	Amorphous or semi-crystalline	Glass transition temperature, T_g(°C)	Melting point, T_m(°C)	Heat deflection temperature, HDT @ 1.82 MPa (°C)	Max. use temperature (°C)	Coefficient of thermal expansion (10^{-6} per °C)
High density polyethylene	Semi-crystalline	–120	137	45–55	80–120	60–110
Polypropylene	Semi-crystalline	–18	176	50–60	105–140	80–100
Polycarbonate	Amorphous	149	–	130–140	120	68
Polymethyl methacrylate	Amorphous	105	–	65–100	85–95	50–90
Acrylonitrile butadiene styrene	Amorphous	88–120	–	85–95	70–95	60–110
Styrene maleic anhydride	Amorphous	114	–	95–130	90–95	80
Polyamide-6 (nylon-6)	Semi-crystalline	58	225	60–70	80–120	80–85
Polyamide-6,6 (nylon-6,6)	Semi-crystalline	65	255	65–105	80–150	80
Polyamide-4,6 (nylon 4,6)	Semi-crystalline	75	295	160	150	110
Polyoxymethylene (acetal)	Semi-crystalline	–50	160–180	110–125	90–100	50–112
Polyethylene terephthalate	Semi-crystalline	69	212–265	55	80	65
Polybutylene terephthalate	Semi-crystalline	52	220–267	85	135	60–95
Polyphenylene sulfide	Semi-crystalline	85	285–290	137	260	27–49

(iii) They exhibit creep and stress relaxation. These are both time and temperature dependent behavior and must be taken into account for long-term application of polymers under load. Creep is manifested by increasing strain with time, even when the stress on the polymer is maintained at a constant value (Fig. 5.2a). Due to creep, a polymer beam under load will show increasing deflection with increasing time. Stress relaxation is manifested by decreasing stress with time even when the strain on the polymer is held constant (Fig. 5.2b). Due to stress relaxation, the joint stress in a bolted joint of a polymer part may reduce with increasing time, making the joint less effective. Both creep and stress relaxation increase with increasing temperature.

(iv) Increasing strain due to creep or decreasing stress due to stress relaxation are, in effect, equivalent to a reduction of the modulus of the polymer with time. When the creep strain exceeds a critical value (which depends on the polymer type as well as temperature), the polymer may fail even though the stress on the polymer is lower than its yield or tensile strength. This phenomenon is known as creep rupture.

(v) Unlike low carbon steels, polymers do not exhibit an endurance limit when subjected to fatigue (cyclic) loading. The fatigue life of a polymer increases with decreasing stress level. In general, the fatigue performance of semi-crystalline polymers is better than that of amorphous polymers. Among the commonly used semi-crystalline polymers, polyoxymethylene (POM or acetals) shows the best fatigue performance, followed by polyamide-6 as well as polyamide-6,6. Another point to note is that, at high frequencies of cycling or at high fatigue stress levels, internal heat generated within the polymer may cause softening of the polymer, which causes a reduction in modulus of the polymer and ultimately, thermal failure.

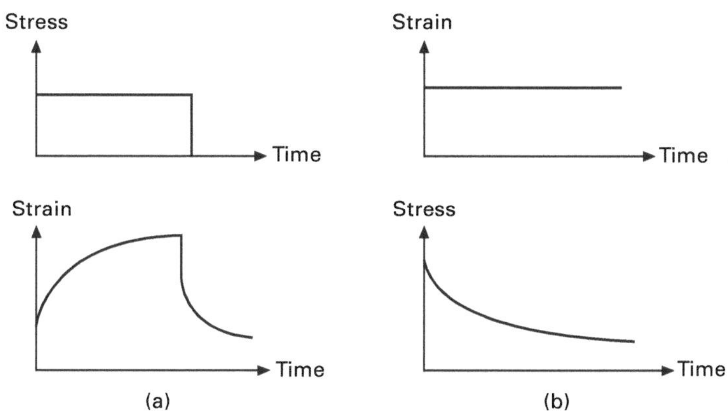

5.2 Schematic representation of (a) creep and (b) stress relaxation.

(vi) In many automotive applications, the polymer parts may be exposed to chemicals such as gasoline, motor oil, antifreeze, paint, cleaning solvents and road salt. In general, semi-crystalline polymers, such as high density polyethylene (HDPE), polypropylene (PP) and polyamides, have much higher chemical resistance than amorphous polymers, such as ABS and polycarbonate. Typically, semi-crystalline polymers are selected for under-the-hood applications, such as windshield washer fluid bottles and radiator end caps.

(vii) The effect of chemicals on polymers is influenced by temperature, and for some polymers, also by the presence of tensile stress. The combination of tensile stress (either applied or residual) and a chemical environment may cause stress cracking in the polymer. The stress crack resistance of a polymer depends on a number of factors, including its chemical structure and molecular weight, tensile stress level, type of chemical to which it is exposed and the time of exposure.

(viii) Many polymers exhibiting relatively high ductility and high impact resistance at room temperature can transform into brittle materials and show low ductility and low impact resistance at temperatures lower than room temperature. For example, polypropylene exhibits up to 200% elongation-at-break at 23 °C, but as the temperature is reduced to 0 °C or lower, it turns into a brittle polymer and exhibits only 1–2% elongation-at-break. If high elongation-at-break is desired at such low temperatures, either an ethylene–propylene copolymer or a rubber-modified polypropylene is selected instead of polypropylene.

(ix) The notch sensitivity of polymers also varies with the polymer type and affects the impact behavior. For example, both polycarbonate and ABS are considered ductile polymers and exhibit high impact strength in standard Izod or Charpy impact tests. However, polycarbonate is more notch sensitive than ABS. Polycarbonate changes from a ductile polymer to a brittle polymer if a sharp notch (e.g. a surface scratch) appears on its surface. The impact resistance of ABS also decreases with notch sharpness, but a transition from ductile to brittle behavior is not observed.

(x) Long-term use of polymers should not only take into account the possibility of creep and stress relaxation, but also the effect of ageing on polymer properties. Ageing occurs because of irreversible degradation of the polymer molecules due to oxidation, reduction in molecular weight, etc., and is aggravated by heat, ultraviolet light, chemical environment, etc. Very often, ageing creates surface embrittlement, but there may also be an overall deterioration of the polymer's properties.

(xi) Polymers have a significantly higher coefficient of thermal expansion than metals. Thus, dimensional changes in polymer parts due to temperature changes are much higher. For a polymer part fitted with a metal part, this can cause interference when the temperature is increased or create gaps when the temperature is decreased. The differential thermal expansion or contraction between a metal part and a polymer part may also create thermal stresses between them.

(xii) The surface finish of polymer parts is controlled mainly by the surface finish of the mold surfaces. Polymers can be molded in a wide variety of colors using colorants or color master batches; however, uniform distribution of color on a large part and color matching with neighboring parts can sometimes be a challenge. Polymer surfaces can also be painted; however, the long-term adhesion between the paint and the polymer surface may be affected by environmental conditions, such as high humidity. Polyethylene and polypropylene surfaces cannot be painted unless they are modified by plasma treatment or flame treatment. The painting process for steel body panels involves exposure to high temperatures (up to 220 °C for 30 minutes or longer). Many thermoplastic polymers will experience heat sagging and creep at such high temperatures, and may not be suitable for painting in paint ovens.

5.2.3 Processing of thermoplastic polymers by injection molding

Injection molding is the principal processing method used for the vast majority of thermoplastic automotive parts. It is capable of producing parts of complex shapes and geometry at high production rates, with good dimensional accuracy and excellent surface finish. In-mold surface decoration and surface texturing can be done in injection molding.

The basic injection molding machine (Fig. 5.3) contains an injection unit and a clamping unit (Rosato and Rosato, 1995). The injection unit consists of a heated cylindrical barrel and a properly designed screw that rotates inside the barrel. Solid polymer pellets or granules are fed into the hopper located on the back side of the injection unit. As the pellets are moved forward by the rotating screw, they start to melt and ultimately turn into a homogeneous liquid polymer, which is collected in a small chamber located in the front section of the injection unit. After the required amount of liquid polymer is collected, the screw rotation is stopped and the screw is now moved forward to act like a piston, forcing the collected liquid polymer to flow into the closed mold cavity located in the clamping unit. Even after the mold is filled, additional liquid polymer is injected into the cavity to compensate for the shrinkage of the polymer as it starts to cool inside the mold. The mold cavity is held closed until the average temperature of the polymer part falls below the solidification temperature of the polymer. The mold is then opened and the part is ejected. Further cooling to room temperature takes place outside the mold.

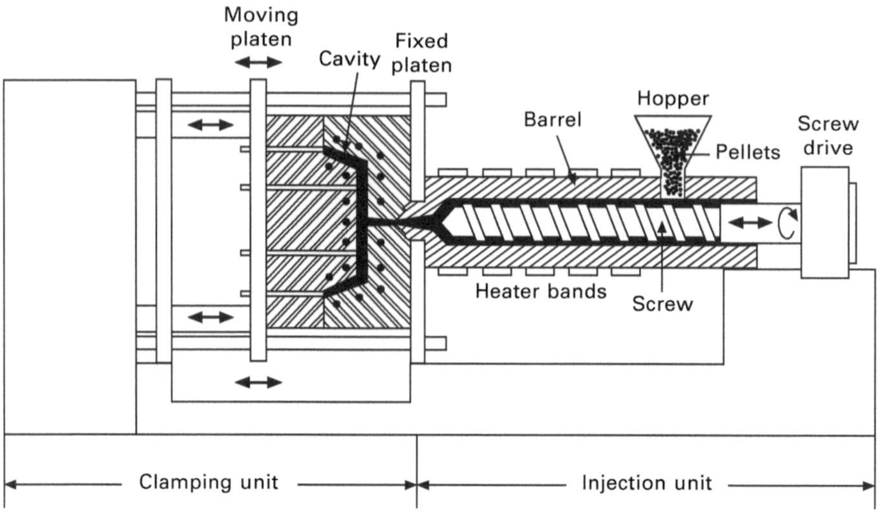

5.3 Injection molding machine.

The cooling time in the mold, which has a major effect on the injection molding cycle time, depends on the part thickness, melt temperature (the temperature of the liquid polymer at the time of injection), mold wall temperature and demolding temperature (the average temperature at which the part is ejected from the mold). For shorter cooling times, the melt and mold temperatures are selected as low as possible and the demolding temperature is selected as high as possible without introducing defects and affecting the quality of the injection molded part.

Depending on the part design, mold design and processing parameters, a variety of molding defects can be observed in injection molded parts. They include unfilled cavity, surface blemishes, internal voids, high shrinkage, warpage, sink marks, weld lines and residual stresses. Many of these defects can be reduced or eliminated by selecting the proper process parameters, such as melt temperature, mold temperature, injection pressure and injection speed. Process parameters also influence molecular orientation, residual stresses and crystallinity (for semi-crystalline polymers), which, in turn, influence the properties and performance of injection molded parts.

A variety of injection molding techniques now exists that has expanded the use of thermoplastics in the automotive industry. They have also created opportunities for adding many new design features and optimizing the use of material (Osswald *et al.*, 2002). Some of these, such as insert molding and outsert molding, are variations of the conventional injection molding process. In insert molding, small components of other materials, such as internally threaded metal inserts, metal pins and electrical strips, are placed in the mold cavity prior to injecting the polymer melt. As the polymer melt is subsequently injected, it surrounds the pre-placed components, making them integral to the molded part. In outsert molding, the polymer melt is injected onto a metal frame placed in the mold cavity. The metal frame contains a number of punched out holes or protruding pins, which serve as the anchoring locations for the polymer. Both processes can lead to considerable savings in assembly operations, assembly time and therefore, assembly cost. A list of some of the other injection molding techniques is given below:

(i) *Lost-core injection molding* to produce hollow injection molded parts with complex internal geometry and undercuts in a single molding operation. (Example: Air intake manifolds)
(ii) *Gas-assisted injection molding* in which compressed nitrogen gas is injected into the liquid polymer after it has partially filled the cavity. The gas creates a hollow channel in the core of the liquid polymer and, as the gas moves forward, it forces the liquid polymer to fill the rest of the cavity. Reductions in part weight, material usage, cycle time, clamping force and surface defects are some of the advantages of gas-assisted injection molding. (Example: Door handles)
(iii) *Structural foam injection molding*, which is used to produce a sandwich structure of thin solid skins and a foamed core, both of the same polymer. The advantage of structural foam molding is that it produces parts with higher stiffness to weight ratio. (Example: Instrument panels)
(iv) *Co-injection* or *sandwich molding*, which involves either sequential or concurrent injection of two dissimilar but compatible polymers, producing a sandwich structure with one polymer forming the skin and the other forming the core. The sandwich structure provides properties or cost advantages not achievable with a single polymer. (Example: Instrument panels)
(v) *Overmolding*, which utilizes sequential injection of the same polymer with different colors to make a multi-colored part. Polymers with different properties can also be overmolded to make multi-functional products. (Example: Multi-colored tail lamp lenses)
(vi) *Microcellular molding*, in which a supercritical gas (usually nitrogen or carbon dioxide) is blended with the liquid polymer to form a single-phase solution. During mold filling, the gas forms highly uniform microscopic cells (0.1 to 10 μm in diameter) in the interior of the part. The pressure created by the foaming process reduces shrinkage and surface defects, such as sink marks. Blending the supercritical gas with the liquid polymer significantly reduces its viscosity, and therefore the part can be molded at lower temperatures and with lower injection pressures. Furthermore, due to a foamed structure, both the part weight and the cycle time are reduced. (Example: Interior trims)
(vii) *Injection-compression molding*, which utilizes a combination of injection

molding and compression molding to manufacture a part. Injection molding is used first to partially fill the cavity with the liquid polymer. Complete filling occurs when one side of the mold is moved forward to compress against the liquid polymer, which spreads it to the rest of the cavity and produces the final thickness. The injection-compression molding process produces more dimensionally stable, residual-stress free parts at much lower pressures and reduced cycle times than the conventional injection molding process. Injection-compression molding is used for making thin-walled parts with high surface accuracy or parts with polymers that cannot be easily injection molded. (Example: Door module carrier)

5.3 Thermoplastic–matrix composites for automobiles

5.3.1 General characteristics

Thermoplastic–matrix composites currently used in the automotive industry are mostly reinforced with E-glass fibers. Although carbon fibers have lower density and can provide higher modulus, E-glass fibers are selected for their low cost. Glass fibers are incorporated in the thermoplastic–matrix in a variety of forms – randomly oriented short fibers, randomly oriented long fibers, randomly oriented continuous fibers, unidirectional continuous fibers and bi-directional fabric. Among these, randomly oriented short glass fiber reinforced thermoplastics are very common, since they can be processed by the traditional injection molding techniques; however, they are mostly used in semi-structural parts (e.g. air intake manifolds and water pump housings) and functional parts (e.g. gears in window cranking mechanisms and electrical switches). Random fiber orientation makes the composite behave as an isotropic material, but produces lower strength and modulus than the modulus and strength of composites with unidirectional fibers. If unidirectional fibers or bi-directional fabrics are added for higher strength and/or modulus, the composite changes from an isotropic to a non-isotropic material.

A number of structural applications of thermoplastic–matrix composites with unidirectional or bi-directional continuous fibers have been developed. They include seat structures, bumper beams and cross members. Continuous fibers are preferred over short fibers for structural applications, since they produce a much higher modulus and strength. However, the use of continuous fiber reinforced thermoplastic–matrix composites lags behind that of thermoset matrix composites. The difficulty encountered in making continuous fiber reinforced thermoplastics is due to the high viscosity of liquid thermoplastics. The continuous fibers are incorporated in thermoset polymers in their low viscosity state before chemical reaction transforms them into a solid polymer.

The viscosity of liquid thermoplastics is several orders of magnitude higher than that of liquid thermosets, which makes it difficult to impregnate the fiber bundles and wet the fibers. The low viscosity of liquid thermosets is a distinct advantage over liquid thermoplastics in manufacturing high performance composites. The advantages of thermoplastic–matrix composites are due to their lower processing time, weldability, higher damage resistance and recyclability.

It is important to note that modulus and strength of continuous fiber reinforced thermoplastics are primarily controlled by the fiber properties, fiber weight fraction, fiber orientation and to some extent, fiber–matrix interfacial bond strength. The matrix and the fiber–matrix interfacial bond strength play much more significant roles in determining the strength and modulus of short fiber reinforced thermoplastics (Mallick, 2008). Since the T_g and T_m of thermoplastic polymers are not affected by the presence of fibers, the selection of the matrix has a large effect on the composite's behavior at different application temperatures. The effect of chemicals (or other environments, such as ultraviolet light) on the properties of thermoplastic–matrix composites is also determined by the chemical (or environmental) resistance of the thermoplastic polymer selected for the matrix.

In this section, various E-glass fiber reinforced thermoplastic–matrix composites are discussed. Among them, injection moldable short and long fiber composites are available with a variety of thermoplastic polymers, such as polypropylene, polyamides, polybutylene terephthalate (PBT) and ABS. Most of the other thermoplastic–matrix composites used today for automotive applications, such as glass mat composites and glass fabric composites, use polypropylene as the matrix. They can be produced with other thermoplastic polymers, including ABS, PET and polyamides; however, polypropylene is selected for its low cost, good mechanical properties and processability.

5.3.2 Short fiber thermoplastics

Short fiber thermoplastics (SFT) contain fibers that are typically less than 1 mm long. They are suitable for injection molding in conventional injection molding machines. The short fibers are compounded with the thermoplastic polymer using a melt-blending process in a compounding extruder, such as a twin screw extruder. The melt-blended material is pelletized and the pellets are then fed into an injection molding machine. The fiber lengths in the pellets are then up to 1 mm long; but due to fiber breakage during injection molding, the fiber length in the injection molded parts becomes much lower than 1 mm. The addition of short fibers to a polymer increases its modulus and heat deflection temperature and decreases its coefficient of thermal expansion and mold shrinkage (Table 5.3). In most cases, yield strength, tensile strength and impact strength are also increased; however,

Table 5.3 Properties of short E-glass fiber reinforced polypropylene and polyamide-6,6

	Polypropylene				Polyamide-6,6 (dry as molded)			
Fiber weight (%)	0	10	20	30	0	13	33	43
Density (g/cm^3)	0.9	0.98	1.04	1.12	1.14	1.22	1.38	1.51
Tensile modulus (GPa)	1.4	2.5	3.73	5.3	1.6	6.2	8.8	13.8
Tensile strength (MPa)	35	43.5	44.8	48.3	83	96.6	179.4	207
Strain-to-failure (%)	150	4	3	2	60	4	3	2
Notched izod impact strength (J/m)	37	42.6	53	58.6	53	53	111.3	138.6
HDT at 1.82 MPa (°C)	54	127	132	135	90	243	252	252
Coeff. of thermal expansion (10^{-6}/°C)	90	46.8	43.2	39.6	70	27	23	22
Mold shrinkage (%)	1.7	0.6	0.6	0.6	1.5	0.7	0.3	0.2

the extent of their increase depends on fiber length and the fiber–matrix interfacial bond strength. Another positive effect of fiber addition is that the creep strain is significantly reduced. On the other hand, the density of the polymer is increased and the strain-to-failure is decreased by the addition of short fibers. Injection molding of SFT may also cause warpage due to the difference in shrinkage in the flow direction and normal to the flow direction, which is caused by the short fibers tending to align more in the flow direction.

The nature of the conventional injection molding process imposes several limitations on the maximum fiber content, fiber length and fiber orientation, all of which are important parameters controlling the mechanical properties of a short fiber reinforced composite. For example, in a short fiber reinforced composite, the fiber length plays a crucial role in determining the strength of the composite, since effective strengthening is achieved only if the fiber length is greater than the critical fiber length. The critical fiber length depends on fiber strength, fiber diameter and fiber–matrix interfacial shear strength (Mallick, 2008). For glass fibers, the interfacial shear strength is controlled by using a chemical coupling agent (most commonly, a silane coupling agent) on the fiber surface. Chemical coupling increases the interfacial shear strength, which, in effect, reduces the critical fiber length.

Fiber content in injection molded SFT is limited because the viscosity of the liquid polymer increases with increasing fiber weight fraction. At high viscosities, it becomes difficult to produce a complete part, since the mold cavity may not be completely filled. High viscosity may also contribute to undesirable fiber orientation in the direction of flow in the mold. For practical purposes, the maximum weight fraction in injection molded SFT is limited to approximately 40 percent.

The limitation on the fiber length arises from fiber breakage, which occurs mainly due to the high-shear mixing and plasticization that take place inside the barrel of an injection molding machine. Additional fiber breakage occurs as the fiber-filled liquid polymer flows through the narrow passages (runners and gates) in the mold before entering the mold cavity. A study conducted to determine the fiber length distribution in injection molded plaques of a short E-glass reinforced polypropylene has shown that the average fiber length in the pellets before injection molding was 0.71 mm. After injection molding, the average fiber length decreased to 0.274 mm. If, due to fiber breakage, the majority of the fibers are shorter than the critical fiber length, the strength of the molded part is reduced.

Under ideal molding conditions, fibers in SFTs are randomly oriented so that the injection molded material can be treated as isotropic. In practice, however, depending on the mold design, part thickness and process conditions used, some of the fibers may be oriented in a preferred direction. In a large complex shaped part, the direction of preferred fiber orientation may vary from location to location. Furthermore, there may be variation in fiber orientation through the thickness of a molded part. Weld lines, where two or more flow fronts meet, are one of the locations where preferred fiber orientation occurs. The weld lines are formed in multiple gated molds or behind core pins used for making molded-in holes (Fig. 5.4). They are often the weakest locations in an injection molded part, since stresses acting normal to a weld line, i.e. transverse to the preferred fiber orientation direction, can initiate cracks and cause failure at relatively low loads.

5.3.3 Long fiber thermoplastics

Long fiber thermoplastics (LFT) contain randomly oriented fibers ranging in length from 5 to 25 mm. The initial fiber length in long fiber thermoplastics ranges from 10 to 50 mm. Because of longer fibers, LFTs exhibit higher tensile modulus, tensile strength and impact strength than SFTs (see Table 5.4). Application of LFTs is found in seat structures, door modules, dashboard carriers, front end modules, bumper beams and spare-wheel wells.

LFT parts can be molded either using pre-compounded pellets or by directly compounding the fibers and other additives in-line with the final part production. The pre-compounded pellets are manufactured by the pultrusion-compounding process, in which continuous fiber rovings are pulled through the heated die of an extruder. As the fiber rovings pass through the die, they

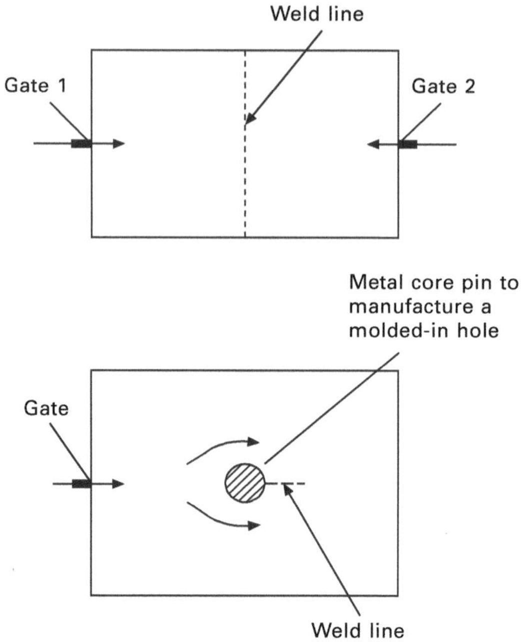

5.4 Weld line formation in injection molded parts. Weld line formed at the center where two flow fronts meet in a double-gated mold. Weld line formed behind the core pin where the two divided flow fronts meet.

Table 5.4 Comparison of short fiber thermoplastics (SFT) and long fiber thermoplastics (LFT) (both with polypropylene matrix)

Fiber wt. %	SFT 30	LFT 30	SFT 40	LFT 40
Density (g/cm^3)	1.12	1.12	1.21	1.21
Tensile modulus (GPa)	6.21	6.89	8.96	8.96
Tensile strength (MPa)	76	100	90	117
Tensile strain at failure (%)	4–5	2.5–3.5	3–4	2–3
Flexural modulus (GPa)	4.83	6.21	6.89	7.58
Flexural strength (MPa)	112	155	131	179
Notched Izod impact strength (J/m)	107	166	107	267
Heat deflection temperature at 1.82 MPa (°C)	141	–	141	154

Adapted from the product data sheet of RTP Co.

are impregnated with the liquid polymer from the extruder. The polymer-impregnated rovings are cooled and then chopped into pellets of the desired length. Standard pellet lengths are 10–12 mm, but lengths up to 40 mm are also used. The pre-compounded pellets are then used to mold the part using injection molding, compression molding or injection-compression molding. Injection molding can be carried out in a conventional injection molding machine; however, to reduce the fiber breakage and maintain the long fiber length, several modifications are needed in injection screw design and mold design, as well as in process parameters. There is very little fiber breakage in compression molding, and compression molding is preferred for LFTs that are difficult to mold by injection molding.

Direct compounding of fibers in-line with the final production of parts eliminates the second heating step needed with the pre-compounded pellets. In this process, the liquid polymer produced by an injection molding machine and continuous fiber rovings are simultaneously fed into either a long coating die (Hawley and Jones, 2005) or a twin-screw extruder (Krause *et al.*, 2003) where fiber impregnation with the liquid polymer takes place. An in-line chopper, located between the fiber impregnation area and the molding machine, cuts the hot polymer-impregnated fiber rovings into desired lengths, which are then directly fed into the molding machine. The material produced by direct compounding is called LFT-D. Since the polymer is not allowed to cool down between the compounding area and the molding machine, it experiences a single melt history, which reduces the possibility of degradation of the polymer. Furthermore, the elimination of the second heating step reduces energy consumption and material handling cost.

5.3.4 Glass mat thermoplastics

Glass mat thermoplastics (GMT) are available in sheet form, typically 3.7 mm in thickness, in which E-glass fiber mats are combined with a thermoplastic polymer. Polypropylene is the most commonly used thermoplastic for GMT; other thermoplastics, such as PET, PBT and polyamides are also used, albeit to a limited extent. The fiber mat usually contains either randomly oriented chopped glass fibers (typically, 25 to 100 mm in length) or randomly oriented continuous glass fibers (Fig. 5.5). GMT is also available with unidirectional continuous glass fibers as well as a bi-directional glass fiber mat. Because of the random fiber orientation, both chopped fiber GMT and continuous fiber GMT exhibit isotropic behavior, i.e. their properties are the same in all directions in the plane of the sheet. For the same reason, the modulus or strength of chopped fiber GMT as well as that of continuous fiber GMT are lower than that of unidirectional or bi-directional continuous fiber composites. Unidirectional continuous glass fibers or bi-directional woven glass fabric can be added to the surfaces or in the middle layers of random fiber mats to improve the modulus and strength in selected directions.

Compression molding (also known as flow molding) is the common manufacturing process used for making GMT parts. In the compression molding process, the GMT sheet is preheated in an infrared oven or a hot

5.5 Glass mat thermoplastics (GMT) with (a) randomly oriented chopped fibers and (b) randomly oriented continuous fibers.

air convection oven to a temperature above the melting point of the polymer matrix. For polypropylene matrix, the preheating temperature is around 220 °C. As the matrix material melts, the mat tends to open up or loft. The lofted mat is transferred to a molding press, where it is placed in the lower half of the mold. Both halves of the mold are preheated to a temperature above the room temperature, but much below the solidification temperature of the polymer matrix. The press is closed at a high speed and, as pressure is applied on the material, it flows outward to fill the mold cavity. The mold is held closed under pressure until the average material temperature reduces to the mold temperature, and then the upper mold half is opened and the part is removed from the mold. The compression molding conditions for GMT with a polypropylene matrix are listed in Table 5.5. Compression molding is used for manufacturing GMT parts with complex shapes and can produce parts with varying thickness.

Thermostamping is also used for manufacturing GMT parts; however, it is limited to produce simple shapes in which the final part thickness is close to the GMT sheet thickness. In this operation, the GMT blank is heated in an infrared oven to 20–60 °C above the melt temperature of the polymer matrix and then transferred to a matched mold held at room temperature. The part shape is formed by the application of pressure as the mold is closed. Very little, if any, flow of material takes place in the mold. For polypropylene GMT, the recommended sheet temperature prior to thermostamping is 200 °C, the mold closing speed is 750–1500 mm/min and the molding pressure is 10–14 MPa (Bigg and Preston, 1989).

PP-based GMT with randomly oriented chopped fibers and randomly oriented continuous fibers have been used in bumper beams, knee bolsters, seat structures, inner door panels, battery trays, compressor brackets, underbody shields, etc. The properties of these two types of GMT are compared in

Table 5.5 Compression molding conditions for GMT with polypropylene matrix

Blank surface temperature (°C)	195–220
Blank heating time (sec)	
(a) in infrared oven	180–240
(b) in hot air convection oven	240–480
Mold temperature (°C)	25–70
Molding pressure (bar)	100–140
Mold closing speed (mm/s)	425
Pressing speed (mm/s)	10–30
Total closing time (s)	< 5
Cooling time in the mold (sec per mm of part thickness)	5–7
Total cycle time (s)	45–90
Shrinkage after molding (%)	0.2–0.4

Table 5.6. As can be seen in this table, continuous fibers provide a slightly higher modulus and strength than the chopped fibers, which is principally due to their greater length. Table 5.6 also gives the properties of PP-based GMT containing unidirectional continuous fibers. The unidirectional GMT has a significantly higher modulus and strength in the longitudinal (L) direction of fibers, but its strength and modulus in the transverse (T) direction are much lower. The transverse strength and modulus are much closer to the strength and modulus of random fiber composites. The unidirectional GMT is useful for beam-type applications, such as bumper beams. In another variation of GMT, called low density GMT, the matrix is aerated using aqueous foam as the dispersing fluid. The low density GMT weighs 0.5 to 2 kg/m^2 compared with 4 to 5 kg/m^2 for a typical GMT. Applications of low density GMT are found in headliners, load floors, door trims, sunshades and parcel shelves (Raghavendran and Haque, 2001).

In advanced GMT composites, random fiber GMT is combined with either woven or stitched fabrics and needled non-woven mats to produce sandwich sheet structures that show significantly higher tensile and impact properties than GMT. The sandwich sheet, produced by a double belt lamination process, can be designed with random fiber GMT either in the core or in the skins. Parts from advanced GMT are produced by compression molding. Properties of two advanced GMTs are given in Table 5.7. Door modules and bumper beams constructed from advanced GMT are reported to exhibit higher crash energy absorption than either GMT or LFT (Törnqvist and Baser, 2002). Advanced GMT is recommended for automotive applications requiring high stiffness, high creep resistance, high fatigue resistance and high crash resistance.

5.3.5 Glass fabric thermoplastics

Glass fabric thermoplastics use commingled rovings of continuous glass filaments and thermoplastic filaments that are woven into a two-dimensional

Table 5.6 Properties of glass mat thermoplastic (GMT) with polypropylene as the matrix

	Chopped fiber GMT		Continuous fiber GMT		Unidirectional fiber GMT	
Fiber weight (%)	32	40	30	40	42	
Density (g/cm^3)	1.14	1.19	1.14	1.21	1.24	
Tensile modulus (GPa)	4.62	5.965	4.3	5.82	10.1 (L)	5.31 (T)
Tensile strength (MPa)	76	90.4	60	108	276 (L)	60 (T)
Strain-to-failure (%)	2.4	2.2	2.2	2.5	2.5 (L)	2.3 (T)
Poisson's ratio	0.317	0.333	–	0.361	0.377 (L)	0.361 (T)
Notched Izod impact strength (J/m)	560	469	–	–	1650 (L)	720 (T)
HDT at 1.82 MPa (°C)	151	160	155	154	158 (L)	143 (T)
Coeff. of thermal expansion (10^{-6}/°C)	36	–	–	43	37 (L)	–
Molding shrinkage (%)	0.1-0.3	0.1–0.3	0.2–0.4	0.1–0.3		0.1–0.3

Adapted from Azdel data sheet.

Table 5.7 Effect of fabric reinforcement on the properties of glass mat thermoplastic (GMT)

	GMT*	GMT* reinforced with fabric		
Fiber weight (%)	40		40	
Density (g/cm^3)	1.22		1.22	
Tensile modulus (GPa)	5.3	9.6 (L)		5.1 (T)
Tensile strength (MPa)	75	200 (L)		84(T)
Strain-to-failure (%)	2.2	2.37 (L)		2.28 (T)
Coeff. of thermal expansion (10^{-6}/°C)	15–25		15–25	

Note: (L) and (T) represent longitudinal and transverse directions.
*Randomly oriented chopped glass fiber mat in polypropylene matrix.
Adapted from Quadrant product data sheet.

fabric (Fig. 5.6). The fabric is highly flexible and can be draped over contoured mold surfaces. Upon heating in the mold, the thermoplastic filaments in the glass fabric thermoplastic melt and form liquid pools around the glass fibers. On the application of mold pressure, the liquid thermoplastic flows in and around the rovings and transforms into a continuous phase. As the material is cooled, the liquid thermoplastic is solidified and becomes the matrix. The advantage of commingling is that the distance the liquid polymer has to flow to coat and wet the glass filaments in the rovings is very small (on the order of 100 μm). This helps in obtaining good fiber impregnation, which is beneficial in achieving good composite properties.

Commercially available glass fabric thermoplastics are based on the Twintex process developed and patented by Vetrotex. In this process, glass

5.6 Commingled fabric.

and thermoplastic filaments (such as polypropylene filaments) are commingled during the glass fiber production process, which assures a uniform distribution of the two types of filaments in the commingled fiber rovings. The glass filament content in the rovings is controlled by commingling different proportions of glass and thermoplastic filaments. The commingled rovings are woven to produce bi-directional (0/90) fabrics of different styles (plain, twill, etc.)

Several automotive parts have been developed using Twintex glass fabric composites with polypropylene matrix. Among them are bumper beams, skid plates, seatback structures, load floors and spare tire covers. Compression molding is used for making these parts. The mold temperature used for compression molding is 200–220 °C and the molding pressure is between 0.45 MPa and 1.38 MPa. Thermostamping is another process that has been used with Twintex fabrics to produce several types of automotive parts. In this process, the fabric is preheated to 200–220 °C in an infrared or hot air convection oven and then quickly transported to the mold. The mold is closed rapidly (at a recommended speed of 60 mm per sec) to ensure that the part shape is formed prior to solidification of the polypropylene matrix. The part is removed from the mold after it has cooled below 100 °C.

The tensile properties of three different Twintex glass fabric composites with polypropylene matrix are shown in Table 5.8. In the composites containing balanced (1/1) fabric (i.e. equal number of commingled rovings in both weft and warp directions), the tensile properties are the same in both weft and warp directions. In the composites containing 4/1 fabric (i.e. four times commingled rovings in the weft direction than in the warp direction), the

Table 5.8 Properties of Twintex commingled glass fabric–polypropylene composites

	1/1 (balanced) fabric	4/1 fabric	1/1 (balanced) fabric
Fiber content (wt%)	60	60	75
Density (g/cm^3)	1.50	1.50	1.75
Tensile modulus (GPa)	15	24/8	21
Tensile strength (MPa)	350	500/180	420
Unnotched Charpy impact strength (kJ/m^2)	220	330/90	300
Heat deflection temperature at 1.82 MPa (°C)	159	159	159

Note: Glass fabric composites with balanced (1/1) fabric has the same properties in the weft and warp directions. Glass fabric composites with 4/1 fabric have higher properties in the weft direction than in the warp direction.

tensile properties are higher in the weft direction than in the warp direction. Due to the high glass fiber content in these composites, both tensile modulus and strength are higher compared with the tensile modulus and strength of glass mat composites.

5.3.6 Laminated thermoplastic composites

A 'prepreg' is a fiber layer pre-impregnated with a polymer. Laminated thermoplastic composite parts are made by stacking several layers of either unidirectional continuous fiber prepregs or bi-directional fabric prepregs, heating the stack either in an infrared oven or in a hot air circulating oven and then thermostamping the heated stack in a press. The principal advantage of laminated composites is that the fiber orientation can be varied from layer to layer (Fig. 5.7), which gives the flexibility of designing the composite to match the design need. Additionally, since continuous fibers are used, the modulus and strength of the laminate are much higher than the modulus and strength of the discontinuous fiber composites described in previous sections. Examples of properties obtained in laminated thermoplastic composites are given in Table 5.9.

Even though laminated thermoplastic composites have a great potential in high performance structural applications, such as front rails, pillars and cross members, they are currently not used in any production parts. There are several reasons for this. First, the cost of thermoplastic prepregs is relatively high, the main reason being the difficulty encountered in coating and wetting the fibers with the high-viscosity thermoplastic melt. Most of the thermoplastic prepregs available today are made by melt impregnation. One such melt impregnation process is called DRIFT (Hartness *et al.*, 2001) in which continuous fiber strands are pulled through a standard extrusion machine to coat and impregnate the fibers with a liquid thermoplastic polymer. It produces

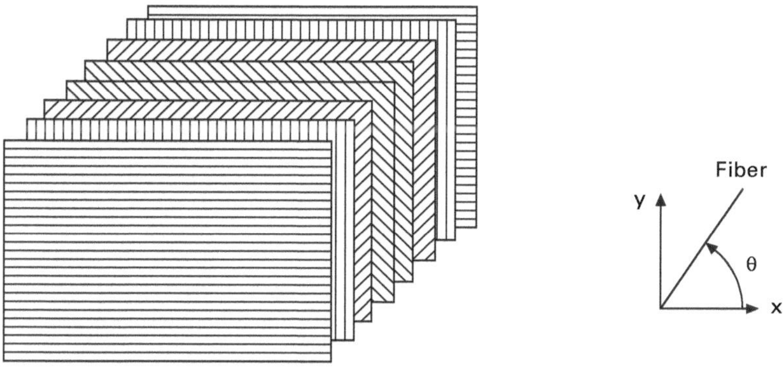

An 8-layered continuous fiber laminate with [0/90/+45/–45/–45/90/0] layers

Fiber oriented angle, θ, which can be varied from layer to layer in a laminated composite

5.7 Laminated composite.

Table 5.9 Properties of laminated thermoplastic composites (carbon fibers in polyamide-6 matrix)

	Unidirectional laminate [0]	Quasi-isotropic laminate $[0/45/90/\text{-}45]_S$
Fiber content (wt%)	65	65
Density (g/cm^3)	1.5	1.5
Tensile modulus (GPa)	120 (L), 7.1(T)	55
Tensile strength (MPa)	1100 (L), 50 (T)	620
Elongation at break (%)	0.92 (L), 0.68 (T)	2
Notched Izod impact strength (J/m)	2400 (L), 6.6 (T)	1700
Coeff. of thermal expansion (10^{-6}/°C)	–0.36 (L)	–0.34

Adapted from Fiberforge data sheet.

excellent fiber impregnation and wetting by the thermoplastic polymer at very high operating speeds. Another process being developed for making thermoplastic prepregs is continuous resin infusion (Wang *et al.*, 1997; Ragone and Mallick, 2008), which is considered a less expensive operation than melt impregnation The second reason is related to the preparation and stacking of layers to form the stack. A process developed by Fiberforge (Burkhart and Cramer, 2005) uses an automated lay up system to produce 'tailored blanks' from pre-impregnated tapes, with highly accurate fiber alignment in each layer of the stack. The lay-up machine can produce these blanks very rapidly with net-shape or near net-shape configurations. Tailored blanks are first consolidated in a press with heated flat platens and then thermostamped in a matched-die mold. For a 14-layer carbon fiber reinforced polyamide-6 stack with $[(0/90)_3/0]_S$ laminate configuration, Burkhart and Cramer (2005) used a consolidation temperature between 220 and 260 °C and a consolidation pressure ranging from 3 to 10 MPa. The preheat temperature of the blank before thermostamping was between 160 and 180 °C and the pressing time was 60–90 seconds.

5.3.7 Natural fiber thermoplastics

Natural fiber thermoplastics use natural fibers, such as jute and sisal, as reinforcement for thermoplastic polymers. Natural fibers are grown as agricultural plants and are commonly used for making ropes, carpet backings, hand bags, etc. They are classified into three groups: (i) bast fibers, such as flax, hemp, jute and kenaf, (ii) leaf fibers, such as sisal, pineapple and banana, and (iii) seed or fruit fibers, such as cotton and coir (which comes from coconut husks) (Drzal *et al.*, 2005). Wood fibers are also used as reinforcement for thermoplastics.

Interest in using natural fibers as reinforcement is for the following reasons:

(i) They are biodegradable and the energy consumption to produce them is very small compared with that for glass or carbon fibers. Therefore, they are more environmentally friendly than glass or carbon fibers.
(ii) The density of natural fibers is in the range of 1.25 to 1.5 g/cm^3 (Table 5.10), which is significantly lower compared to 2.54 g/cm^3 for E-glass fibers and 1.8 to 2.1 g/cm^3 for carbon fibers.
(iii) The modulus-to-density ratios of several natural fibers are greater than that of E-glass fibers, which means that they can be very competitive with E-glass fibers in stiffness-critical designs.
(iv) Natural fiber composites have a higher acoustic damping value than E-glass or carbon fiber composites, and are therefore selected for noise attenuation, an increasingly important design requirement for interior automotive components.
(v) Natural fibers are much less expensive than both E-glass and carbon fibers.

There are several drawbacks to natural fibers. Their tensile strength is low compared with that of either E-glass or carbon fibers. Among other drawbacks

Table 5.10 Properties of selected natural fibers

Property	Hemp	Flax	Kenaf	Coir	Sisal	Jute
Density (g/cm^3)	1.48	1.4	1.45	1.25	1.33	1.46
Modulus (GPa)	70	60–80	53	6	38	10–30
Tensile strength (MPa)	550–900	800–1500	930	220	600–700	400–800
Strain-to-failure (%)	1.6	1.2–1.6	1.6	15–25	2–3	1.8
Moisture absorption (%)	8	7	–	10	11	12

are their low decomposition temperature and tendency to absorb moisture. At temperatures higher than 200 °C, most natural fibers start to degrade, which leads to odor, discoloration, release of volatiles, and deterioration of mechanical properties. The low decomposition temperature of natural fibers also limits the processing temperature and the selection of thermoplastic polymers that can be used for producing natural fiber thermoplastics. Most of the natural fiber thermoplastics utilize either polypropylene or polyethylene as the matrix, since their melting point is lower than that of polyamides and many other thermoplastics. Even then, to keep the degradation of natural fibers to a minimum, the processing temperature and thermal exposure should be as low as possible. There is also poor adhesion between the natural fibers and the polymer matrix and, unless the adhesion is improved by coupling the two together, the mechanical properties, particularly strength and impact resistance, of natural fiber thermoplastics may not be high. Coupling is done by either chemically treating the fiber surface with a coupling agent or by adding an adhesion-promoting grafting agent to the polymer (Malkapuram *et al.*, 2009)

Automotive applications of natural fiber reinforced polypropylene include inner door panels, headliners, package trays, trunk floors, seatbacks and dashboards (Holbery and Houston, 2006). The basic form of material used for making these parts is needle-punched non-woven mat, which is produced by carding or air-laying a mixture of natural fibers and polypropylene fibers. The production of parts starts with cutting a blank from the needled mat, preheating it above the melting point of the polymer (lower than 200 °C) for a short time, transferring the heated blank to a room-temperature mold, and pressing it into shape. The mold pressure is low, usually in the range of 0.05 to 0.5 MPa. Cycle time in the press is typically less than 1 minute. The mechanical properties obtained in one such composite are shown in Table 5.11.

Table 5.11 Properties of natural fiber reinforced polypropylene (fully coupled)

		Natural fiber		
	None	Jute	Sisal	Wood
Fiber content (wt%)	0	50	50	50
Tensile modulus (GPa)	1.7	8.5	6	5.5
Tensile strength (MPa)	32	73	60	39
Flexural modulus (GPa)	1.4	7.5	5.1	5.3
Flexural strength (MPa)	41	100	85	68
Notched Izod impact strength (J/m)	24	40	55	23
Unnotched Izod impact strength (J/m)	620	210	190	90

Source: Global Resource Technologies.

5.3.8 Self-reinforced thermoplastics

Self-reinforced thermoplastics (SRT) are single polymer composites in which the materials for the reinforcing fibers and the matrix are of the same thermoplastic polymer type; for example, polyethylene fibers in a polyethylene matrix or polypropylene fibers in a polypropylene matrix. Thus, unlike the glass or carbon fiber reinforced thermoplastics, the self-reinforced thermoplastics are completely recycalable. Since the fibers and the matrix are of the same chemical structure, a strong interfacial bond exists between the two, which helps in achieving high tensile strength for the composite.

The polymer molecules in the reinforcing fibers of self-reinforced thermoplastics are highly oriented in their length direction, which gives them the high modulus and strength needed to serve as the reinforcement for the matrix. There are three methods for making self-reinforced thermoplastics: (i) hot compaction, (ii) co-extrusion and (iii) film stacking. In hot compaction, a woven fabric made of closely spaced polymer fibers is heated to high enough temperature to melt the skin of each fiber (Ward and Hines, 2004). On cooling, the melted skin recrysatllizes to form the matrix. In co-extrusion, a thin layer of slightly lower melting polymer matrix is co-extruded with highly oriented closely spaced fibers of the same polymer. In film stacking, thin films of the matrix polymer are stacked on the polymer fiber layers and the two are consolidated by hot pressing. The key issue in making these composites is to find a suitable temperature range so that the polymer fibers do not melt, but the surrounding polymer matrix melts and wets out the fibers completely.

Self-reinforced polypropylene is the only self-reinforced thermoplastic currently marketed. It is available in both woven and sheet forms. The sheets, which range in thickness from 0.3 to 3 mm, are prepared by hot-compacting several layers of the fabric under heat and pressure. The temperature used for hot compaction is between 165 and 190 °C and the pressure ranges from 2.8 to 7 MPa. The final part shape from these sheets is produced by pressure thermoforming. In this process, the sheet is preheated to 140–160 °C, either in an infrared oven, or in a hot air circulating oven and then pressed in a matched metal mold to form the shape. The pressure during thermoforming is between 1 to 2 MPa, which is low enough to use an aluminum mold instead of a steel mold. The sheet must be clamped around the edges in a metal frame during heating and pressing to prevent shrinkage.

The density of self-reinforced polypropylene is significantly lower than that of glass fiber reinforced polypropylene (Table 5.12). One outstanding property of self-reinforced polypropylene is its high impact strength (also shown in Table 5.12), which is nearly three times higher than that of continuous glass mat reinforced polypropylene. Self-reinforced polypropylene is currently being investigated for door panels (Hayes, 2008).

Table 5.12 Comparison of self-reinforced polypropylene with GMT

Property	Hot compacted self-reinforced polypropylene (polypropylene fabric in polypropylene matrix) Tegris*	Curv†	Polypropylene (unreinforced)	Random short glass fiber mat–polypropylene GMT (with 40 wt% fibers)
Density (kg/m^3)	780	920	900	1185
Tensile modulus (GPa)	5–6	4.2	1.2	3.5–5.8
Tensile strength (MPa)	200	120	27	99
Notched Izod impact strength (J/m)	4800	4750	200	672
Heat deflection temperature @455 kPa (°C)	130	160	110	154
Co-eff. of thermal expansion (10^{-6}/°C)	6	68	96	27

*Milliken & Co.
†Propex Fabrics.

5.3.9 Thermoplastic nanocomposites

Thermoplastic nanocomposites contain nanometer (10^{-9}m) size reinforcements, such as nanoclay, carbon nanofibers and carbon nanotubes. The properties of these nano-reinforcements are considerably higher than those of conventional reinforcing fibers, such as glass and carbon fibers. Furthermore, their surface area to volume ratio is very high, which provides a greater interfacial interaction with the thermoplastic–matrix. These composites show not only high modulus and strength, but also excellent thermal, electrical, optical and other properties, and in general, at relatively low reinforcement content.

Nanoclay is a platelet-type smectite clay mineral containing several layers of silicates. Each silicate layer is 1 nm thick and has a surface area of 100 nm^2 or higher. The most common smectite used in nanocomposites is called the montmorillonite. To be an effective reinforcement, the silicate layers have to be exfoliated so that they are completely separated from each other, and uniformly dispersed in the polymer matrix. The clay particles are chemically treated to promote the exfoliated dispersion. A variety of techniques are available to mix nanoclay particles with a thermoplastic polymer. Melt mixing in an extruder or an injection molding machine is one of these techniques.

The ability of montmorillonite to significantly improve modulus and strength of polyamide-6 was first reported by Toyota in 1987 (Okada and

Usuki, 2007). The composite was prepared by the *in-situ* polymerization method. With the addition of only 4.2 wt% of exfoliated montmorillonite, the tensile modulus of polyamide-6 was nearly doubled and its tensile strength increased by more than 50%. The heat deflection temperature was increased by 80 °C compared with polyamide-6. The first automotive application of this material was the timing belt cover in a Toyota car. Since then, nanoclay reinforced thermoplastics have been used in engine covers, body side moldings, cargo floors and seat backs (Wang and Xiao, 2008). Polyamide-6 reinforced with only 2 wt% nanoclay has five times the resistance to gasoline permeation compared with polyamide-6; this has prompted the use of the material in fuel lines.

Apart from nanoclay, a great deal of research is currently being conducted on developing carbon nanofiber as well as carbon nanotube reinforced thermoplastics. Both types of reinforcement significantly increase modulus and strength and decrease the coefficient of thermal expansion of the thermoplastic. The other major benefit is an increase in electrical conductivity of the thermoplastic (Harris, 2004), which helps in dissipating static electricity build-up in electronic components and fuel lines, and during on-line painting of thermoplastic body panels. At present, the use of these nano-reinforcements is relatively few, mainly because of their high cost and low availability. Their use in automotive applications is at an early stage of exploration and requires further development.

5.4 Joining of thermoplastic–matrix composites

Thermoplastic and thermoplastic–matrix composite parts can be joined together by mechanical joining, adhesive bonding and welding or fusion bonding (Rotheiser, 2004). While mechanical joining and adhesive bonding are the only options for joining thermoplastics or thermoplastic–matrix composite parts to metal parts, welding can be used for joining one thermoplastic or thermoplastic–matrix composite part to another. Mechanical joining in thermoplastics and thermoplastic–matrix composites not only includes bolted joints and threaded joints using metal inserts and screws, but also snap fits. Holes required for bolted joints can be molded-in during injection molding or compression molding, or drilled afterward. If the weld lines formed near the molded-in holes cause early failure in injection molded or compression molded parts, it is recommended that they should be drilled in post-molding operations. Metal inserts with internal threads can be molded-in or heat-staked in a drilled hole after molding. Heat staking requires heating the metal insert and pressing it into a slightly undersized hole in the thermoplastic part. Snap fits are very common in thermoplastic parts, since they can be engaged and disengaged quickly with very little force.

One of the advantages of mechanical joints is that they can be disassembled repeatedly for repairs or replacements. Adhesive joints, on the other hand, are permanent joints and they cannot be disassembled without destroying or damaging the substrates. Two types of adhesives are available for making adhesive joints: (i) two-component adhesives, such as epoxies, urethanes and acrylics, and (ii) one-component adhesives, such as cyanoacrylates. The two-component adhesives take a longer time to cure and set. Surface preparation, such as removal of surface contamination, degreasing and surface treatment with a primer, is recommended for improved adhesive joint strength. Polypropylene, which is one of the common thermoplastics used in the automotive industry, is difficult to join using adhesives. Polypropylene is a non-polar polymer and its surface energy is relatively low among the thermoplastic polymers. The surface polarity of polypropylene can be increased by plasma treatment, fluorination and flame treatment.

One great advantage of thermoplastics over thermosets is that they can be welded or fusion bonded. Welding between two thermoplastic parts requires melting them locally and pressing them together to make a permanent bond (Stokes, 1989). There is a variety of welding techniques available for thermoplastics and thermoplastic–matrix composites (Table 5.13). The three most common welding techniques used in the automotive industry are vibration welding, spin welding and ultrasonic welding. They are briefly described below.

In vibration welding, also called linear friction welding, the two surfaces to be joined are held under pressure (typically 0.5 to 10 MPa) and frictional heat is generated at their interface by vibrating one surface relative to the other at 120–240 Hz frequency and less than 0.5 mm amplitude (Fig. 5.8). The vibration is continued until the surfaces start to melt. The molten film created at the interface is allowed to cool and solidify under pressure, resulting in a welded joint. The welding time is typically between 1 and 10 seconds.

Spin welding is a variation of vibration welding and is applicable to parts with circular cross sections. It can also be used for welding tubular parts to flat surfaces. In this process, one part surface is rotated relative to the other with a circular spinning motion while they are held under pressure.

Table 5.13 Various welding techniques used with thermoplastics and thermoplastic matrix composites

Thermal	Mechanical	Electromagnetic
• Hot gas welding	• Vibration welding	• Resistance welding
• Hot plate welding	• Spin welding	• Induction welding
• Extrusion welding	• Ultrasonic welding	• Dielectric welding
• Infrared welding	• Friction stir welding	• Microwave welding
• Laser welding		

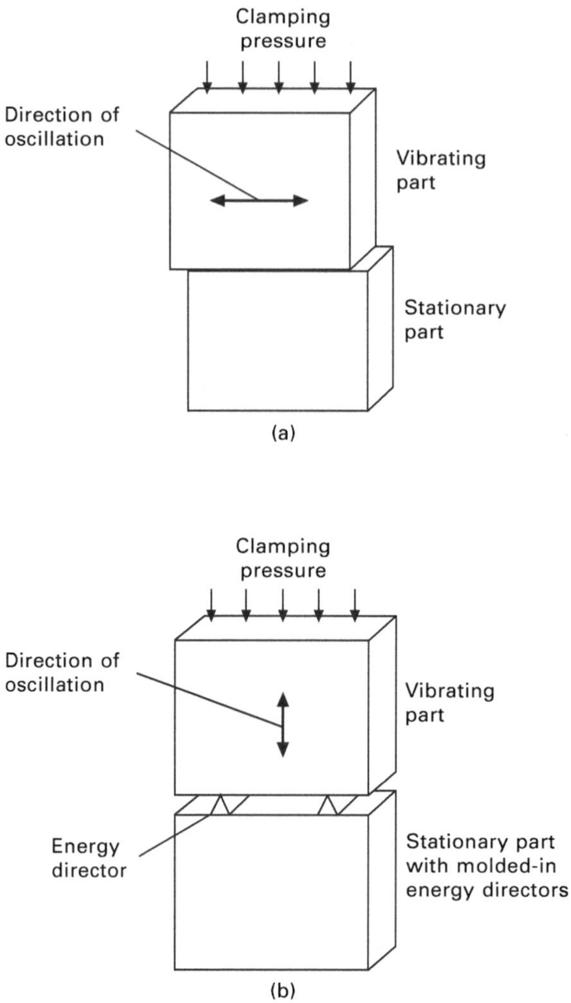

5.8 (a) Vibration welding and (b) ultrasonic welding.

Ultrasonic welding utilizes high-frequency (20 to 50 kHz), low amplitude (15 to 60 μm) vibration to cause localized melting. The parts to be joined are held together under pressure (typically 1–10 MPa) and then subjected to the vibratory motion, either normal to or parallel to the interface between the two parts. For normal-to-the-interface vibration, one of the surfaces must contain energy directors, which are molded-in wedge-shaped protrusions on the surface (Fig. 5.8). Contact between the two surfaces is maintained through these protrusions. Vibrational energy concentrated at the protrusions generates heat which melts them. The liquid polymer created by melting of the protrusions flows under the hold pressure and wets the interface.

Vibration welding is particularly suitable for joining large parts along flat seams and seams with small out-of-plane curvature. Automotive applications of vibration welding include bumper beams, instrument panel assemblies, headlight and intake manifolds. Ultrasonic welding is used for welding over small areas and can be used for spot welding of thin sections. It is a fast process with cycle times, typically on the order of 1 second, and it produces welds with very little flash. In the automotive industry, ultrasonic welding is used for joining glove box doors, instrument clusters, valves, electrical connectors, electrical terminals and switch components.

Both vibration welding and ultrasonic welding have been applied to join thermoplastic–matrix composite parts (Yousefpour *et al.*, 2004). However, the strength data available to date are limited to mostly short-term properties, such as static lap shear strength. Several studies have shown that these properties are influenced by the welding parameters. For both types of welding, the main welding parameters are vibration frequency, amplitude, clamping pressure and hold time. In addition, for ultrasonic welding the design of the energy directors and the ultrasonic head play important roles. More work needs to be done to determine their influence on long-term properties of welded joints under both static and dynamic loading conditions before these or other welding techniques can become widely accepted.

5.5 Conclusion

This chapter has presented an overview of thermoplastics and thermoplastic–matrix composites for automotive applications. Besides weight saving (due to their low density), they provide opportunity for parts consolidation, styling and design options, modularization, energy saving, cost reduction and easier recycling compared to other materials. The vast majority of their applications in today's automobiles are in the interior of the vehicle, where they provide significant advantages over other materials, especially in terms of component consolidation, modular design and aesthetic styling. These include dashboards, consoles, climate control ducts, inside door panels, instrument clusters, electrical components and various trims. Exterior applications of thermoplastics and thermoplastic–matrix composites are fewer. The loading, environmental, appearance and assembly requirements for exterior applications are more demanding and even though there are many thermoplastics that can meet such requirements, the cost-conscious automotive industry will remain slow to accept them in the near future. For body panel applications, the major technical issues related to thermoplastics are heat stability during the paint baking cycle, gaps and flushes due to differences in thermal expansion, long-term weather resistance, low-impact damage resistance and class-A surface finish. Several thermoplastic–matrix composites are currently used for exterior body applications, such as bumper beams, fenders, spoilers and spare wheel wells. More structurally demanding applications in body structures, such as front rails, cross members, etc., will require higher strength and modulus that can be obtained only with continuous fibers. The advantages of using thermoplastic–matrix composites in these applications are not only in significant weight saving and parts consolidation, but also in manufacturing cost reduction due to shorter processing time and lower tooling cost.

5.6 References

Bigg D M and Preston J R (1989), 'Stamping of thermoplastic–matrix composites', *Polymer Composites*, 10, 261–268.

Burkhart A and Cramer D (2005), 'Feasibility of continuous-fiber reinforced thermoplastic tailored blanks for automotive applications', *2005 SPE Automotive Composites Conference & Exposition*, Troy, MI.

Drzal L, Mohanty A and Misra M (2005), *Natural Fibers, Biopolymers and their Composites*, Boca Raton, CRC Press.

Harris P J F (2004), 'Carbon nanotube composites', *International Materials Reviews*, 49, 1, 31–42.

Hartness T, Husman G, Koenig J and Dyksterhouse J (2001), 'The characterization of low cost fiber reinforced thermoplastic composites produced by the *DRIFT*™ process', *Composites: Part A*, 32, 1155–1160.

Hawley R C and Jones R F (2005), 'In-line compounding of long-fiber thermoplastics for injection molding', *J. Thermoplastic Composite Matls.*, 18, 459–462.

Hayes H (2008), 'Self-reinforced polypropylene composites – a new class of material for the motorsports industry', *2008 SAE World Congress*, Paper No. 2008-01-2947, Warrendale, Society of Automotive Engineers.

Holbery J and Houston D (2006), 'Natural-fiber-reinforced polymer composites in automotive applications', *JOM*, November, 80–86.

Krause W, Henning F, Tröster S, Geiger O and Eyerer P (2003), 'LFT-D – a process technology for large scale production of fiber reinforced thermoplastic composites', *J. Thermoplastic Composite Matls.*, 16, 289–302.

Lange W (2003), 'Polymers in automobile applications', in *Plastics and the Environment* (ed. Andrady A L), John Wiley & Sons.

Malkapuram R, Kumar V and Negi Y S (2008), 'Recent development in natural fiber reinforced polypropylene composites', *J. Reinforced Plastics and Composites*, Vol. 28, pp 1169–1189, 2009.

Mallick P K (2008), *Fiber Reinforced Composites*, Boca Raton, CRC Press.

Maxwell J (1994), *Plastics in the Automotive Industry*, Cambridge, Woodhead Publishing.

Okada A and Usuki A (2007), 'Twenty-year review of polymer–clay nanocomposites at Toyota Central R & D Labs., Inc.', *2007 SAE World Congress*, Paper No. 2007-01-1017, Warrendale, Society of Automotive Engineers.

Osswald T A, Turng L-S and Gramann P J (eds.) (2002), *Injection Molding Handbook*, Munich, Hanser.

Raghavendran V and Haque E (2001), 'Development of low density glass mat thermoplastic composites for structural applications', *2001 SAE World Congress*, SAE Paper No. 2001-01-0100, Warrendale, Society of Automotive Engineers.

Ragone C and Mallick P K (2008), 'Development of a thermoplastic prepreg manufacturing process by continuous resin infusion' *2008 SAMPE Fall Conference*, September, Covina, Soc. for the Advancement of Materials and Process Engineering.

Rosato D V and Rosato D V (1995), *Injection Molding Handbook: The Complete Molding Operation Technology, Performance, Economics, Second Edition*, New York, Chapman & Hall.

Rotheiser J I (2004), *Joining of Plastics*, Munich, Hanser.

Stokes V K (1989), 'Joining methods for plastics and plastic composites', *Polymer Engineering and Science*, 29, 1310–1324.

Törnqvist R and Baser B (2002), 'Structural modules with improved crash performance using thermoplastic composites', *2002 SAE World Congress*, Paper No. 2002-01-1038, Warrendale, Society of Automotive Engineers.

Wang X, Mayer C and Neitzel M (1997), 'Some issues on impregnation in manufacturing thermoplastic composites by using a double belt process', *Polymer Composites*, 18, 701–710.

Wang Z and Xiao H (2008), 'Nanocomposites: Recent development and potential automotive applications', *2008 SAE World Congress*, Paper No. 2008-01-1263, Warendale, Society of Automotive Engineers.

Ward I M and Hines P J (2004), 'The science and technology of hot compaction', *Polymer*, 45, 1423–1437.

Yousefpour A, Hojjati M and Immarigeon J-P (2004), 'Fusion-bonding/welding of thermoplastic composites', *J. Thermoplastic Composite Materials*, 17, 303–340.

6
Thermoset–matrix composites for lightweight automotive structures

P. K. MALLICK, University of Michigan-Dearborn, USA

Abstract: This chapter presents an overview of the materials, properties and manufacturing processes for thermoset–matrix composites used for automotive applications. It also reviews the applications of carbon fiber reinforced thermoset–matrix composites in automobiles.

Key words: thermoset polymers, compression molding, SMC, RTM, SRIM, filament winding, vacuum bag molding, carbon fiber reinforced thermoset composites.

6.1 Introduction

The majority of the long or continuous fiber reinforced polymer matrix composites used for automotive body and chassis applications in today's automobiles is based on thermoset polymers. The principal reason for using thermosetting polymers instead of thermoplastic polymers is their much lower viscosity, which makes it much easier to combine them with long and continuous fibers and produce structural composites with high strength and modulus. Thermoset polymers are prepared by chemically reacting one or more liquid pre-polymers in the presence of a catalyst or a curing agent. The polymerization reaction, called cross-linking or curing, transforms the liquid pre-polymers to a solid polymer. The fibers are combined with the thermosetting polymers in the pre-cured low-viscosity liquid state. In the cured state, thermoset polymers have higher heat and chemical resistance than most thermoplastic polymers, and they do not exhibit as much creep deformation. However, there are also a few limitations of using thermoset polymers as the matrix. First, depending on the curing system and temperature used, the curing reaction may take several minutes to several hours to complete. This makes the processing time for thermoset polymers significantly longer than that for thermoplastic polymers. Another limitation is that thermoset polymers cannot be re-melted or returned to liquid state, and therefore cannot be directly reused or recycled.

Applications of thermoset–matrix composites in current automobiles include radiator supports, bumper beams, fenders, hoods, roof panels, deck lids, and a number other exterior and interior body components. There are also a few chassis, suspension and powertrain components where thermoset–matrix composites are used, such as leaf springs, cross members and drive shafts. In most of the current applications of thermoset–matrix composites, E-glass fiber is used as the reinforcement. The reason for using E-glass fibers is their much lower price compared with carbon fibers. However, carbon fibers, on account of their greater modulus-to-density and strength-to-density ratios, can produce higher stiffness and higher load carrying capacity than glass fibers at significantly lower weight. Thus, carbon fibers are much more effective than glass fibers for component weight reduction and are expected to see much greater use in future vehicles, particularly as vehicle weight reduction becomes an important priority for improving fuel efficiency.

This chapter presents an overview of the current state of the art of thermoset–matrix composites for automotive applications. It starts with a section on materials and material selection considerations. The major manufacturing processes for making structural automotive parts using thermoset–matrix composites are then described. The application of carbon fiber reinforced thermoset–matrix composites in current vehicles is also reviewed.

6.2 Materials

6.2.1 Material selection considerations

Fiber reinforced polymers are prepared by combining fibers with a polymer matrix. Each of these constituents has its own role. The fiber, by virtue of its high modulus, serves as the principal load carrying member. The polymer matrix serves as the binder that not only keeps the fibers in place, but also provides the load transfer mechanism between the fibers. Other constituents that may also be found in fiber reinforced polymers are coupling agents, coatings, and fillers. Coupling agents and coatings are applied on the fiber surface to improve the wetting between the fibers and the matrix and to promote bonding across the fiber/matrix interface, both of which are essential for making the fiber and matrix interact with each other to provide a good composite action. Fillers are used primarily to reduce cost, but they may also improve modulus and dimensional stability of the polymer matrix.

Fibers in a polymer matrix composite can be used in continuous lengths or discontinuous lengths. Continuous fibers are much more effective than discontinuous fibers in carrying the load and in providing both high strength and high modulus. For a unidirectional continuous fiber composite in which all of the fibers are oriented in the loading direction (Fig. 6.1), the longitudinal modulus and strength of the composite in the fiber direction are given by

$$E_L \approx E_f v_f \qquad [6.1]$$

$$S_L \approx S_f v_f \qquad [6.2]$$

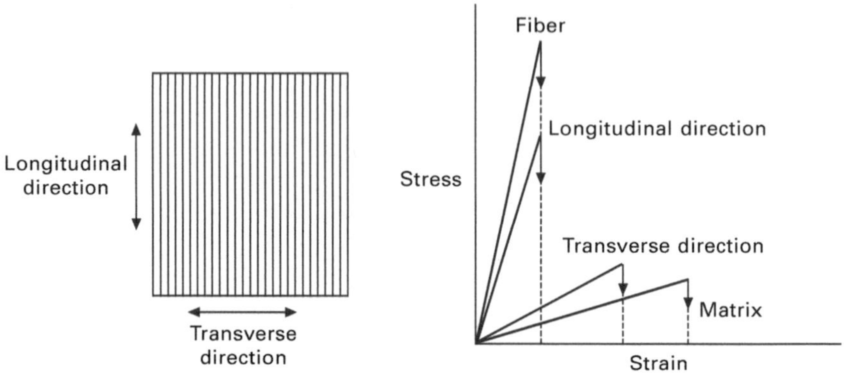

6.1 Tensile stress–strain characteristics of a unidirectional continuous fiber composite in the longitudinal and transverse directions of fibers.

where, E_L and S_L are the longitudinal modulus and longitudinal strength of the unidirectional continuous fiber composite, E_f and S_f are the modulus and strength of the selected fiber, and v_f is the fiber volume fraction in the composite.

Equation [6.1] indicates that, to obtain a high longitudinal modulus for the composite, a high modulus fiber should be selected. Carbon fibers have much higher modulus than glass fibers, and therefore, are much more desirable for high stiffness applications. Similar to Equation [6.1], Equation [6.2] also indicates the importance of selecting a high-strength fiber if high longitudinal strength for the composite is desired. The tensile strength of glass fibers is comparable to that of many carbon fibers, but the strength of glass fibers deteriorates relatively easily with abrasion, contact and moisture absorption, and the full benefit of their high strength may not be achievable in the composite form. This, combined with carbon fibers' lower density, gives carbon fibers an advantage over glass fibers in strength-critical applications. Another parameter appearing in both Equations [6.1] and [6.2] is the fiber volume fraction, which should be as high as possible for obtaining both high modulus and strength. In practice, the fiber volume fraction is limited to 60%, since above this value it becomes difficult to wet the fibers with the polymer matrix, which is required for good composite properties.

The properties of a unidirectional continuous fiber composite depend on the fiber orientation angle with the loading direction (Fig. 6.2). The modulus and strength are the highest when the fibers are aligned in the loading direction. Such a material is called an orthotropic material. In contrast, for an isotropic material, such as steel or aluminum alloy, properties are the same in all directions.

The fiber orientation in a fiber reinforced composite can be varied to achieve the desired stiffness or the load carrying capacity. For designing slender beams or tie rods which experience uniaxial stress condition, fibers

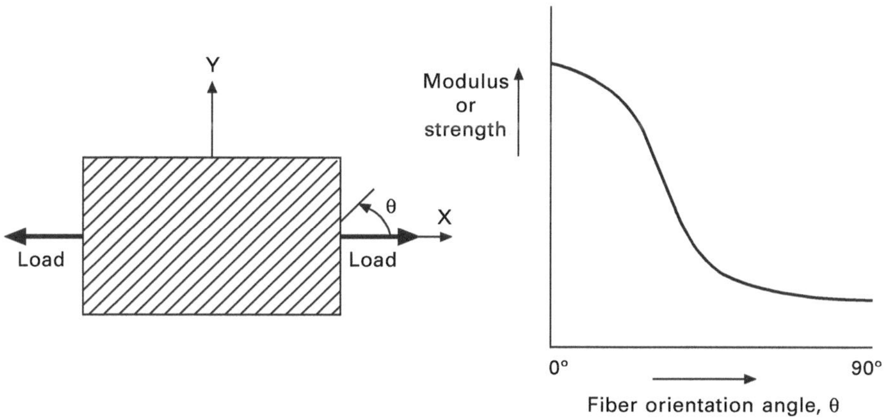

6.2 Variation of tensile modulus and tensile strength of a unidirectional. continuous fiber composite with fiber orientation angle θ.

are aligned along their lengths so that fiber direction and the stress direction coincide. But for other applications where biaxial stress conditions exist or stress directions may change during the application of the material, it may be necessary to use a laminate containing several layers, each layer containing unidirectional fibers, but the fiber orientation angle varying from layer to layer. Such a laminate can be formed by stacking several unidirectional continuous fiber layers in a specified sequence of fiber orientations; however, it should be noted that the properties derived from the laminate, the laminate failure mode and the order of failure in various layers in the laminate depend on the fiber orientation angle in each layer and the difference in fiber orientation angles in successive layers in the laminate. In general, a symmetric laminate is preferred, since such a laminate will be free from bending or torsional deformations when an in-plane load is applied on it. A common laminate which produces equal modulus in all directions in the plane of the laminate is a symmetric quasi-isotropic laminate, such as a [0/90/45/-45/-45/45/90/0] laminate (Fig. 6.3a). For maximum bending stiffness, the 0° layers are placed on the outside of such a laminate. For a tube subjected to torsion, the principal stresses are at 45 and -45° angles to the length direction of the tube and, in such an application, a symmetric [45/-45/-45/45] laminate is recommended. The construction of such a laminate is shown in Fig. 6.3b.

If discontinuous fibers are used instead of continuous fibers, fiber length becomes an additional important parameter in determining the modulus and strength of the composite. Both longitudinal modulus and strength of a unidirectional discontinuous fiber composite are lower than the longitudinal modulus and strength of a unidirectional continuous fiber composite containing the same fiber volume fraction. Discontinuous fiber composites are used with random fiber orientation, since they are much easier to produce than

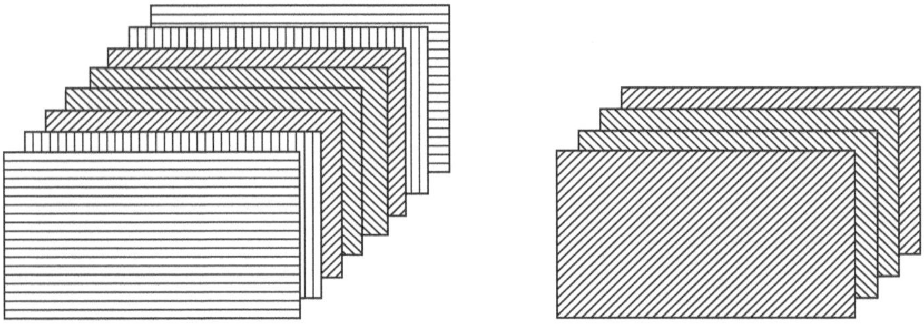

6.3 Stacking of unidirectional continuous fiber layers in two commonly used symmetric laminates.

unidirectional discontinuous fiber composites. With random fiber orientations, the modulus and strength become even lower, but the random fiber composite behaves as an isotropic material with equal properties in all directions.

For applications involving fatigue loading, carbon fibers are recommended over glass fibers, since carbon fiber reinforced composites exhibit significantly higher fatigue strength than glass fiber composites. For unidirectional continuous fiber composites loaded in the fiber direction, the fatigue strength at 10^6 cycles is between 80 and 90 percent of the static tensile strength if the fibers are high modulus carbon fibers, but only between 50–60 percent if the fibers are glass. An important point to note is that the fatigue failure mode and damage tolerance of fiber reinforced composites are very different from the ones observed in metals. In general, fiber reinforced composites exhibit slow and progressive damage growth under fatigue loading, which often accompanies a continuous decrease in stiffness of the structure (Mallick, 2008). In metals, the fatigue crack initiation phase can be long; however, as the initiated crack reaches its critical length, the failure is fast and often catastrophic.

Fiber reinforced composites are also being considered in crush resistant structures for which high impact energy absorption in a controlled fashion is the primary design consideration. In terms of specific energy absorption (SEA) or energy absorption per unit mass, carbon fiber reinforced composites are better than glass fiber reinforced composites (Caliskan, 2000). In general, thermoset–matrix composites show lower SEA than thermoplastic matrix composites. Interfacial bond strength between the fibers and the matrix, and interlaminar strength between the various layers in the composite also play major roles, since they control the failure modes, such as local microbuckling, fiber kinking, fiber fracture and delamination, during the crushing of the fiber reinforced composite structure. Fiber architecture, geometry (round cross section vs. rectangular or square cross section) and crush triggering mechanism are also among the factors to consider for obtaining progressive crush and improved SEA of the composite structure.

6.2.2 Fibers

Fibers are the principal load carrying members in a fiber reinforced polymer. Proper selection of the fiber type, fiber volume fraction, fiber length and fiber orientation angle are all important in obtaining the desired performance of a fiber reinforced composite, since they influence the modulus, strength, fatigue resistance, and electrical and thermal properties of the composite. Fiber selection also influences the density and cost of the composite.

Table 6.1 lists properties of several fibers that are commercially available and are considered for automotive applications. E-glass is the most common reinforcing fiber used today. The principal advantages of E-glass fibers are their low cost, high tensile strength, high chemical resistance and excellent insulating properties. However, they have higher density, low static fatigue resistance, lower tensile modulus and lower fatigue strength than carbon fibers. Another drawback of E-glass fibers is their high sensitivity to moisture absorption and abrasion, which reduce their tensile strength from 3.45 GPa in the as-produced condition to approximately 1.72 GPa after incorporating

Table 6.1 Properties of selected reinforcing fibers

Fiber	Density (g/cm^3)	Tensile modulus (GPa)	Tensile strength (GPa)	Strain at failure (%)	Coefficient of thermal expansion ($10^{-6}/°C$)	Poisson's ratio
Glass fiber						
E-glass	2.54	72.4	3.45	4.8	5	0.2
S-glass	2.49	86.9	4.30	5.0	2.9	0.22
PAN-based carbon fiber						
T-300	1.76	230	3.53	1.5	–0.6 (L)*	0.2
T-600	1.79	230	4.31	1.89		
AS–4	1.78	228	4.28	1.87		
IM-7	1.78	276	5.18	1.87		
UHM	1.87	441	3.45	0.8		
Pitch-based carbon fiber						
P-55	2.0	380	1.90	0.5	–1.3 (L)	
P-100	2.15	758	2.41	0.32	–1.45 (L)	
Aramid fiber						
Kevlar 49	1.45	131	3.62	2.8	–2 (L)	0.35
Extended chain polyethylene fiber						
Spectra 900	0.97	117	2.59	3.5		

*L represents the longitudinal direction of fibers.

them in a matrix to make a composite. Another type of glass fibers are S-glass fibers, which were originally developed for aircraft components and missile casings. They have the highest tensile strength of all reinforcing fibers. However, the compositional difference and higher manufacturing cost make them more expensive than E-glass fibers. A lower cost version of S-glass, called S-2 glass, is also available. Although S-2 glass is manufactured with less stringent nonmilitary specifications, its tensile strength and modulus are similar to those of S-glass.

Carbon fibers are available with a variety of tensile modulus, ranging from 207 GPa on the low side to 1035 GPa on the high side. In general, the lower modulus carbon fibers have lower density, lower cost, higher tensile strength and higher tensile strain-to-failure than higher modulus carbon fibers. The principal advantages of carbon fibers are their exceptionally high modulus-to-density ratio and strength-to-density ratio, very low co-efficient of thermal expansion, high fatigue strength and high thermal conductivity. The disadvantages are their low strain-to-failure, low impact strength and high electrical conductivity (which may cause electrical shorting in unprotected electrical machinery in production places where carbon fibers are used).

There are several other fibers that find limited applications in the automotive industry. One of them is aramid fiber, which is an organic polymer fiber; the most common trade name of these fibers is Kevlar. Aramid fibers have the lowest density and highest tensile strength-to-density ratio of currently available reinforcing fibers. They are also capable of high energy absorption under impact loading. Among their limitations are (i) high cost, (ii) high moisture absorption and (iii) low compressive strength. Aramid fibers are not recommended in applications involving high compressive stresses. Another organic polymer fiber is the extended chain polyethylene fiber which has the lowest density of all reinforcement fibers. These fibers also have high elongation to failure. Both aramid and extended chain polyethylene fibers can be used to provide high impact resistance to glass or carbon fiber composites when they are combined to make a hybrid laminate.

Natural fibers, such as jute and kenaf, are also used with thermoset polymers, but the applications of natural fiber reinforced thermoset polymers are limited to interior trim components. Natural fiber properties are given in Chapter 5 and are not discussed in this chapter.

6.2.3 Fiber architecture

Man-made fibers are produced as very small diameter continuous filaments, with the filament diameter typically ranging between 2 to 20 μm. Since filaments of this small diameter are extremely fragile and are difficult to handle, fibers are supplied as untwisted or twisted bundles of filaments. Untwisted bundles contain multiple parallel filaments. They are called rovings in the glass fiber industry and tows in the carbon fiber industry. Twisted bundles are called yarns. In a typical glass fiber roving, there may be 200 to 400 glass filaments. Similarly, carbon fiber tows may contain 1000 to 160 000 filaments. Carbon fibers used in the aerospace industry typically contains 6000 to 12 000 filaments. High filament counts in the rovings and tows are desirable for achieving higher productivity in continuous manufacturing operations, such as filament winding; however, if the filament count is very high, it becomes increasingly difficult to coat the filaments with the matrix and the resulting composite properties may be lower than the expected values.

The three basic fiber forms, namely rovings, tows and yarns, can be arranged in a variety of two-dimensional architectures, such as non-woven continuous fiber mats, woven fabrics, knitted fabrics and tubular braids (Fig. 6.4). In addition, they can also be cut into small lengths and supplied as chopped fibers, which can then be matted to produce chopped strand mat (CSM) containing randomly oriented fibers. Another form of two-dimensional random mat is called continuous fiber mat (CFM) in which the fibers are continuous, but laid out in random swirl patterns. Using these different fiber architectures, a variety of mechanical and thermal properties can be generated for the composite. Selection of the fiber architecture for an application will depend on the performance requirements for the part and the processing method to be used to make the part. During processing, the liquid thermoset polymer must be able to flow through the open spacing between the fibers, wet the filaments in each fiber bundle and displace the air between them in order to obtain void-free composite parts with uniform matrix distribution.

The most common two-dimensional fiber architecture with continuous fibers is a bidirectional fabric in which the fiber bundles are woven in two mutually perpendicular directions, called warp and fill, representing 0° and 90° directions, respectively. Fiber counts in the warp and fill directions can be varied. Depending on the design and processing requirements, the weave pattern can also be varied between plain weave, basket weave and satin weave. Multi-directional fabrics with fiber bundles woven in more than two directions are also available. An example of a two-dimensional multi-directional fabric is the 0/45/-45 fabric that is shown in Fig. 6.4. Since in a woven fabric, the fiber bundles are crimped as they pass over and under one another to form an interlacing pattern, the tensile strength of a fabric reinforced composite is generally lower than a laminated composite containing straight continuous fibers oriented in different directions in different layers.

Two-dimensional architecture can also be prepared using knitting and braiding processes. In a knitted fabric, the fiber bundles are inter-looped instead of interlaced. Knitted fabrics are more flexible than woven fabrics and are more suitable for making part shapes with tight corners. Braided fabrics are produced on braiding machines by intertwining two sets of continuous

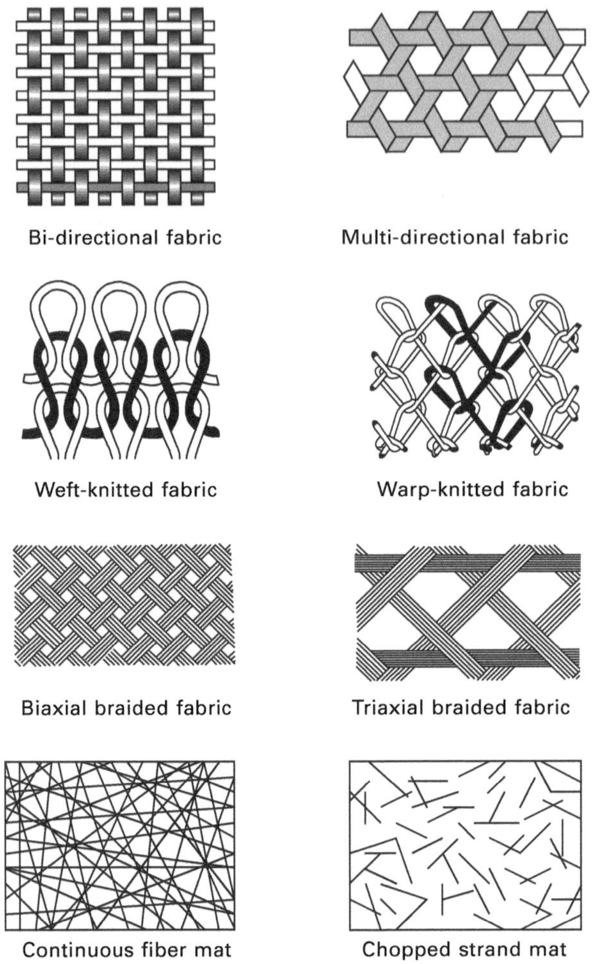

6.4 Two-dimensional fiber architectures.

fiber bundles, one in the positive angle direction and the other in the negative angle direction relative to the braiding axis. Braided construction is most suitable for making tubular parts.

6.2.4 Dry fiber preform

A dry fiber preform is an assembly of dry fiber layers that have been pre-shaped to the form of the desired product and bonded together using a binder resin. Dry fiber preforms are placed in the mold as a single piece. They are used with liquid injection molding processes, such as resin transfer molding and structural reaction injection molding. The use of preforms improves moldability for parts with complex shapes (particularly with deep draws) and geometric features, such as corners and ribs. Since preforms are near-net shape constructions, trimming after molding the part may not be necessary.

There are various methods of producing preforms. For random fiber preforms, either a spray-up process or a water slurry process is commonly used. In the spray-up process, continuous fiber rovings are cut into 10 to 80 mm lengths using a chopping gun and the chopped fibers are sprayed onto a screen which is pre-shaped in the form of the part. Vacuum applied to the rear side of the screen holds the fibers securely on the screen. A thermosetting binder resin sprayed with the fibers keeps them in place and maintains the preformed shape. In a recently developed preforming process, called Programmable Powdered Preforming Process or simply P4, chopped fibers are sprayed along with a thermoplastic binder onto the preform screen using a programmable robot. Positive airflow through the screen holds the chopped fibers on the screen surface. After the spraying is complete, the thermoplastic binder is melted by drawing hot air through the screen. Two advantages of the robotic control in the P4 process are that the fiber direction in the preform can be more precisely controlled and the preform thickness can be varied from location to location. In the water slurry process, a slurry of chopped glass or carbon fibers and thermoplastic binder fibers in water is first created by mixing them in a slurry tank. The preform is produced by rapidly removing the water from the slurry through a preform screen and then heating it in an air circulating oven to remove the remaining water and melt the thermoplastic binder fibers.

Continuous strand mat preforms containing long random fibers are produced by a stamping operation using a simple press and a pre-shaped die. Both thermoplastic and thermoset binders are available for retaining the formed shape after stamping. With woven fabrics containing bidirectional fibers, a 'cut and sew' method is used in which various patterns are first cut from the fabric and then stitched together by polyester, glass, or Kevlar sewing threads into the shape of the part being produced. Braiding and textile weaving processes are used to produce two- or three-dimensional preforms. Braiding is particularly suitable for producing tubular preforms, which can be subsequently used in the resin transfer molding process to manufacture the frame rails and the front end structure of a vehicle.

6.2.5 Thermosetting polymers

The matrix in a polymer matrix composite does not carry much load; but, as mentioned before, one of its major roles is to transfer load from one fiber to the next. It influences the compressive and shear strengths of the composite and contributes to the fracture resistance and energy absorption capability of the composite. Furthermore, it protects the fibers from environmental conditions and controls the following characteristics of a polymer matrix composite.

- Chemical resistance
- Maximum use temperature
- Amount of moisture absorption
- Tolerance to fluids, which in the automotive environment can range from gasoline to battery acid

The selection of matrix also controls the processing characteristics of a polymer–matrix composite. For a thermoset–matrix composite, the important processing characteristics for the matrix resin are

- Low initial viscosity for good resin flow through the fiber layers, which is important not only for impregnating the fiber bundles, but also for removing air, solvents or reaction by-products from the composite prior to the completion of the curing reaction.
- Fast cure for high productivity
- Fiber surface wettabilty, which can be improved by using fiber surface treatment
- Low curing shrinkage for dimensional control

Thermoset polymers commonly used in the automotive industry are either polyester or vinyl ester resins. Epoxy resins have better mechanical properties than polyesters and vinyl esters, and are the principal thermoset polymer used as matrix in aerospace composites. However, they are more expensive and the curing time for epoxies is in the range of several minutes to several hours and is considered too long for automotive applications. Because of these two reasons, epoxies are not considered as the major matrix material for automotive composites. Properties of polyester, vinyl ester and epoxy resins are given in Table 6.2.

Polyesters are selected for many automotive applications because of their low cost, ease of processing due to their low viscosity, and fast cure time. Vinyl esters are similar to polyesters in terms of their processing characteristics, but they are more expensive than polyesters. Their mechanical properties

Table 6.2 Properties of thermoset polymers commonly used in thermoset-matrix composites

Polymer	Density (g/cm^3)	Tensile modulus (GPa)	Tensile strength (MPa)	Elongation at failure (%)	Coeff. of thermal expansion ($10^{-6}/°C$)	Shrinkage (%)
Polyesters	1.1–1.43	2.1–3.45	34.5–103.5	1–5	–	5–12
Vinyl esters	1.12–1.32	3–3.5	73–81	3.5–5.5	–	5.4–10.3
Epoxies	1.2–1.3	2.75–4.10	55–130	–	55–80	1–5

are also better than the mechanical properties of polyesters. Both polyesters and vinyl esters exhibit higher mold shrinkage than epoxies and both have excellent moisture resistance compared to epoxies. Vinyl ester resins are usually preferred over polyesters for parts subjected to fatigue and impact loading.

Polyesters are available in two chemical forms: orthophthalic and isophthalic. Orthophthalic polyester is more commonly used than isophthalic polyester, but the latter has higher strength, crack resistance and chemical resistance. Polyester resins come mixed with styrene monomer, which acts as both a diluent and a reactant. As a diluent, it reduces the initial viscosity of the polyester resin, and as a reactant, it chemically reacts with the polyester molecules and forms cross links between them, which transforms the liquid polyester to a solid polymer. The curing reaction needs the addition of a catalyst and an elevated temperature. Accelerators and reaction inhibitors can be varied in the resin mix to control the curing reaction rate of the resin.

The curing of vinyl ester resins is very similar to that of polyester resins. They are also mixed with styrene monomer, which has the dual role of acting as a diluent and a reactant. The curing of epoxy resins, on the other hand, is quite different. They are mixed with a stoichiometric amount of a hardener, which acts as the curing reaction initiator. For many epoxy resins, curing reaction is initiated by raising the temperature; however, after the curing reaction has started, it can be also slowed by lowering the reaction temperature to create a partially cured or B-staged condition. The B-staged resin can be transformed into a hard, solid mass by completing the curing reaction at a later time.

6.3 Manufacturing processes

Manufacturing of thermoset polymer matrix parts involves curing of the uncured or partially cured thermoset resin at elevated temperatures. High cure temperatures are required to initiate and sustain the chemical reaction that transforms the uncured or partially cured material into a fully cured solid. External pressure is used during the manufacturing process to provide the force needed for the flow of the resin or fiber–resin mixture in the mold, wetting the fiber surface with the resin, expelling air from the mold and consolidating individual layers into a bonded laminate. The magnitudes of these two important process parameters, as well as their duration, significantly affect the quality and performance of the molded product. The time required to properly cure a part is called the cure cycle. Since the cure cycle determines the production rate for a part, it is desirable to achieve the proper cure in the shortest amount of time. It should be noted that the cure cycle depends on a number of factors, including resin chemistry, catalyst reactivity, cure temperature, and the presence of inhibitors or accelerators.

Hand lay-up technique is one of the early manufacturing processes used for making structural parts with fiber reinforced polymers. Although hand lay-up is a reliable process and is still used, it is by nature a very slow and labor intensive process. In recent years, partly due to the automotive industry's interest in fiber reinforced polymers, several different manufacturing methods have been developed that can support high production rates. Compression molding and filament winding represent two such manufacturing processes. Liquid injection molding methods, such as resin transfer molding (RTM) and structural reaction injection molding (SRIM) also have the ability to produce composite parts with complex shapes at relatively high production rates. Many of these processes can be highly automated.

6.3.1 Compression molding

Compression molding is currently the most common manufacturing process for producing thermoset–matrix composite parts in the automotive industry, mainly because of its high production rate, ability to produce large size parts with complex shapes and automation. Examples of compression molded parts are exterior and interior body panels, seats, trunk covers, lift gates, cargo boxes, bumper beams, radiator supports, wheels and cross members. Compression molding requires the use of matched metal dies, and if proper material and process conditions are used, it is capable of producing the Class A surface appearance required for exterior body panels, such as outer hood panels and front fenders. The tooling cost for compression molding is lower than the tooling cost for stamping similar size steel parts. Because of this, compression molding is considered competitive with stamping up to a production volume of 150 000–200 000 parts per year.

The compression molding process uses sheet molding compound (SMC) as the starting material (Mallick and Newman, 1990). Sheet molding compound (SMC) is a thin ready-to-mold continuous sheet containing fibers dispersed in a thermosetting resin. The resin in SMC is in a highly viscous, but uncured state. Curing of the resin takes place during the compression molding operation. The common thermosetting resins for SMC sheets are polyesters and vinyl esters. Longer cure times for epoxies have limited their use in SMC. Recently, thermoset resin derived from soybean oil has also been developed for SMC applications (Lu and Wool, 2007).

The majority of the sheet molding compounds in current use contains randomly oriented discontinuous fibers, typically 25 mm long. They are designated as SMC-Rxx, where xx is a two-digit number representing the nominal weight percentage of fibers in the SMC. For example, the nominal fiber content in SMC-R50 is 50% by weight. The higher the fiber content, the higher are the modulus and strength of SMC-R (Table 6.3). Typically, higher fiber content is selected for load-bearing structural parts (for example, radiator supports), whereas lower fiber content is used for semi-structural or non-structural parts (for example, grille opening panels). SMC sheet can also be prepared with continuous fibers which provide high modulus and strength in the fiber direction (Table 6.3). Continuous fiber SMC, which also contains small percentages of randomly oriented discontinuous fibers, is used in structural applications, such as cross members. The continuous fiber SMC is designated as SMC-Cxx Ryy, where xx and yy are two-digit numbers representing the weight percentages of continuous and random fibers, respectively. For example, in an SMC-C20 R30, the continuous fiber content is 20 percent by weight and the random fiber content is 30 percent by weight. In some applications, continuous fiber SMC is combined with random fiber SMC to locally stiffen or strengthen the compression molded part.

The starting material in an SMC is a resin paste which is combined with either randomly dispersed chopped fibers or a combination of continuous and chopped random fibers in an SMC machine to make a continuous roll of sheet material. The resin paste contains the following basic ingredients and is prepared using a high shear mixer just prior to making the sheet.

- Liquid resin premixed in styrene monomer
- Filler, typically calcium carbonate, which is used to reduce the material cost as well as control curing shrinkage
- Low profile additive, such as polyethylene or polystyrene powder, to control curing shrinkage and surface finish
- Thickener, usually magnesium oxide or magnesium hydroxide, to increase viscosity without causing any curing reaction

Table 6.3 Properties of various sheet molding compound (SMC) composites

Property	SMC-R25	SMC-R50	SMC-R65	SMC-C20R30
Density (g/cm^3)	1.83	1.87	1.82	1.81
Tensile modulus (GPa)	13.2	15.8	14.8	21.4 (L) 12.4 (T)
Tensile strength (MPa)	82.4	164	227	289 (L) 84 (T)
Strain at failure (%)	1.34	1.73	1.63	1.73 (L) 1.58 (T)
Poisson's ratio	0.25	0.31	0.26	0.30 (LT) 0.18 (TL)

Note: (i) SMC-R25, SMC-R50 and SMC-R65 contain 25, 50 and 65 wt% randomly oriented E-glass fibers, respectively.
(ii) SMC-C20R30 contains 20 wt.% unidirectional continuous E-glass fibers and 30 wt% randomly oriented E-glass fibers. L and T directions are the longitudinal and transverse directions, respectively.

- Catalyst to initiate the curing reaction when heat is applied
- Inhibitor to prevent premature curing reaction
- Accelerator to increase the reaction rate
- Mold release agent which acts as an internal lubricant and prevents sticking of the cured part to the mold surface

Thickener is an important ingredient in an SMC formulation since it increases the viscosity of the compound without curing the resin, and thus makes it easier to handle the SMC sheet prior to molding. However, the thickening reaction should be sufficiently slow to allow proper wet-out and impregnation of fibers with the resin. At the end of the thickening reaction, the sheet becomes dry, non-tacky, and easy to cut and shape. At the end of the sheet manufacturing process, the SMC roll is stored at room temperature for 1 to 7 days to 'mature' (thicken or increase in viscosity) The matured sheet can be either compression molded immediately after the proper level of maturation is achieved or stored at a sub-zero temperature for future use. With the application of heat in the mold during compression molding, the thickening reaction is reversed and the resin paste becomes sufficiently liquid-like to flow in the mold.

Compression molding is carried out at an elevated mold temperature, which is typically about 150 °C for polyester and vinyl ester resins, using a pressure of 2 to 25 MPa. The compression molding operation starts by placing a pre-cut assembly of SMC sheets on the bottom half of mold, which is already preheated to the mold temperature. Normally, about 60–70% of the mold surface is covered by the SMC sheet assembly. As the mold is closed and the SMC material is heated up, it starts to flow and fill the mold cavity. The mold is kept closed for the material to cure under pressure for about 1 to 3 minutes, depending on the part thickness. The molded part is removed from the mold, often with the aid of ejector pins, after the desired level of curing is achieved.

Many new developments have taken place in the last several years that have made SMC a more viable material for automotive body panels. A few examples are given below.

- Toughened polyester resins which have helped reduce the microcracking that usually develops during molding and is the principal reason for paint 'pops' on SMC parts when they pass through paint baking ovens
- Low density SMC which contains hollow glass spheres or other low density fillers instead of calcium carbonate
- In-mold coating that helps achieve Class A surface
- Vacuum assisted compression molding to reduce porosity in the molded part
- In-mold cure monitoring sensors that detect the cure completion and

automatically send signals to open the mold, thus reducing part-to-part cure variability
- Laser surface analyzers to measure the surface quality for Class A surfaces, such as the Laser Optical Reflective Index Analyzer (LORIA)

6.3.2 Resin transfer molding

Resin transfer molding (RTM) is a liquid injection molding process that utilizes liquid resin injection under pressure through either a stack of dry fiber layers or a dry fiber preform placed in a closed mold cavity. The resin in RTM is either a polyester, a vinyl ester or an epoxy premixed with the catalyst or curing agent and other ingredients, such as fillers, solvents (which is used for decreasing the initial viscosity), etc. Table 6.4 shows the typical composition of a polyester resin mix used in RTM. The resin mix is injected through one or more ports into the closed mold cavity containing the dry fiber stack or the dry fiber preform. The injection pressure used in RTM is in the range of 0.4 to 1 MPa. As the uncured liquid resin flows and spreads throughout the mold, it fills the space between the fibers, displaces the entrapped air through the air vents in the mold and coats the fibers. Depending on the type of the resin-catalyst system used, curing is performed either at room temperature or at an elevated temperature in an air-circulating oven. After the cured part is pulled out of the mold, it may sometimes be necessary to trim the part at the outer edges to achieve the final dimensions.

Compared to the compression molding process (which requires significantly higher pressure and temperature as shown in Table 6.5), RTM has a very low tooling cost and simple mold clamping requirements. In some cases, a ratchet clamp or a series of nuts and bolts can be used to hold the two mold halves together. RTM is also a low-pressure process, and therefore parts can be resin transfer molded in low-tonnage presses. A second advantage of the RTM process is its ability to encapsulate metal inserts, stiffeners, washers, etc. within the molded laminate. It is also possible to make a sandwich structure by including a foam core between the top and bottom preforms

Table 6.4 Typical resin composition in resin transfer molding

	Weight %
Polyester resin in styrene	48–68
Calcium carbonate (filler)	28–48
Cobalt octoate (accelerator)	0.2
Dimethylaniline (accelerator)	0.1
Inhibitor	<0.1
Internal mold release	0.1-1
MEK peroxide (added just prior to injection)	2

Table 6.5 Temperature and pressure requirements for various thermoset-matrix composite processing methods

Processing method	Processing temperature (°C)	Pressure (MPa)
Compression molding	130–175	2–25
Resin transfer molding	20–120	0.4–1
Reaction injection molding	60–120	0.5–1.5
Vacuum bag molding	130–175	0.5–1

of a preform assembly. This adds stiffness to the structure without adding significant weight and allows molding of complex three-dimensional shapes in one piece. RTM offers a cost-saving alternative to the labor-intensive bag molding process or the capital-intensive compression molding process. It is particularly suitable for producing low- to mid-volume parts, say 5000–50 000 parts a year.

With the current RTM resin systems, the curing time may vary from several minutes to several hours. The RTM cycle time can be speeded up using a rapid injection system, which requires the initial resin viscosity to be as low as possible and the use of multiple injection ports. However, the most time-consuming part of the production cycle is the cure cycle, for which fast curing resin needs to be developed.

Vacuum assisted resin transfer molding (VARTM) is a variation of the RTM process in which vacuum is used, in addition to the resin injection system, to pull the liquid resin into the preform. Another variation of the RTM process is SCRIMP, which stands for Seemann's Composite Resin Infusion Molding Process, a patented process named after its inventor William Seemann. Vacuum is also used in SCRIMP to pull the liquid resin into the dry fiber preform, but in this process, a porous layer is placed on the preform to distribute the resin uniformly throughout the preform. The porous layer is selected such that it has a very low resistance to flow and it provides the liquid resin an easy flow path into the preform. In both VARTM and SCRIMP, a single sided hard mold is used. The preform is placed on the hard mold surface and covered with a soft vacuum bag.

The quality of resin transfer molded composite parts depends on resin flow through the dry fiber preform, since this determines mold filling, fiber surface wetting and void content in the molded part. The principal molding problems observed are incomplete filling, dry spots, non-uniform resin distribution, void formation, non-uniform cure and low degree of cure. The main source of void formation is the air entrapped in the preform. Good resin flow and mold venting are essential in reducing the void content in the composite. There may also be fiber displacement and preform distortion as the liquid resin moves though the fiber preform, especially if the viscosity increases rapidly before the mold filling is complete.

6.3.3 Structural reaction injection molding

Structural reaction injection molding (SRIM) is also a liquid injection molding process and is currently used in the automotive industry for manufacturing large external body parts, such as a pick-up truck cargo box. SRIM is based on the reaction injection molding (RIM) technology in which two highly chemically reactive, low-viscosity liquid chemicals are impinged on each other at very high speeds in a mixing chamber immediately before injecting the liquid mix into a closed mold cavity (Fig. 6.5). In the SRIM process, either a dry fiber stack or a preform is placed in the mold prior to injecting the liquid mix. Commercial RIM resins are mostly based on the polyurethane or polyurea chemistry. The reaction rate for the resins formed in RIM or SRIM is much faster than the cure rate of the epoxy, polyester or vinyl ester resins that are commonly used for compression molding or other liquid injection molding processes. Polyurea is selected instead of polyurethane if the molded part needs to exhibit higher dimensional stability at elevated temperatures during its application phase. However, both polyurethane and polyurea have lower modulus and hardness than the other thermoset resins, which is a drawback for the SRIM process.

The processing temperature for SRIM is in the range of 60 to 120 °C and the molding pressure is in the range of 0.5 to 1.5 MPa. Thus, in many respects, the SRIM process is similar to the RTM process. The principal advantage of SRIM is the processing speed. A typical reaction time for the SRIM resin systems is of the order of 30 seconds, so that a typical mold closing to mold opening cycle is around 1 minute compared to 5 minutes or more for RTM. The viscosity of the SRIM resin is usually an order of magnitude lower than that of the RTM resin. Because of their high reactivity, the mixing of the two parts of the SRIM resin requires a very fast, high pressure impingement mixing, whereas the RTM resin parts are usually mixed using a low pressure mixing process. Other differences in RTM and SRIM processes are listed in Table 6.6.

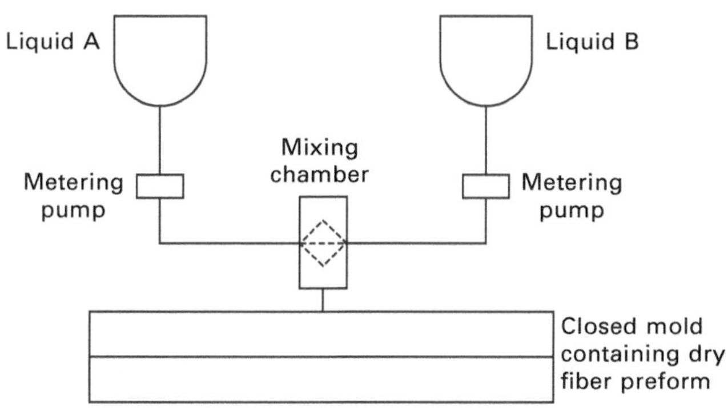

6.5 Schematic of the structural reaction injection molding (SRIM) process.

Table 6.6 Resin transfer molding (RTM) versus structural reaction injection molding (SRIM)

	RTM	SRIM
Mold pressure (MPa)	0.3	2.4
Mold temperature (°C)	25–40	95
Flow rate (kg/min)	2.3	55
Mixing	Static mixing	Impingement mixing
Cycle time (min)	10–60	2–6
Void content (vol%)	0.1–0.5	0.5–2
Mold material	Epoxy	Steel
Equipment cost	$30,000	$500,000

Adapted from Davies (2003).

6.3.4 Filament winding

The basic filament winding operation involves pulling a band of continuous fibers through a pre-catalyzed liquid thermosetting resin and wrapping it around a rotating mandrel to produce hollow parts. The resin-wetted fiber band is moved back and forth, parallel to the rotating mandrel. Depending on the ratio of the rotational speed of the mandrel and the translational speed of the fiber band, a variety of winding patterns can be created. After winding a number of layers of fiber bands to obtain the desired part thickness, the filament wound part is cured in an oven or in an autoclave, and then the mandrel is either removed or left inside the filament wound part to act as a permanent liner (e.g. in gas tanks). For some applications, such as high pressure gas tanks, a non-removable hollow metal mandrel is used, which acts as a leakage barrier for the gas inside the tank. The removable mandrels can be made of inflatable rubber, metal skin with a collapsible internal structure, low melting alloy, Plaster of Paris or a soluble salt.

Filament winding can be used to manufacture parts as simple as axisymmetric hollow tubes or as complex as an aircraft fuselage or an automobile steering wheel. The complex and non-axisymmetric parts require numerically controlled multi-axis filament winding machines, which are now commercially available. In the automotive industry, filament winding is used for manufacturing drive shafts, leaf springs, compressed natural gas (CNG) tanks and hydrogen storage tanks for fuel cell vehicles. More complex parts, such as frame rails and roof rails, can also be manufactured using the filament winding process.

The advantages of filament winding are its low tooling and operating costs and excellent reproducibility. The winding speed is limited to about 60–120 m/min due to the inability of the resin to wet the filament surfaces at higher winding speeds.

6.3.5 Vacuum bag molding

Vacuum bag molding is the primary composite manufacturing process for making laminated structures and is very common in the aerospace industry. It has limited use in the automotive industry since it is a labor-intensive process and may involve long cycle times. It has been used primarily for making prototype automotive composite parts, but may find use in the future for making large components, such as roof panels and floor pans, especially if a laminated construction is used for these components and the production volume is relatively small.

The starting material for the vacuum bag molding process is a prepreg that contains either unidirectional continuous fibers or a bidirectional fabric in a partially cured (also called B-staged) thermosetting polymer, typically an epoxy. Plies of desired shape, size and fiber orientation are cut from the prepreg roll and are stacked according to the design requirement, either by hand or by using a numerically controlled tape-laying machine. The stack is placed on the mold surface and is covered with a flexible polymer film, which serves as the vacuum bag (Fig. 6.6). After sealing the vacuum bag around the edges, the assembly is placed either in an air-circulating oven or in an autoclave for curing. The application of a vacuum removes air from the vacuum bag. The temperature in the oven or the autoclave is then raised at a controlled rate until the preset curing temperature is reached. If autoclave curing is used, a positive pressure is applied on the assembly to consolidate the separate plies in the stack into a solid laminate. If oven curing is used, the consolidation takes place at one atmospheric pressure. Depending on the resin used, the curing process may take anywhere from 30 minutes to several hours. Since autoclave molding utilizes both vacuum and positive pressure, it produces better consolidation and lower void content in the cured

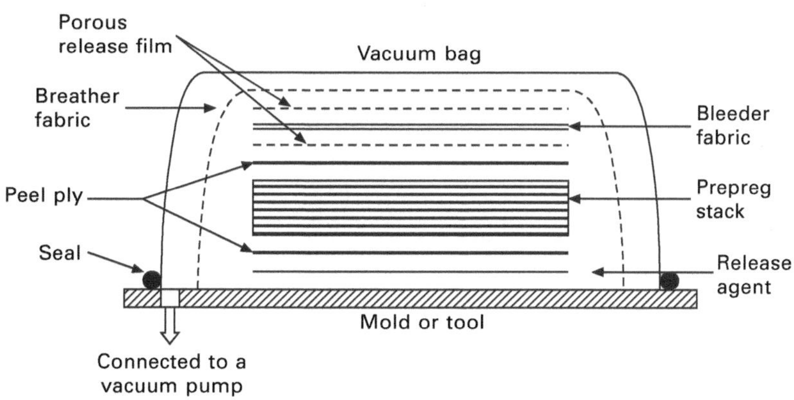

6.6 Schematic of a vacuum bag molding set-up.

laminate than oven curing and is generally recommended for manufacturing thick laminates.

6.4 Carbon fiber reinforced thermoset–matrix composites

Carbon fiber reinforced thermoset–matrix composites, although very expensive relative to other materials used in today's vehicles, continue to raise interest among vehicle manufacturers due their great weight saving potential. However, so far they have found only limited applications in production vehicles, mostly in low-volume sports cars for which customers are willing to pay higher price for improved acceleration and performance. Since the mid-1970s, several concept cars or prototype cars have been built to prove the weight saving potential with carbon fiber reinforced composite. They were produced using the proven aerospace technology, such as carbon fiber reinforced epoxy prepreg and vacuum bag molding process. The earliest known of these concept cars was developed in 1979 by Ford Motor Co. to make a direct weight comparison with a production steel car. As shown in Table 6.7 , the total weight saving was 33%, with the maximum weight saving provided by the body-in-white. The ride quality and vehicle dynamics of the composite vehicle were judged to be at least equal to those of the steel vehicle (Beardmore and Johnson, 1986). There are also examples where only the body-in-white or the chassis structure was either built or designed using carbon fiber reinforced epoxy or vinyl ester resin and RTM as the key manufacturing process. A recent design study (Boeman and Johnson, 2002) shows that a carbon fiber reinforced composite body-in-white is capable of 60 percent weight reduction compared to the steel baseline while meeting the stiffness targets. Another example is found in the design of the carbon fiber/

Table 6.7 Weight comparison of steel and carbon fiber/epoxy components in a prototype composite car built by Ford in 1972

	Weight (kg)		
Component	Steel	Carbon fiber/epoxy	Weight saving (%)
Body-in-white	191.6	72.5	62.2
Front end	43.0	13.6	68.4
Frame	128.2	93.3	27.2
Wheels (5)	41.5	22.2	46.5
Hood	22.2	7.8	64.9
Decklid	19.4	6.5	66.5
Doors (4)	63.9	25.1	60.7
Bumpers (2)	55.7	19.9	64.3
Driveshaft	9.6	6.7	30.2
Total vehicle	1698.7	1134.3	33

epoxy clamshell floor pan/frame architecture of a pick-up truck (Rastogi, 2004). It was a laminated structure containing unidirectional continuous fibers, 45/-45 woven fabrics and 0/90 woven fabrics in various layers of the laminate. There was a 44 percent weight saving with the carbon fiber/epoxy structure over the baseline steel design.

In recent years, several vehicle manufacturers have started introducing carbon fiber reinforced composite parts in their production vehicles. Although these are niche and high-priced vehicles, the experience gained in designing and manufacturing these composite parts and evaluating their on-vehicle performance will be extremely valuable. The most noted among the vehicles in which these parts are found are the BMW M6 coupe and the Corvette Z06. The roof panel in the M6, for example, was made of a carbon fabric reinforced epoxy and was manufactured using RTM. Resin injection and curing took approximately 30 minutes. The exterior hood panel on the 2004 Corvette Z06 used unidirectional carbon fiber reinforced epoxy prepreg and was vacuum bag molded in autoclaves. The thickness of the carbon fiber/epoxy hood was 1.2 mm compared to 2.5 mm for glass fiber reinforced SMC-R and it was approximately 60 percent lighter. The successful application of the carbon fiber/epoxy hood led to further development of carbon fiber/epoxy front fenders for the 2006 Corvette Z06. The material used for the front fenders was also unidirectional carbon fiber reinforced epoxy which was vacuum bag molded in an autoclave (Remy *et al.*, 2005a). Other carbon fiber reinforced thermoset–matrix parts for the 2006 Corvette Z06 (Remy *et al.*, 2005b) were the floor panel and the front wheelhouses (both inner and outer). Compression molding was the process used for manufacturing all these panels. The floor panel was of sandwich construction, consisting of chopped carbon fiber reinforced polyester in the skins and a balsawood core. The wheelhouse was also made of chopped carbon fiber reinforced polyester, but in this case hollow glass microspheres were added as filler to reduce the density of the material. The Dodge Viper, first introduced in 1992, also used compression- molded chopped carbon fiber reinforced SMC to reduce weight in many of its structural components. A carbon fiber/epoxy driveshaft was introduced in Ford's Econoline vans in the mid 1980s and GM's full-size trucks in the mid-1990s. This one-piece driveshaft was significantly lighter than the two-piece steel driveshaft used in truck applications. Replacing the two-piece driveshaft with a one-piece driveshaft, it was also possible to eliminate the center support bearing, which contributed to further weight saving. The body shell, doors, hood, chassis and seats in the 2004 Mercedes-Benz SLR McLaren were all made of carbon fiber reinforced composites (Birch, 2004). As part of the SLR's front structure, two conical carbon fiber reinforced composite longitudinal members were utilized to absorb impact energy in a front-end crash. The primary structure of the SLR was about 30 percent lighter than the conventional steel structure of a comparable front/mid-engine vehicle. Other notable applications of carbon fiber/epoxy composites in production vehicles, albeit in small volumes, are the chassis of the Porsche Carrera GT, front end structure of the Dodge Viper, and rear deck inner panel of the Ford GT.

Despite all the successful applications mentioned above and the promises of weight saving potential, the future of carbon fiber reinforced composites in production vehicles remains unsettled, primarily for two reasons, namely high cost and limited availability. There are also a number of technical issues that need to be considered. Among them are:

- Joining and assembly techniques
- Long-term performance in the automotive use environment
- Durability and damage tolerance
- Repair strategy
- Production cycle time
- Quality assurance techniques

6.5 Conclusion

The use of fiber reinforced thermoset–matrix composites in automobiles has grown quite significantly, largely due to the application of glass fiber reinforced sheet molding compounds in a variety of exterior and interior body applications. Despite the great weight saving potential of carbon fiber reinforced composites over other structural materials, their use has not occurred on a large scale, primarily due to the high cost and limited availability of carbon fibers. There are also the problems of long cycle times and lack of design information specific to the automotive environment, which have slowed further use of fiber reinforced composites in the automotive industry. Developments of more efficient structural analysis and design tools, faster curing resins, automated processes and reliable non-destructive testing tools are needed for further growth in the use of fiber reinforced composites in future automobiles.

6.6 References

Beardmore P and Johnson C F (1986), 'The potential for composites in structural automotive applications', *Composites Sci. Tech.*, 26, 251–281.

Birch S (2004), 'Mercedes-Benz SLR McLaren', *Automotive Engineering International*, February, 158–161.

Boeman R G and Johnson N L (2002), 'Development of a cost effective competitive, composite intensive, body-in-white', *2002 SAE World Congress*, Paper No. 2002-01-1905, Warrendale, Society of Automotive Engineers.

Caliskan A G (2000), 'Crashworthiness of composite materials and structures for vehicle applications', *2000 SAE World Congress*, Paper No. 2000-01-3536, Warrendale, Society of Automotive Engineers.

Davies G (2003), *Materials for Automobile Bodies*, Burlington, Elsevier.
Kim P, Schuh T and Wittig W (2004), 'CFRP – from motor sports into automotive series: Challenges and opportunities facing the technology transfer', *Plastics in Automotive Engineering*, Düsseldorf, VDI Verlag GMbH.
Lu J and Wool R P (2007), 'Sheet molding compound resins from soybean oil: Thickening behavior and mechanical properties', *Polym. Eng. Sci.*, 47, 1469–1479.
Mallick P K (2008), *Fiber Reinforced Composites*, Boca Raton, CRC Press.
Mallick PK and Newman S (1990), *Composite Materials Technology: Processes and Properties*, Munich, Hanser Publishers.
Rastogi N (2004), 'Stress analysis and lay-up optimization of an all composite pick-up truck chassis structure', *2004 SAE World Congress*, Paper No. 2004-01-1519, Warrendale, Society of Automotive Engineers.
Remy J, Voss M, Blackwell D and Di Natale C (2005a), '2006 Corvette Z06 carbon fiber fender – engineering, design and material selection considerations', *2005 SAE World Congress*, Paper No. 2005-01-0468, Warrendale, Society of Automotive Engineers.
Remy J, Hamilton D, Moss E D, Pukalo B, Zuk G and Johnson R (2005b), '2006 Corvette Z06 carbon fiber structural composite panels – design, manufacturing and material development considerations', *2005 SAE World Congress*, Paper No. 2005-01-0469, Warrendale, Society of Automotive Engineers.

Part II

Manufacturing and design of lightweight automotive structures

7 Manufacturing processes for light alloys

G. T. KRIDLI, University of Michigan–Dearborn, USA,
P. A. FRIEDMAN, and J. M. BOILEAU,
Ford Research and Innovation Center, USA

Abstract: This chapter highlights the manufacturing processes used to fabricate lightweight automotive parts. The chapter provides an overview of aluminum and magnesium alloys; describes design issues and manufacturing challenges for light alloys; highlights commonly used metal casting and metal forming processes; identifies enablers to significantly increase the use of light alloys in the production of automotive parts; and describes some of the promising metal forming technologies for light alloys.

Key words: metal forming processes, metal casting processes.

7.1 Choosing light alloys

In the past three decades, weight reduction in automotive components has become a key focus area. This has been due to the need to meet increased customer expectations of vehicle safety, performance, and reduced emissions while complying with regulations on fuel economy. One important method of weight reduction involves redesigning existing components to use lightweight materials for vehicle body construction.

The key to the effective use of these lightweight materials is to tailor the material and the processes to the parts of the vehicle structure. Typically, sheet is used to create two- and (simpler) three-dimensional shapes such as hoods, decklids, and body panels. More complex three-dimensional forms with reinforcing webs are more commonly created from castings. In the selection of the fabrication method, cost and functionality must be analyzed to create both an efficient and cost-effective solution. Under certain conditions, an assembly created from multiple stamped parts welded together will be the most effective; under other conditions, a single large-scale casting that integrates multiple features may be optimum.

7.2 Materials of interest

7.2.1 Aluminum

Aluminum is a material which is compatible with existing manufacturing processes and has attractive properties such as low density, good mechanical properties, and high corrosion resistance. In one study, it was estimated that replacing steel with aluminum for body-in-white (BIW) structures and closures could result in weight savings as high as 55% (Powers, 2001). Currently (2008), aluminum is commonly used for automotive components in both the wrought (sheet, extrusions, and forgings) and cast forms.

Aluminum alloys are classified using the Aluminum Association System, which assigns a numerical code that uniquely identifies each alloy; the numerical code is of the general form 'XYYY' for wrought and 'XYY' for castings; often, an alphanumeric temper (or heat-treatment code) is added at the end. The system divides the 240 major alloy compositions into 8 families, based upon the major alloying elements present; Tables 7.1 and 7.2 present the classifications for wrought and cast aluminum alloys, respectively; Table 7.3 presents the temper codes.

Alloying is important, as it strengthens the material through the metallurgical mechanisms of solid solution strengthening and precipitation hardening. Solid solution strengthening is characterized by the dissolution of the alloy element atoms into the aluminum on the atomic level. Thus, the alloy element atoms are incorporated into the crystallographic lattices of the aluminum. This causes the lattices to be distorted, strengthening the resulting alloy. The strength of the alloy is a function of the type of alloying element, percentage of the element present, and the relative difference in the atomic diameters of the aluminum and alloy element atoms; in general, solid solution strengthening is most effective when the relative difference in diameters is less than 15%. Since this strengthening mechanism is not thermally activated, these alloys are often referred to as 'non-heat treatable alloys'. Cold work is the only practical method for increasing the strength in these alloys.

Precipitation hardening is characterized by thermally activated diffusion of elements within the lattice of the aluminum alloy. Precipitation hardening requires a heat-treatment process; this consists of any combination of the following three steps: a solutionization, a rapid quench, and an aging treatment. The solutionization involves heating the component to a high (> 490 °C) temperature to allow the precipitating alloy atoms to be uniformly dispersed throughout the aluminum matrix. The quench is a rapid cooling process that 'freezes' the precipitating alloy atoms in place, creating a supersaturated solution. The aging process controls the rate at which the precipitating alloy atoms in the supersaturated solution are allowed to come out of the solution to form the precipitates. Since the structure and morphology of the precipitates are controlled by the times and temperatures used in the heat-treatment sequences, this allows for considerable optimization of the strength, ductility, and toughness of the aluminum alloys. Additionally, the heat-treatment may be performed at any stage in the manufacturing process, including post-forming.

Table 7.1 Classifications of wrought aluminum alloys (Hatch, 1984)

Series	Major alloying elements	Heat-treatable?	Attributes	Applications (typical alloys)
1xxx	None (>99% Al)	No	Excellent corrosion resistance, high thermal and electrical conductivity, and excellent formability. Low strengths.	Electrical (1350) and chemical (1100)
2xxx	Cu	Yes	High strength, good formability, but poor corrosion resistance. Requires solution heat treatment for optimal properties.	Aircraft (2024/2124/2219) and automotive
3xxx	Mn	No	Moderate strength and good formability; excellent drawing characteristics.	Beverage cans bodies (3003)
4xxx	Si	No	Lower melting point relative to other Al alloys	Welding (4043) and brazing (4345)
5xxx	Mg	No	High strength, good formability and good welding characteristics; some alloys have excellent bright finished surfaces. Potential stress corrosion cracking at Mg levels > 3.5%.	Beverage can tops/tabs (5182/5042) and automotive sheet (5082)
6xxx	Si, Mg	Yes	Moderate to high strength, good formability, and good corrosion resistance. Requires heat treatment for optimal properties.	Automotive sheet (6010) and general use alloy (6061/6063)
7xxx	Zn	Yes	Very high strength, moderate formability and crack-growth resistance. Requires heat treatment for optimal properties	Aircraft structures (7075/7175/7475)
8xxx	Varies	Varies	Miscellaneous alloys not otherwise categorized in the above classes; includes high temperature/pressure corrosion-resistant alloys, high-performance bearing alloys, and low-density, medium strength Al alloys.	Nuclear reactors (8001) and automotive bearings (8280)

Table 7.2 Classifications of cast aluminum alloys (Hatch, 1984)

Series	Major alloying elements	Heat-treatable?	Attributes	Applications (typical alloys)
1xx	None (>99% Al)	No	Excellent corrosion resistance, high thermal and electrical conductivity. Soft and ductile.	Large collector rings and conductor bars (150)
2xx	Cu	Yes	Excellent hot and room-temperature strengths, and good machinability. Poor corrosion resistance and castability (very vulnerable to hot tearing.) Requires heat treatment for high strength properties.	Aircraft gear housings (201/206) and motorcycle heads (242)
3xx	Si + Cu and/or Mg	Yes	Very good castability and fluidity. Good strength, ductility, toughness, machinability, and pressure-tightness. Significant strengthening can be obtained via heat treatment. Most commonly used alloy series	Automotive engine blocks and cylinder heads (319/356/380/383/390), wheels (356), and structural nodes (356).
4xx	Si	No	High corrosion resistance, good weldability, and lower density. Very good fluidity and pressure-tightness for very intricate shapes. Moderate ductility with limited strengths.	Cast-in lettering/high definition surfaces (413) and tire molds (443).
5xx	Mg	No	Very high corrosion resistance, good machinability, and good weldability. Very polished surface finishes possible. Castings are costly due to low fluidity/high oxidizing tendency of Mg. Alloys with >10% Mg can self–age at room temperatures.	Ornamental hardware/architectural fittings (513, 520) and escalator parts (518).
6xx	–	–	Designation no longer exists	–
7xx	Zn	No	High melting point, high strengths with moderate ductility, good machinability and brazability.	Components assembled by brazing (710, 712). Trailer parts and mining equipment components (713).
8xx	Varies	Varies	Very good load-bearing strength (especially compressive), good corrosion resistance in oils, and excellent self-lubricity. Large solidification range causes issues with macrosegregation and hot-cracking.	Nuclear reactors (8001) and automotive bearings (8280)

Table 7.3 Classifications of heat treatments

Designation		Definition
F		As-cast or as-fabricated (no heat-treatment)
O		Annealed or recrystallized (wrought products only)
H		Strain-hardened (wrought products only)
	H1	Strain-hardened only
	H2	Strain-hardened and then partially annealed
	H3	Strain-hardened and then stabilized
T		Thermally treated to produce stable tempers other than F, O, or H
	T4	Solution-treated and naturally-aged to a substantially stable state
	T5	Cooled after casting; no solution treatment; artificially aged at elevated temperature to produce a stabilized microstructure
	T6	Solution-treated and artificially aged at elevated temperature to produce peak strength properties
	T7	Solution-treated and artificially aged at elevated temperature to produce a stabilized microstructure
	T8	Solution-treated, cold worked, and artificially aged (wrought products only)

Wrought alloys

In terms of non-heat treatable alloys, the primary aluminum alloys of interest for automotive body construction are the Al–Mg alloys (AA5xxx). These materials are used extensively in vehicle structures and closure inners because they can be processed to have a variety of strengths. However, these alloys are not the typical choice for outer closure panels due to their relatively low strength, and the possibility of stretch marks forming in some alloys.

Heat treatable alloys (such as AA6xxx) are ideal for outer closure panels. The design requirements for these panels are strength driven so that dent resistance can be achieved. However, the strength required for dent resistance would normally make these panels difficult to form. Therefore, they are formed from alloys in the naturally aged (low-strength, high-ductility) condition. After attachment to the vehicle, these panels go through the elevated temperature paint–bake cycle. This cycle not only cures the paint, but acts as an artificial aging treatment that causes precipitation to occur. The strength is thus raised to the level required for high dent resistance. Heat treatment thus allows for a wider forming process window. Additionally, it allows thinner gauges of aluminum to be used. As an example, an aluminum hood manufactured from a typical AA5xxx alloy would need to be approximately 20% thicker than an AA6xxx alloy to achieve the same functionality.

It should be noted that, while heat-treatable alloys can offer very high strengths in the T6 and T7 tempers (artificially aged), these alloys are difficult to form after heat treatment since they often retain little ductility in the final hardened temper. There are thermal forming techniques that can be employed after heat treating. However, these techniques often reduce or eliminate the strengthening and can introduce problems such as part distortion, which then requires fixing through an expensive and often manual calibration process to return the part to the intended shape. This can result in significant issues in mass production because the required solutions involve increased costs.

Cast alloys

The majority of cast aluminum components used in the auto industry use the Al–Si– (Cu and/or Mg) alloys (AA3xx). The AA356 (Al–Si–Mg) alloy has been the most commonly used alloy for structural applications to date. The ease of processing of the alloy, combined with its ability to be heat-treated to high strength while retaining moderate ductility and toughness, has made it the primary choice for control arms, engine cradles/crossmembers, structural nodes, and wheels. There are also a number of specialty 3XX-series alloys (such as C448™ and Silafont™) that have been developed to have high strengths with improved ductilities and elongations relative to AA356. These alloys have been used in applications where the need for increased mechanical properties supersedes their increased cost; for example, the C448™ has been used for the shock towers in the Jaguar XJ.

7.2.2 Magnesium

A second lightweight material that shows promise in enabling significant reductions in vehicle mass is magnesium. With a density less than 25% of steel, Mg has the potential to remove significant mass from an existing steel vehicle.

In the US, magnesium alloys are classified using the ASTM System, which assigns an alphanumerical code that uniquely identifies each alloy; the code is of the general form 'XY##Z – Temper Code'. In this code, the 'X' refers to the primary alloying element and 'Y' to the secondary alloying element; the '#' gives the approximate percentage of that alloying element present in the alloy. Further, the 'Z' refers to a ranking of the purity of the alloy, with the greater the letter, the higher the purity. The temper code is the same as that for aluminum. Table 7.4 presents the classifications for selected wrought and cast magnesium alloys.

Like aluminum alloys, certain specific magnesium alloys can be heat-treatable; these alloys also use precipitation to strengthen the material. Other magnesium alloys rely upon solid solution strengthening for improved mechanical properties. Generally, magnesium alloys cast using high-pressure die-casting processes are not heat-treated, even if the alloy is capable of being heat-treated.

Currently, automotive structural components created from Mg are primarily

Table 7.4 Classifications of selected magnesium alloys

Series	Major alloying elements	Heat-treatable?	Attributes	Applications (typical alloys)
AZ	Al & Zn	Yes	Low cost, low density, good room to elevated (< 120 °C) temperature properties, moderate to good corrosion resistance (with high-purity versions), moderate to good creep resistance (with alloying), and easily die-cast. Mechanical properties are sensitive to part thickness.	Transmission cases, gearboxes, engine brackets, and component covers (AZ91); luggage, ladders, and aircraft racks/shelves (AZ31).
AM	Al & Mn	Yes	Low cost, moderate strength, excellent ductility, and easily die-cast. Mechanical properties are sensitive to part thickness.	Automobile instrument panels, wheels, steering wheels, grill opening reinforcements, and door inner panels (AM50/60)
AS	Al & Si	No	Low cost, moderate strength, moderate ductility, good elevated (< 150 °C) creep properties, and easily die-cast.	Automotive engine blocks and transmissions (AS41)
'X'E or E'X'	Rare earth element-based alloys	Specific alloys	Optimized for high temperature mechanical properties, including creep resistance (up to 240 °C for EZ alloys) and strength (up to 200 °C for QE alloys); moderate to expensive cost. Certain alloys have good to excellent castability and pressure tightness for thin-walled castings.	Military and aerospace components; experimental Mg engine block (USCAR). block (USCAR).
ZK	Zn & Zr	Yes	Very highly grain-refined microstructure yields moderate to very high (>350 MPa) strengths with high (>10%) ductilities; poor to moderate weldability	Military components; tent poles and tennis racquets (ZK60).

limited to cast applications. This has been due to the relatively high cost of converting Mg into wrought products. The majority of the applications have used high-pressure die cast AM50 and AM60 Mg for components such as instrument panels and grille opening reinforcements. In these applications, the ease of processing of the alloy allows for a high degree of part consolidation relative to stamped and welded steel components. Further, these alloys have moderate strength, high ductility, and good toughness, which make them attractive for components where crash response and energy absorption are important. Additionally, the aluminum/rare-earth alloy AE44 was developed for the Corvette Z-6 engine cradle; this alloy provides a high-strength, high creep-resistant material that allows the integration of the cradle into a single unit.

There are few automotive applications of wrought Mg alloys; currently, only the Ford GT has an Mg extrusion (AZ31 rail). However, recent developments in the area of continuous casting of Mg sheet hold promise to reduce production costs significantly (Hunt and Herling, 2005). This has driven renewed interest and research into the forming of magnesium sheet for high volume automotive components.

7.3 Vehicle architecture design and manufacturing

7.3.1 Wrought materials

While there are several technical challenges to engineering and manufacturing a vehicle from lightweight alloys such as aluminum, these can be overcome. Indeed, the recent successes by several automotive companies in producing aluminum-intensive production vehicles (such as the Audi A2 and Jaguar XJ) have shown that aluminum is a viable material for vehicle body construction.

Compared to other common sheet materials, aluminum alloys have sufficient formability to produce most automotive panels. However, the formability is lower than typical drawing quality steel; this can result in restrictions in outer body panel geometry which, in turn, can limit automotive design. Further, the limited formability can require that parts be split into two or more separate, easier-to-form panels; this will increase both cost and complexity. For example, the inner panel of a door is often stamped in one piece from steel; in aluminum, this same panel may require up to four separate stamped pieces. This adds to cost because: (i) additional dies are required to form the separate pieces, (ii) fixtures are required to hold the pieces for assembly, and (iii) a joining process must be added to create a finished panel ready for assembly into a car. This additional assembly process may also introduce additional dimensional variability as compared to one-piece construction.

Wrought magnesium alloys are still in a developmental phase in the automotive industry; however, aerospace applications of magnesium have shown that this material has great potential. The key issue is that, at room temperature, magnesium has limited ductility due to its hexagonal closed packed (HCP) crystallographic structure. However, when magnesium sheet alloys (such as AZ31) are heated to 200 °C, thermal activation of pyramidal and prismatic slip planes occurs (The Dow Company, 1984; Ullmann's Encyclopedia of Industrial Chemistry, 1990); this has been experimentally demonstrated to substantially improve formability (Behrens *et al.*, 2004). The enhanced ductility permits the forming of more complex 3-dimensional shapes with magnesium sheet. Further, the enhanced slip has also increased interest in using technologies such as warm (~0.5T_{melt}) drawing and hot (~0.75T_{melt}) gas forming to form parts for the automotive industry.

7.3.2 Vehicle architecture design and manufacturing

The automotive industry utilizes a wide spectrum of manufacturing processes to effectively and economically accommodate different vehicle production volumes. In order to develop the most cost-effective production strategy, investment in tooling (fixed cost) must be balanced with variable cost for a given volume. For very low-volume production, minimization of investment is paramount. This is usual in the specialty car industry, where the annual production volumes may be less than 1000 units. In these cases, more expensive materials and somewhat longer cycle times can be tolerated to minimize investment in tooling and equipment. For more medium-volume production (in the order of 25 000 units per year), the use of automation is the key enabler in achieving competitive costs and cycle times. For high-volume production (such as vehicles where the total volume over a six-year cycle can be several million vehicles), the fixed cost is spread over the millions of vehicles produced. Therefore, minimizing variable costs are critical; lower-cost materials and rapid cycle times are crucial to the economic success of the product.

Thus, the need to manage investment in tooling and equipment with variable cost has a significant effect on how a vehicle body is designed. In response to this, several approaches to vehicle design exist, including body-on-frame, monocoque, space frame, and unibody. The original method for vehicle body design, body-on-frame, is mostly used today for trucks. It is well suited to high-volume production and is based on a body structure attached to (but independent of) the chassis system.

In monocoque construction, a vehicle structure is made from extrusions and relatively simple stampings or break-forms. This creates a very simple, 'tub'-like structure which offers attachment points for the chassis, the front end, and the closure panels. Because this method minimizes the number of stampings and castings, the number of expensive dies for forming and casting is also minimized; thus, monocoque construction is a very low-investment method for body design.

In the space frame approach, frame members are typically fabricated from extrusions. These extrusions are usually joined together by castings called 'nodes.' This frame is then used to carry the loads and offer attachment points for the outer panels. This approach is well designed for low- to medium-volume vehicle production runs.

In the unibody approach, the vehicle body is composed mostly of stampings joined together to integrate the body and the chassis into a single structure. Today, this method has become the standard method for producing high-volume cars. In the unibody approach, the required stiffness and strength is usually achieved by creating boxed sections comprised of two separate stampings. In recent years, there has been an increase in the use of tubular hydroformed sections to replace some of these stamped and welded sections. The unibody design requires a significant investment in tools and dies to fabricate the complex sheet metal stampings. However, it is well suited for high-volume production because of the relatively low variable cost of stampings. The unibody approach is also suitable for high volumes, due to the fast cycle times present in stamping and body assembly operations.

7.3.3 Lightweight material design issues

While the yield stress (YS) of most aluminum alloys used in vehicle construction is similar to that of drawing-quality steels, the elastic modulus is roughly one-third (70 GPa vs 210 GPa). Given that the majority of the body components on a vehicle are stiffness limited rather than strength limited, a significant change is needed in the structural design of an aluminum vehicle. However, with proper design, an aluminum body-in-white and closures can achieve a mass saving of approximately 50% over a similar-sized steel vehicle. To achieve this mass saving, a combination of new panel design and selective use of metal upgauging is required. The new panel design is required because the elastic modulus of aluminum cannot be changed or improved by either cold work or heat treatment. Thus, some of the reduced stiffness of aluminum can be offset by using design approaches where the geometry of the panel is altered to offer more 'geometric' stiffness. For example, adding ribs and bosses can significantly stiffen a panel, while use of a reinforcement part to create a box section can greatly increase structural stiffness.

For components that are strength limited, such as closure panels (for dent resistance) and some structural parts, using aluminum alloys with increased yield stress can enable reductions in material gauge. However, to accomplish this reduction, the panel stiffness requirement must be satisfied as well; methods to accomplish this include changes to panel geometry and the application of reinforcement supports.

Safety-critical parts can typically be classified into 'bending collapse' parts and 'axial collapse' parts, depending on how they are designed to fail. For components designed for bending collapse, the critical failure mode is due to buckling instability; in this case, material modulus, material gauge, and part geometry are key factors. The amount of energy absorbed during bending collapse depends upon the YS and the ultimate tensile strength (UTS). Improving the material yield stress and workhardening capability can be used to down-gauge components used as crash members. However, higher strength can only be used to down-gauge panels up to a certain point; beyond this, the panel stability will be degraded due to a lack of stiffness. For components designed for axial collapse, the yield stress, UTS, cross-sectional area, and material gauge are key design parameters. In this case, improving the work hardening during plastic deformation will increase the initial yield stress, again allowing components to be down-gauged to the point where stiffness limitations become a significant issue.

7.3.4 Manufacturing challenges

There are several challenges with the manufacture of light alloys, including areas such as forming, joining, painting, mechanical properties, and repairability. Some of these challenges stem from the fact that most existing technologies were developed for steel. Thus, these techniques require modifications to meet the needs of manufacturing products out of light alloys. For example, the repairability of steel is well understood due to the many decades of use in automotive manufacturing. Now that aluminum-intensive vehicles are being manufactured, understanding the repairability of aluminum and developing guidelines and practices are needed. However, it must be stated that many of the manufacturing processes used with light alloys are consistent with many of our current manufacturing processes.

7.3.5 Factors influencing the manufacturing cost of metal products

The successful production of light alloy parts brings together key elements of alloy, tool design (die, mold, etc.), quality, and process. In general, a successful component is one that meets the design specifications at the minimum cost. For cast alloys, a very thorough discussion of these factors, as well as a recommendation on design and process guidelines to be followed for high-quality castings, is presented in Campbell (1988).

Alloy

The alloy that is chosen has a significant effect upon the cost. Certain alloying elements can be expensive due to their limited availability (such as Ni or rare-earth elements). The presence or absence of impurity elements also affects the cost. Primary alloys made from reduced aluminum ore alloyed with pure

Table 7.5 Macroscale defects observed in castings

Defect	Characteristics	Causes
Misrun	Metal solidifies before completely filling the cavity.	Reduced fluidity in molten metal due to pouring temperature being too low, pouring rate being too slow, or the mold cross-section being too thin.
Cold shut	Two separate metal flows meet but do not fuse together, leaving a seam.	
Shrinkage cavity	A large pipe or internal cavity exists	Inability to continuously feed molten metal to an area during solidification.
Hot tear	Large tear or crack in casting	Metal is constrained during cooling and does not pull away from mold wall. High tensile stresses arise internally and pull the metal apart.
Gas blow	Large circular cavity/ opening in casting	Arises from mold gases that are generated after an area is filled. Can be an issue when high resin concentrations are used in sand cores.
Penetration	Metal penetrates into sand mold walls	Metal fluidity may be too high and/or sand not be bonded properly.

Table 7.6 Microscale defects observed in castings

Defect	Characteristics	Causes
Gas porosity	Circular and/or hemispherical cavities within the casting.	Hydrogen gas (often caused by moisture/water vapor) or entrapped air.
Microshrinkage porosity	Tortuous, irregularly-shaped voids or cavities within the casting; individual metal dendrites are observable in the void interior	Natural volumetric contraction and an inability to continuously feed molten metal to an area during solidification. Issue tends to increase as the alloy freezing range increases
Intermetallics	Brittle, hard, or weak features in the microstructure that fracture easily	Undesirable second-phases that arise due to metal impurity and/or improper alloy element concentrations. Fe-based and Cr-based intermetallics are often very deleterious
Inclusions	Brittle, weak features in the microstructure that fracture easily	Oxides result from entrainment and/or turbulence in the molten metal. Sand entrapment can be due to low resin concentrations or poor mold cleaning. Dirt and organics can come from contaminated scrap, poor molten metal equipment practices, or too thick of a mold coating in die-casting.

elements, are expensive due to their use of virgin material. Secondary alloys, since they are made from recycled aluminum alloys, are less expensive; however, they often contain significant amounts of impurities (such as Fe or Mn) that can degrade certain properties if their presence is not taken into account. The prescribed pre-forming characteristics of the material also influence cost. These include alloys tempers, coating, and surface treatment. They also include the cost of processing needed for grain size refinement required by some forming processes, such as superplastic forming.

Certain elements can make a metallic alloy more difficult to melt and pour. As an example, low silicon aluminum alloys (such as 2xx and 5xx) have low fluidity. Thus, casting with these alloys may require larger gating systems/risers and/or higher temperatures to assure complete filling. Larger gates and risers decrease the amount of metal in the component to the metal poured (called the 'casting yield'). As the casting yield decreases, the energy and labor costs per casting increase. Similarly, as temperatures are increased, energy costs increase as well. Additionally, the amount of metal lost due to oxidation increases unless a protective environment is used; this will raise metal costs or equipment costs. The need to heat treat a casting also influences the alloy choice; in certain cases, it is less expensive to use a heat-treatable alloy (such as 319) in the un-heat treated condition simply because it is more common and less expensive compared with other non-heat treatable alloys.

Tooling

The cost of tooling is affected by the processing conditions and the production volumes. For example, a complete conventional stamping tool set for an automotive door (including forming, trimming and flanging tools) can cost upwards of a million dollars. This cost can be amortized over the life of the product and the production volume. For products with low production volume (a few hundred to a few thousand units per year), the cost of tooling per vehicle is too high if conventional stamping is used. Therefore, alternative forming processes, such as superplastic forming, are employed.

For cast products, the design of the mold has a significant effect on the cost. In general, the size, weight, and complexity are key factors as they directly influence the type of casting process chosen, as well as the material and labor required to produce a casting. In sand castings, the amount of sand per mold may equal 10–20 times the weight of cast metal poured into the mold. Molds with internal cores are generally more expensive because of the additional costs associated with the core equipment, the placement of the cores, and their removal following casting. In permanent mold castings, the tooling used to create the metal dies is the key economic factor. These castings have very high initial costs and lead times due to the need to purchase large-capacity equipment and the need to make the tooling from high quality tool steels. Production volumes also need to be closely calculated, as the fixed costs are distributed over the number of parts manufactured. When possible, multiple-cavity dies are preferred, as they allow for the production of more parts per production cycle. However, the increased size and complexity of the die will raise costs. These dies are more difficult to create and will require a higher-tonnage capacity machine for operation.

Quality requirements

The quality requirements also affect the cost. In general, quality indicates how well a manufactured part meets the customer's requirements and how consistently these requirements are met in production quantities. Premium quality requirements increase costs by requiring more complex production methods, greater numbers of more highly skilled workers, more expensive materials/production equipment, and additional finishing and inspection operations. These requirements will increase the number of rejected parts.

For formed parts, the key quality measures are fit and finish specifications. For castings, the common quality measures include the chemical and mechanical properties of the metal, the soundness of the casting, and the accuracy and consistency of dimensions. However, it should be stressed that decreased costs are *never* a valid reason for purchasing substandard castings; off-dimension or poor-finish castings will increase machining and finishing costs.

Process

Finally, the process exerts a strong influence on the final component costs. In the case of wrought metal forming processes, some have relatively long production cycle times, some require elevated temperatures, some require low strain (deformation) rates, and some require expensive control systems. For metal forming applications with lightweight alloys, one or more of these factors can be present. One example is aluminum tube hydroforming, which has a cycle time that is several times longer than conventional stamping and a required forming temperature that needs to be elevated to produce the desired level of elongation.

In order to cost-effectively form lightweight alloys, a holistic approach is needed to realize many of the benefits. The lower forming stresses required by the lightweight alloys makes it possible to form multiple parts in a single operation. This can often offset the costs associated with additional heating cycles. It can also offset the apparent increase in forming cycle time when the comparison is made with a single-cavity die. Additionally, tooling cost may be reduced by selecting forming processes that allow for part consolidation or by choosing processes that do not require matched die sets (such as sheet hydroforming and superplastic forming).

For casting, the focus of the process is on the minimization of casting defects. In general, defects are classified into two categories: macro defects (those visible to the eye); and micro defects (those visible using optical microscopy). Tables 7.5 and 7.6 detail the typical casting defects and their causes. Thus, the metal melting and holding system significantly influences the cleanliness of the metal; the higher the cleanliness, the lower the costs due to fewer rejected castings. The metal delivery system (such as gravity or pressure-assisted) strongly influences the quantity, type and location of defects. A system that minimizes turbulence and/or entrapped air will again lower costs due to fewer rejected castings. The mold material (such as sand or metal) has a strong influence on the solidification rate, affecting the coarseness or fineness of the microstructure as well as the segregation of second phases and porosity.

7.4 Forming of structural components

Metal forming includes a wide range of processes that effectively use the plastic deformation inherent within a metallic alloy. In general, the purpose is to produce parts out of wrought alloys that possess sufficient ductility under the processing temperature and strain rate. These processes use a specifically created forming tool that reflects the desired product geometry to purposely deform sheet and bulk material. The type of tool used is related to the forming process. Parts can be shaped by pushing material into a die cavity, by draping material over a punch, or by using a matched tool set (such as a punch and die). Several common processes include stamping, where matched die sets form metal, and sheet hydroforming, where the metal is shaped using either a die or a punch. Forming can be performed at room (cold), warm, or hot forming temperatures. The use of elevated temperatures can enhance the ductility of the material, reduce the required forming stress, or both. The strain rate (i.e. deformation speed) also plays a role in metal forming, especially when elevated temperatures are used. Table 7.7 presents select room temperature properties of some automotive alloys. (It should be noted that, at present, wrought magnesium sheets are only available in limited widths.) The UTS is directly proportional to the press tonnage requirement to form the material into parts, and the elongation to fracture is proportional to the material ductility and thus to the complexity of the geometry that may be formed.

Commercial metal forming processes can be classified as either sheet forming processes or bulk forming processes. Sheet forming processes involve techniques where a sheet metal blank is formed into the desired shape under predominantly tensile (stretching) loads. Bulk forming processes generally

Table 7.7 Examples of room temperature properties of select alloys

Alloy	UTS (MPa)	Elongation to fracture
Aluminum 5182-O	275	21%
Aluminum 5454-O	250	22%
Aluminum 6063-T4	170	22%
Magnesium AZ31B	170	15%
Magnesium AZ91D-F	230	3%
Magnesium annealed sheet	160–200	3%–15%
DDS-cold rolled	280	45%
HSLA 410	500	27%
DP 590/600	640	24%
TRIP 590/600	625	30%

encompass techniques where metal plates, slugs and billets are formed into shape under compressive loading. Shearing is often used to cut the blanks, billets, or slugs prior to metal forming; it is also used to trim and pierce the part during certain forming operations. Bending is employed in operations such as flanging (where part edges are bent to allow for the joining to other panels) and in operations such as hemming.

There are several advantages to using metal forming processes. First, the relatively short cycle time (compared with other manufacturing methods) is a key advantage. Second, forming at room temperature can improve the strength of the manufactured part through work-hardening. Third, bulk forming processes (rolling, forging, and extrusion) have high levels of material utilization and can be near net-shape or even net-shape processes for some products. Finally, for many automotive applications, sheet metal forming processes are the only type of metal manufacturing processes that are capable of generating the class-A surfaces required on closures and parts that are visible to the customer.

The limitations of metal forming processes can be categorized based on material characteristics and process capabilities. As discussed in Section 7.3.1, the available slip systems for some alloys at room temperature may be limited based on the unit cell structure of the alloy. This will limit the cold formability of the alloy. For example, the hexagonal close-packed microstructure of Mg alloys limits their elongation to only a few percent at room temperature. Therefore, Mg alloys are often restricted to elevated-temperature forming processes. In terms of processing limitations, contact between the alloy and the tooling affects the quality of the finished product, and the generated interface friction leads to an increase in the required forming stress. In sheet metal applications, friction can cause thickness variations within the formed part; excessive thinning can result in localized failure (splits). Moreover, the strain path during sheet metal deformation is also of importance. For example, deforming regions within the part experiencing plane strain conditions need to be identified and properly managed to prevent localized failure in these regions. Another limitation for many metal forming processes, is the cost of the forming tools and presses – which makes these processes feasible for annual production volumes of 25 000 units or more.

7.4.1 Sheet metal forming processes

Stamping

Room temperature stamp forming is a commonly used method for manufacturing automotive sheet products made of such ductile metals as sheet steels and aluminum alloys. Stamping refers to a series of manufacturing steps performed using a press, including shearing (blanking, piercing, trimming), drawing and bending, that transform a sheet metal blank into a useful product. Stamping operations utilize a matched die set (punch and die) and a binder that supports the sheet over the die shoulder during the stamping operation and controls metal flow into the die (Fig. 7.1). For large panels of complex geometries (i.e. automotive body side panels), draw beads are used, at pre-determined locations, to control material flow into the die and prevent wrinkling. Additionally, the pressure applied on the binder and the draw beads may be varied around the periphery of the die in order to optimize material flow during the process.

Stamping can be performed using a progressive press line or a transfer press line. In progressive stamping, the desired geometry is formed over a series of stamping operations performed at different stations within the press. The part being formed remains connected to the coil through a 'carrying web' until the final station in the progressive line, which is a cutoff station used to separate the part from the carrying web. A transfer press is typically used for larger part sizes, such as automotive closures and structural elements, and uses robotic arms to transfer the metal from one station to the next in a series of two to five stations. The stations include one or two drawing

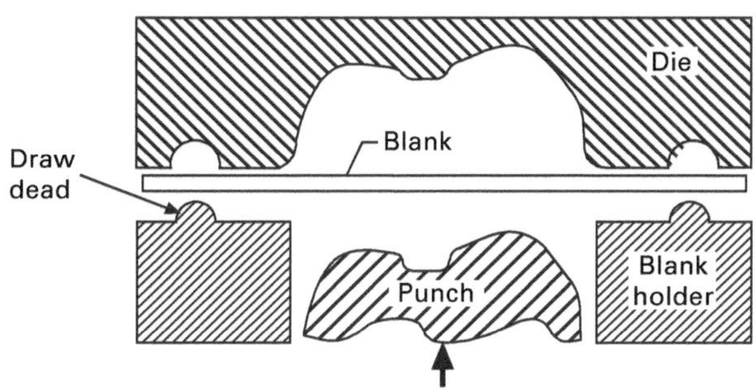

7.1 Schematic of a stamping die, showing the use of draw beads to control the flow of material into the cavity.

stations followed by one to two piercing and trimming stations to remove excess metal, then a flanging station to generate joining surfaces for the part. The number of required stations depends on the part geometry, and the arrangement of the finishing stations (trimming, piercing, and flanging).

Stamping can be performed on flat sheets as well as tailor-welded blanks (TWBs). Tailor-welded blanks are made by joining two or more sheets of similar or dissimilar gauge and/or mechanical properties in order to replace heavier welded structures in automotive components. In addition to weight savings, TWBs can result in cost savings and improved stiffness. On the other hand, difficulties have been experienced when forming TWBs due to preferential straining of the thinner sheet, which can lead to premature failure (Fig. 7.2), wrinkling, and shifting of the weld line (Kridli *et al.*, 2001). The weld line shift results from preferential straining where a higher level of strain is experienced by the lower strength (or lower thickness) side of the TWB; thus shifting the weld line into the higher strength (higher thickness) side of the TWB.

The challenge of stamping parts from light alloys is mainly due to their limited formability. Mg alloys do not possess sufficient room temperature ductility to form them into automotive parts. Al alloys can be stamped at room temperature; however, parts with complex geometries may often require dividing the part into sections to be stamped individually, then assembling (joining) the stampings to create the part. The use of multiple stampings increases tooling cost and poses assembly challenges due to the form and fit of each stamping. Also, special consideration should be given to springback after room temperature forming of light alloys. Due to their lower elastic moduli compared with drawing quality steel, a higher level of springback

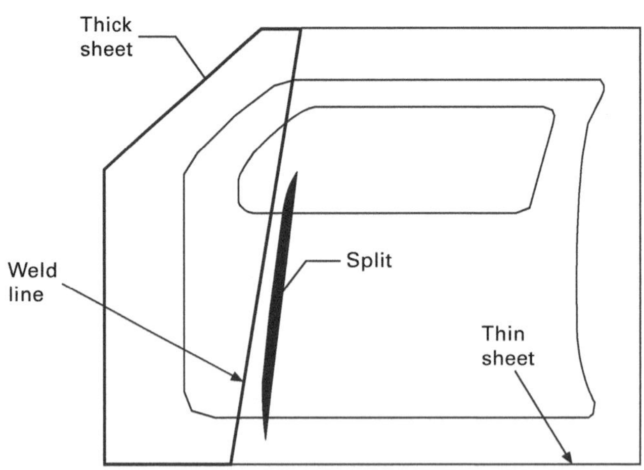

7.2 Schematic of a door inner TWB showing the most likely location of a split near the weld line.

compensation is required for parts made of light alloys. Additionally, large parts such as hoods and roofs require more elaborate inner support structure to reduce the width of any unsupported section in order to stiffen the parts (The Aluminum Association, Inc., 1998a).

Trimming is an important aspect of automotive manufacturing due to its influence on part quality. In shearing, stress is applied locally until a shear zone develops and the material fractures. In general, the quality of the sheared edge is affected by several factors, including the material being sheared, the shearing process, the shearing machine, and the shearing tool. Each of these factors contributes to dimensional inaccuracy, shape error, and/or positional error along the sheared edge (Hilditch and Hodgson, 2005). The quality of the sheared surface has a significant effect on any post-trimming operations that involve material deformation, such as bending, hemming, and flanging.

To evaluate the quality of a sheared edge, the burr height is typically used. It is most desirable to have a burr-free edge so that the need for a secondary de-burring operation is eliminated. Figure 7.3 shows a schematic of the shearing process, and Fig. 7.4 is a schematic showing the different regions of a sheared edge. The four zones shown in Fig. 7.4 represent the shape error in the shearing process. The positional error is caused by the accuracy of the shearing machine; the shape and condition (wear) of the shearing tool; and the shearing process parameters (clearance, speed, etc.). Finally, dimensional error is also caused by the processing parameters and the condition of the shearing tool. The influence of the sheared material on the quality of the sheared edge is attributed to the sheet thickness; the fracture mode of the material (which is affected by its microstructure) also significantly

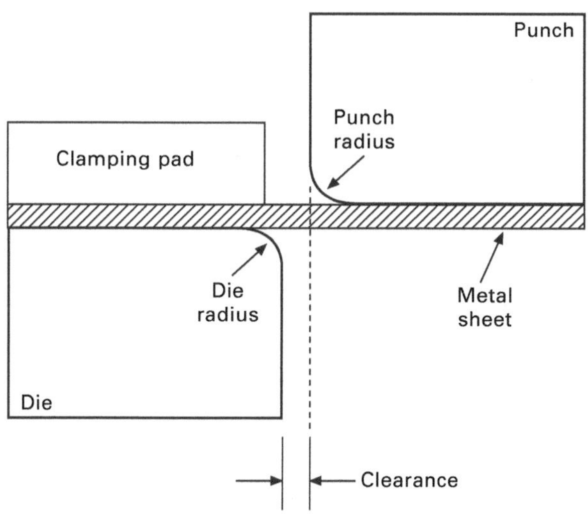

7.3 Schematic of the shearing process.

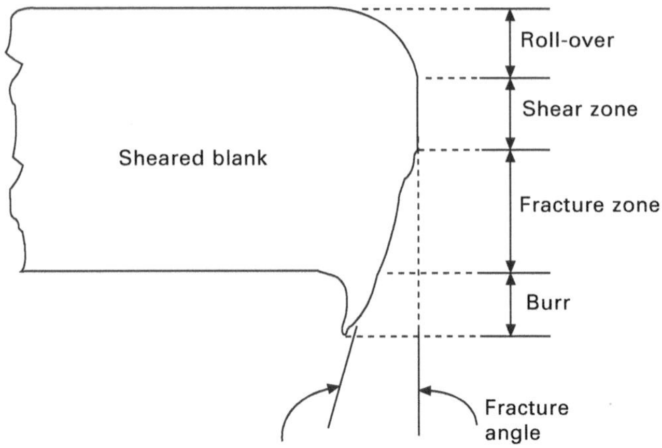

7.4 Schematic of the sheared edge.

affects the quality of the sheared edge. The quality is also affected by any prior thermo-mechanical processing of the sheet, the phases present in the microstructure as well as their ratio, void fraction in the microstructure, the grain size, and the degree of grain anisotropy in the material.

Compared with drawing-quality steels, aluminum and magnesium alloys exhibit a different sheared-edge behavior. For example, due to their lower levels of ductility, the depth of the rollover zone to material thickness is significantly lower for Al and Mg alloys compared with drawing quality steels. Additionally, burr formation for Al and Mg alloys starts at a lower level of tool clearance between the die and the shearing punch. This indicates that tighter control of the shearing operation is required in lightweight alloys. Additionally, the shape of the fracture zone differs for different types of alloys. Moreover, slivers (fragments of metal from the sheared edge) are observed when shearing aluminum alloys, which affects the quality of the sheared edge.

Hemming is a widely used process for constructing vehicle closure panels, where the outer material is first trimmed larger than the inner panel; then, the overlapping edge of the outer panel can be bent around the inner. Hemming can be accomplished with a number of different machines; however, all of the machines have the same basic approach, shown schematically in Fig. 7.5. The typical process begins with a flange on the edge of the outer sheet of approximately 90 degrees. The hem is then produced with two separate processes. First the flange is bent to an intermediate angle (nominally 45 degrees). This is followed by another process which closes the hem. The final step is typically performed with a flat tool face that bends the flange until it is parallel to the inner piece.

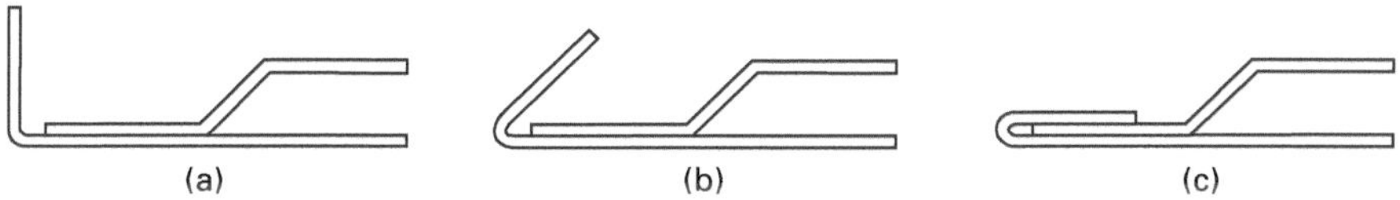

7.5 Typical hemming sequence of (a) flanging, (b) pre-hemming, and (c) final hemming.

Sheet hydroforming

The term 'hydroforming' refers to the metal forming processes in which hydraulic pressure is utilized to deform the material blank. Direct contact between the pressurized fluid and the blank material poses challenges in terms of sealing, pressurization schedule, anti-contamination in the pressure system, and subsequent cleaning of the workpiece and the equipment. Depending on the process arrangement and material thickness, the required hydroforming pressure can exceed 350 MPa in some cases. Based on the blank shape, hydroforming processes are classified as either sheet metal hydroforming or tube hydroforming processes. Since tube hydroforming of aluminum alloys is one of the promising manufacturing processes for light alloys, it is described in Section 7.7.2. Sheet hydroforming is a potential method to improve the formability of aluminum alloy sheets at room temperatures. The process has been used in the commercial manufacture of automotive steel parts; however, the technology is also ready for the production of aluminum parts. At present, only one aluminum part is known to have been manufactured by this process; the component is a Toyota part hydroformed in Japan (Koganti and Weishaar, 2008).

The Amino process is a form of sheet hydroforming that employs a water-based fluid to manufacture automotive closure panels. This process uses a punch, a water chamber, and a process control system. To form parts, a blank is held between the water chamber and the blankholder. Then, a punch moves toward the blank and begins the forming process. As the sheet moves into the water chamber, the hydrostatic pressure generated by the water increases. The hydrostatic pressure (or back pressure) delays the onset of necking and failure, and allows for deeper draws (i.e. more detailed part features to be formed that cannot be made with conventional stamping). As the process continues, the hydrostatic pressure is controlled and adjusted by draining the fluid chamber until the part is completely formed. A schematic of the process is shown in Fig. 7.6.

Like all processes, there are several advantages and disadvantages associated with sheet hydroforming. A major advantage of the process is its ability to extend material formability due to both the back pressure and the reduction in friction. This results in a better thickness distribution in the formed part. Additional advantages of sheet hydroforming include a reduction in tooling

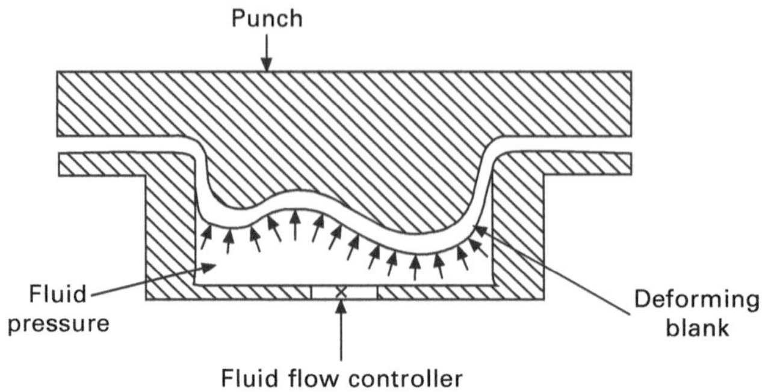

7.6 Schematic of the sheet hydroforming process.

cost (since only a punch is needed) and an improvement in the formed part surface (since the surface away from the punch does not contact the hard tool). However, the size of the part that can be manufactured by this process is limited by the availability of presses with sufficient tonnage to generate the required part forming loads. Additionally, while the forming tool cost is reduced, the cost of the press is relatively high due to the required pressure and control levels. Further, pressure limitations can reduce the ability of this process to generate tight radii. And, finally, the cycle time of sheet hydroforming is slower than stamping.

Superplastic forming

Superplastic forming (SPF) and its automotive-driven processes, such as Ford's Superplastic Forming Technology (FAST) and GM's Quick Plastic Forming (QPF), are processes that have recently been implemented in production. These processes typically require alloys with a grain size that is smaller than 10 μm. They are performed at temperatures above the recrystallization temperature of the alloy (typically between 350 °C and 500 °C) and at strain rates on the order of 10^{-3}mm/mm/s. For automotive applications, AA5083 is the material of choice for aluminum parts; however, the process has also been demonstrated for commodity alloys such as AA5182, and magnesium alloy AZ31.

The SPF process uses gas pressure to form a pre-heated blank into a die cavity or drape it over a punch. This eliminates the need for matched tooling sets and hence reduces the tooling cost per unit. Furthermore, due to the low forming stresses resulting from the high forming temperatures, multiple parts can potentially be formed simultaneously in the press, thus reducing the cycle time per part. A schematic of a superplastic forming tool is shown in Fig. 7.7.

The advantages of the process include part consolidation, extended material formability, low forming stresses, and elimination of springback. However,

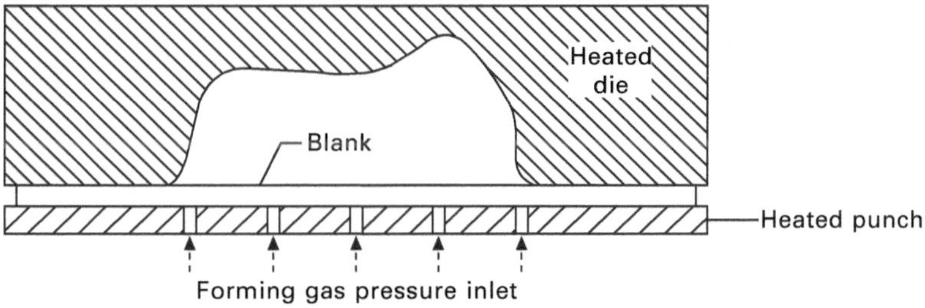

7.7 **Superplastic forming tool showing the blank and formed part.**

the process limitations include high cycle times (2 to 10 minutes), which limit the economic feasibility to low- to mid-volume production levels (under 100 000 units per year). Additionally, the quality of the generated surface is sensitive to, and can be significantly affected by, the exposure to heat; this will require additional post-process steps, such as polishing/buffing, in order to obtain a Class-A surface. Finally, the process requires the use of a sophisticated press that is capable of controlling and adjusting the forming pressure to maintain the desired strain rate.

7.4.2 Bulk metal forming processes

Forging

Forging is a manufacturing method in which compressive forces are used to shape a metal slug into a near-net shape product with the use of a tool. The method is adaptable for a range of shapes, from simple to the very complex; for intricate geometries, forging is performed using several tool sets (multi-step process). Forgings are typically classified according to the type of tooling used in the forging process; these include open-die forging, closed-die forging, and impression-die forging.

For automotive manufacturing, forging is a well-established process used in the manufacture of powertrain components such as connecting rods, as well as chassis system components such as control arms. Depending on the material and the forming stress requirements, forging can be performed at temperatures ranging from room temperature to hot temperatures (>0.5 T_{melt}). Compared with castings, forged products typically have much higher strength, toughness, and ductility; this is due to the refinement that forging performs on the wrought microstructure of the forged alloys. Currently, automotive aluminum forgings can be made of 5xxx, 6xxx, or 7xxx series alloys. Mg alloys in the AZ and ZK alloy families are suitable for automotive forging, with forged wheels constituting the largest automotive application for magnesium.

A newer variant on forging is Precision Forging, a near net-shape flashless hot forging process that is used to form parts in a closed die. The process uses a billet of approximately the same volume of material as the final product; this reduces post-forging processing and material waste (Dröder and Janssen, 1999). It is not a new process, but rather a refinement of conventional forging methods. The advantages of precision forging over conventional forging techniques include a reduced draft angle of 0° to 1°, the ability to forge thin sections, the ability to achieve tight tolerances, the ability to produce better surface finish, and the ability to forge tighter radii. Precision forged parts have good fatigue life and good intergranular corrosion resistance; this is a result of the process creating improved grain alignment within the part. Typically, machining of precision forged parts is needed only for drilling and tapping. Like conventional forging, precision forging techniques can be used for the manufacture of wheels, gears, crankshafts, connecting rods, and other components in single or multi-step sequences of operation (such as forming crankshafts). However, since the cost of precision forging is higher than conventional forging, cost-effective process selection must be made on a part-by-part basis.

The challenge in forging aluminum and magnesium alloys is that the forging temperature window for many of these alloys is more limited than that for steel alloys. Below the lower end of the working temperature range (~200 °C for Mg and ~425 °C for Al), the alloys do not have sufficient formability and cracking may occur. Above the higher end of the range (~300 °C for Mg and ~550 °C for Al), microstructural damage can occur due to hot shortness. Within the last five years, significant research has been focused on the development of AZ and ZK magnesium alloys with reduced alloying elements to enhance their ability to be forged.

Extrusion

Extrusion is a discrete manufacturing process in which a billet of material is pushed through a die orifice using a ram, and can be performed at room temperature or at elevated temperatures. The material exiting the die takes the shape of the die orifice. The extrusion dies are thus simple in shape and the process can be used to produce tubes, channels, bars, and other shapes of a constant cross-sectional geometry. A schematic of the forward extrusion process used for extruding a variety of commercial products is shown in Fig. 7.8.

With the increasing use of light alloys in automotive applications, the use of extruded sections has also been increasing due to the high extrudabilities of these alloys, especially aluminum alloys (it should be noted that steels have significantly lower extrudabilities compared with aluminum and magnesium alloys, thus limiting the use of extrusion as a process to fabricate steel

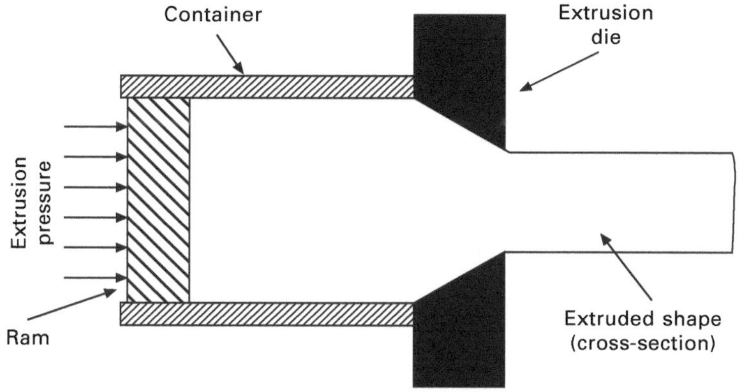

7.8 Schematic of the forward extrusion process.

for automotive applications). Aluminum extrusions are used in a range of automotive components including drive shafts, body structures, suspension components, intake manifolds, bumpers, seats, and doors. These automotive-based extruded products are typically made of 6xxx-series aluminum alloys, which have higher extrudability compared with other aluminum alloy series. Magnesium alloys in the AZ family (such as AZ31 and AZ80), as well as the ZK family, can also be extruded; they are being investigated as potential candidates for use in automotive applications.

Extrusions, in general, offer the ability to improve structural geometric stiffness over stampings due to the ability to extrude a wide range of cross-sectional shapes. Hence, the shapes of extruded components can be optimized to meet the local design requirements. Accordingly, the use of extrusions can result in significant weight savings, thus impacting upon vehicle stability, ride and handling (The Aluminum Association, Inc., 1998b). The main challenge for using aluminum and magnesium alloy extrusions in automotive body structures is energy absorption under crash conditions. With the selection of a suitable alloy along with proper design for crash energy management, light alloy extrusions can be used in the construction of automotive structures that meet occupant safety requirements.

A further challenge for extruding magnesium alloys, in particular, is that extrudeability issues need to be resolved to reduce part cost. This is caused by the narrow temperature forming window for extruding magnesium alloys (similar to that in forging), as well as the speeds at which these alloys can be extruded. For example, the extrusion speed for magnesium alloys ranges from 5 to 10 times slower than that for aluminum alloys. Preliminary research on the use of hydrostatic extrusion (Fig. 7.9) for magnesium sections showed promising results in terms of extending the operating temperature window to temperatures as low as 100 °C, as well as increasing the extrusion speed (Agnew, 2004).

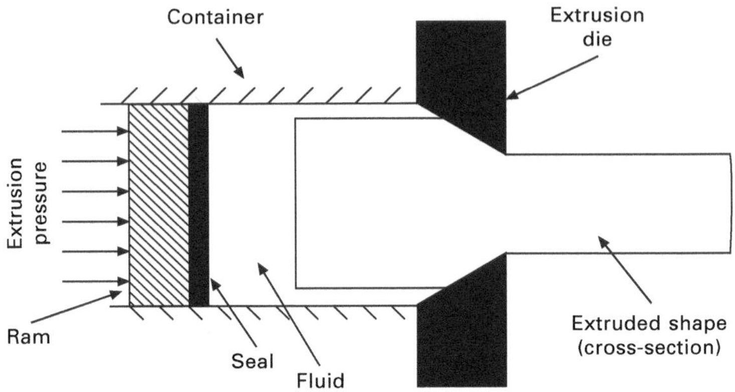

7.9 Schematic of hydrostatic extrusion.

7.5 Cast structural components

7.5.1 Advantages and limitations

Casting is a process in which a molten material flows into a mold and is allowed to solidify, taking the shape of the mold. Casting can be done simply (such as gravity-pouring metal into an open mold), or it can be part of a complex process (such as forming a semi-solid slurry that is injected into a closed metal mold). The process of casting offers a number of advantages over other metal forming processes; these include reducing component cost through creating parts with complex internal and external geometries, reducing the amount of assembly required ('part consolidation'), and creating near–net shape components (reducing the amount of material wasted and the amount of machining required). Casting is a very flexible process. A wide variety of casting processes exist, allowing production rates from job (<10–100 units) to mass (> 1 000 000 units) over a very wide size range of components; for example, castings are used for dental crowns (grams) as well as hydroelectric pump housings (tons).

Like any process, there are limitations associated with casting. In general, the mechanical properties are lower relative to wrought metals; this can be due to the presence of large-scale porosity and/or inclusions in the solidified metal. Further, dimensional tolerances usually need to be greater than those in wrought processes. There are significant safety concerns that need to be addressed, including molten metal processing and handling practices; this is key in magnesium castings, where the melt must be covered with a protective gas during the melting and pouring steps. In addition, there are environmental issues associated with the smoke and fumes due to the interaction of the melt with organics, fluxes, and degassing agents. The use of certain molding materials such as polymer-based sand binders and polystyrene-based cores requires appropriate ventilation and air-handling equipment.

7.5.2 Key design factors

In general, the same design factors that were discussed for the wrought materials apply to the design of structural castings. Both must be designed to withstand peak static and dynamic loads while driving the vehicle, have a high stiffness to provide transverse rigidity, and be sufficiently ductile and tough to rapidly absorb large amounts of energy in crash situations.

7.5.3 Casting as a process of choice

The choice of process is critical to creating the most optimal component; thus, the 'best' product is one that meets the required functions at the lowest cost possible. Castings are important because the process yields two major functional advantages: wide design freedom and part consolidation.

The wide design freedom comes from the use of molten metal to directly form the component. Thus, casting allows for the independent forming of differing internal and external features at the same time. Metal can be placed in differing amounts in differing locations depending upon specific property requirements such as wear resistance, strength, and stiffness. Further, the use of molten metal allows a complex component to be created as a single unit. This reduces the number of joints and the potential for misalignment and leakage associated with joining technologies such as welding and brazing. Part consolidation is a key attribute of the casting process. Properly designed, a casting can allow the number of individual parts to be combined into one unit. This simplifies production, since fewer parts need to be manufactured, tracked, and inventoried. Additionally, it reduces direct assembly costs, as less equipment and/or equipment of fewer varying types is required; often, the number of jigs, dies, and fixtures can be substantially reduced. Production costs are also minimized through the use of castings; the fewer the number of parts, the less likely a failure will occur and stop production. Also, the lower the number of parts, the lower the number of production flow, control, and inspection operations.

7.6 Casting processes

7.6.1 Mold categories

Casting processes are often subdivided into 'expendable mold' and 'permanent mold' processes, based upon the ability of a mold to be reused. With expendable molds, the mold is used only once; following casting and solidification, the mold is usually destroyed to remove the solidified casting. This is a very scalable process where the production rate is often a function of the mold creation rate. It is also a very flexible process that allows the casting of a wide variety of part geometries, weights, and quantities. With the advent of rapid prototyping equipment, the computer aided engineering (CAE) models of a component can be used to generate polymer-based or wood-based tooling that allows the creation of a small number of parts in a very rapid (< 4–6 week) timeframe.

Permanent molds are molds that are fabricated out of metal (usually tool steel) and are re-used repeatedly. In these molds, the production rate is a function of how fast the components can be cast and removed from the mold. Thus, to offset the high costs in equipment and mold tooling, this process is usually used for high-volume, rapid production cycle components that do not have a high degree of complexity. This type of mold generally has higher cooling rates than sand casting, allowing a finer microstructure to arise. Because the surface finish and dimensional tolerances are usually superior to sand casting, permanent mold castings often require less machining prior to use.

7.6.2 Selected expendable mold processes

Sand casting is the most commonly used expendable mold process. In this process, sand is bound together (as detailed in Table 7.8) and shaped to conform to the negative impression of the desired component. Molten metal is then poured into the sand mold; upon solidification, the metal has the geometry of the desired component.

Advantages of this process include the fact that nearly all cast alloys can be poured into sand, it is the only method for high-melting-point alloys, thin-wall sections (< 4 mm) can be obtained, castings can be made in a wide range of weights (grams to tons) and production volumes (one to millions), and it is the most easily scalable of the casting processes.

Some limitations of sand casting include the fact that high volume production requires a large investment in automated mold/core making and assembly equipment as well as floor space. Additionally, casting yields are often < 50%, requiring a larger amount of machining and finishing prior to use. Metallurgically, cooling rates in sand castings are often slow, especially in thick sections; this often yields lower mechanical properties and higher amounts of porosity. Further, sand issues are important considerations, including the consistency and quality of the sand, the volume of the sand used (up to 10 – 20x the weight of part), and sand removal, disposal, and/or recycling.

Sand casting is the process of choice for many large powertrain components, including blocks, heads, and intake manifold. Sand castings are less common

Table 7.8 Binder systems used for creating sand molds

General classifications		Components	Procedure
Green sand		Sand, bentonite clay, coal, and water	Mixture is placed in a core box and squeezed against the pattern to form a mold. Mold can be used immediately for casting. For very large or heavy castings, mold is often baked/dried in an oven to improve mold strength and rigidity. Sand is usually physically removed after casting by shaking or hammering ('rapping') to fracture the mold into small sections
Resin-bonded	Air-set ('No bake' or 'Cold box')	Sand and a resin system based upon esters, furans, phenolic urethanes, soldium silicates, or phenolics	Sand and resin are blended to create a semi-solid mixture which is quickly deposited into a core box containing the pattern. The resin reacts with the moisture in the air and hardens the mixture; the working time is a function of the type and amount of resin used. After hardening, the mold is removed from the core box and can be used immediately for casting. After casting, sand is removed either by physical methods (shaking/rapping) or via thermal operations that pyrolize the binder, leaving the sand free to flow out of the casting and be reclaimed
	Gas-cured ('Cold box')	Sand and a gas-reactive resin system based upon phenolic urethanes, acrylics, or sodium silicates; gases include amines (TEA/DMEA), SO_2, and CO_2	Sand is first mixed with the gas-reactive resin system and deposited into a core box. The mixture is instantly hardened by blowing a specific gas through the core box; thus, working time is flexible as it is a function of the delay between molding and gassing. The mold can be used immediately for casting. Sand is removed either by physical methods (shaking/rapping) or via thermal operations that pyrolize the binder
	Heat-cured ('Hot box')	Sand and a heat-curable resin system based upon furans, phenolics, or urea formaldehydes	Sand is first mixed with the heat-reactive resin system and deposited into a heated pattern (200–260 °C); working time is flexible as mixture is stable until heated. The mixture rapidly harden once heated and is immediately available for use. Sand is usually removed using physical methods. Commonly used for 'shell molding' process

in structural applications, due to their slower cooling rate and generally coarser microstructure. However, in certain applications, such as the cast 356-T6 nodes for the Ford GT spaceframe, need for specific geometries at low production volumes requires the use of sand castings. Other expendable mold processes, such as investment casting and lost foam casting will not be discussed, since their use for structural components is limited to special, small-volume production components.

7.6.3 Selected permanent mold processes

High-pressure die casting

High-pressure die casting (HPDC) is a very commonly used process for creating structural components, especially in Mg. In this process, a metal die having a cavity with the negative geometry of the part is created; simple dies usually consist of two matching halves, while more complex dies can add sliding features that create holes and undercut areas. The die is mounted onto a machine capable of injecting molten metal at high velocities. The die cavity is closed, molten metal is poured into a shot sleeve, the sleeve opening is closed, and a ram moves forward to force the metal into the die in a very short time (10–100 ms), generating high levels of applied pressure. Following this, the ram pressure is maintained for a short time; often, active cooling occurs as internal water passages in the die are activated. Then, the pressure is released and the ram is withdrawn; the die opens and ejector pins push out part. The process cycle for the HPDC process is usually very rapid; for example, a current 110 lb HPDC V6 engine block has a process cycle time of approximately 90 seconds. The rapid speed of the HPDC yields a number of advantages. In terms of costs, HPDC creates castings with a high casting yield (up to 95%), high to very high near net shape, high levels of surface finish, and high dimensional control; all of these factors minimize the amount of post-casting machining and finishing that must be performed.

Additionally, depending on the size and geometry of the component, multiple cavity dies can be created; this allows for the production of multiple parts within each individual process cycle. Further, there are no sand/binder costs or issues associated with HPDC. Metallurgically, the rapid injection and cooling rates tend to yield a very refined microstructure with improved levels of mechanical properties. The use of internal cooling passages within the die means that the solidification of the molten metal (and, hence, the microstructure) can be controlled; this allows more optimized properties to be obtained in specific locations within the casting.

The HPDC process also has several limitations. The most serious limitation is related to the need to have a die that opens and closes along a specific dimension ('the parting line'). This means that the desired component cannot be made with complex internal geometries: while a major limitation for some components (such as cylinder heads), this is less of an issue with structural components due to their simpler three-dimensional geometries. The up-front costs associated with HPDC are high; metal dies and tooling are required along with the hydraulically-actuated die-casting machine. Another issue is that this process is not easily scalable. The HPDC machines have a finite capacity for production; if the production requirements exceed this capacity, another machine must be purchased along with the dies and tooling. Metallurgically, the high injection velocities often entrap air in casting, allowing porosity to arise. Additionally, the rapid cooling rates can concentrate porosity and inclusions along the centerline of a casting. Both of these factors can reduce the mechanical properties of the casting. Also, the presence of the trapped air often prevents these castings from being heat-treated; thus, the maximum strength properties obtainable are often much lower in HPDC components compared with other permanent mold and sand-cast processes. Finally, the non-uniformity of properties tends to increase as the size and/or thickness of the HPDC component increases.

HPDC has been used to create a large number of magnesium automotive structural components. The grill opening reinforcement (GOR) in the current Ford F150 is a HPDC AM60 alloy structure. This structure consolidated eleven steel stampings into a single Mg casting while reducing the weight from 69 to 40 pounds. Instrument panels are further examples of high-volume (>4 million) HPDC components cast from AM50/60 alloys. Examples include the Ford Expedition (14 steel stampings integrated into 1 Mg part with a 12 lb weight saving) and the Cadillac CTS and STS (20 steel pieces integrated into 2 HPDC Mg parts.)

Low pressure casting systems

Low pressure (LP) casting systems are the ones that rely upon pressurization levels of up to 0.8 bar to feed the molten metal into the mold; usually, the mold is at, or above, the level of the metal being poured. LP systems generally fall into three categories: unsealed, sealed, and vacuum-assisted. In unsealed LP systems, the metal is transferred from a furnace through the use of a pump; an example of this is the Cosworth® process, where an electromagnetic pump is used to lift the metal up and into the mold. The sealed LP systems use a furnace that can be sealed and pressurized; in this furnace, one or more hollow tubes connect to the top of the furnace and extend down into the molten metal. A mold is placed on top of the furnace, the melt in the furnace is pressured and the molten metal is forced up through hollow tubes into the mold. The VRC/PRC® process developed by Alcoa and CMI is an example of this system. Vacuum-assisted LP systems are similar to the sealed LP systems, except that a vacuum is applied to the casting prior to/at the same time as pressurization of the melt. Counter-pressure and pressure-counter-pressure systems such as Intermet's CPC/PCPC® use this approach to enhance the traditional LP system.

There are a number of advantages of low pressure casting systems. The molten metal quality is often very high, having low levels of inclusions and hydrogen gas. This is especially true when the melting and degassing of the melt is done in the same furnace that will be pressurized for filling. Metal quality is also high because the molten metal used for casting comes from the middle of the furnace; oxides and inclusions tend to settle to the bottom or rise to the surface. Pressurizing the furnace means that the metal is never transferred from the furnace to a ladle, a process that involves breaking through the melt surface and stirring some of the surface oxides back into the melt. Casting is accomplished by feeding metal through the bottom of the casting and up through the casting. Compared with gravity casting, this allows for more control over the fill rate, minimizing turbulence and the formation of oxides. Additionally, the bottom feeding means that solidification occurs from the top of the casting to the bottom; the bottom of the casting contains pressurized molten metal, which is available to help properly feed the casting and reduce the amount of shrinkage present. The process is very flexible, and can be used with sand molds, metal dies, or combinations of both. A number of 2.0L–3.0L aluminum cylinder heads have been cast using a LP semi-permanent mold system where a metal die forms the outer portion of the cylinder head and sand cores are used to create the internal passages.

Metallurgically, the LP system can produce very sound castings. The use of a controlled, non-turbulent fill almost eliminates air entrapment; additionally, the ready supply of molten metal in the direction of solidification reduces the shrinkage associated with solidification. Both of these factors can significantly reduce the porosity present in a component. Because of this, many LP-cast components can be heat treated to optimize their properties. Further, the use of a metal die with integral cooling passages allows for controlled cooling of the casting; this can further improve mechanical properties through refinement of the microstructure.

The majority of the limitations associated with LP systems are related to the equipment. Like HPDC, many LP systems require specialized machines; additionally, in LP systems, an integral melting furnace capable of being sealed and pressurized must be present. Many of the LP systems used for structural aluminum components also require metal tooling and dies for the mold; thus, initial costs, projected production runs, and timing must be accounted for prior to beginning production. Metallurgically, care needs to be taken with the change in the level of the molten metal in the feeder tubes between castings; it has been observed that too much change in height can create oxides that will enter the next mold to be cast. Additionally, the design of the feeder system is such that it is difficult to put molten metal filters in the system; this also increases the chance that an oxide might enter a mold during the casting process.

A large number of automotive structural components have been cast in aluminum using LP systems. The majority of cast aluminum wheels in service today are manufactured using LP systems and permanent molds (Hatch, 1984). An engine cradle for the Cadillac CTS was cast using the 356 Al alloy and the VRC/PRC® process. This component is, at 33.5 pounds, 40% lighter than its steel counterpart; additionally, the casting consolidated 32 parts into a single casting (Hadley *et al.*, 1999). In the late 1990s, a study was conducted to redesign a Chrysler minivan liftgate inner panel from an assembly of 11 steel stampings to one casting. The results of the study indicated that a 21% weight savings could be obtained for a cost penalty of $1–$2 per pound saved (Meyer *et al.*, 1999).

Gravity permanent mold systems

Gravity permanent mold casting processes use the metal die in combination with a gravity-fed metal delivery system. This combination lowers the overall cost of the process while taking advantage of the higher cooling rate of the metal die. The advantages and limitations of this system are similar to those of the LP permanent mold systems.

There are several different ways of using this casting process. The simplest is pouring the metal into the top of a metal mold and allowing it to solidify. While this is relatively inexpensive (the only major cost is the die), there are issues with metal turbulence, oxide creation and entrapment, solidification shrinkage, and microshrinkage porosity.

Other processes have been developed to try to minimize the turbulence created. One method, called Tilt-Pour®, involves placing the metal die at one end of a long beam on a fulcrum-type device and the molten metal at the other end. Casting begins as the end with the molten metal is rotated and raised upwards; when this end is level with the die, the metal begins to slowly run down the long beam axis and into the die. The process is complete when the empty metal end is raised to the maximum position and the die is at the minimum position. A variation on this process involves angling the die slightly and feeding the metal from the side of the mold. The metal is pumped up from the furnace, fed into the side of the mold, where gravity fills the mold. This variation was used to create the front lower control arms for the Ford F150. A 356 aluminum alloy was used to minimize solidification shrinkage, while a water-cooled mold was used to keep the microstructural features as fine as possible. Following casting, the arm was heat treated to a T6 condition to increase the strength and toughness.

Squeeze casting

Squeeze casting is a process that combines gravity and pressurized casting. In general, molten metal is poured into a pre-heated die. When filling is complete, a ram is used to slowly apply high pressure to the molten metal head. This pressurization helps ensure that metal flows throughout the solidifying casting, minimizing shrinkage and microshrinkage porosity. When combined with a controlled cooling step, this process can produce a very fine microstructure. In certain cases, mechanical properties approaching those of forged parts can be obtained.

There are several advantages to this process. In general, the equipment required for this process is less expensive than in HPDC or forging. Because sand cores can be used with this process, parts of great internal detail can be formed. The low porosity and fine microstructure greatly improve the mechanical properties and pressure tightness of the casting.

The main limitations of this process are in part size and cycle time. Because of the need to pressurize the metal head, the maximum size of the component strongly influences the machine size and capacity. Additionally, the combination of a gravity pouring plus a pressurized solidification will greatly increase the cycle time compared with HPDC. There can also be issues with proper die venting to avoid entrapping air and with oxide inclusions being forced into the casting from the pressurization. In thin-walled castings, this process is vulnerable to misruns unless the section thicknesses are increased relative to a HPDC casting.

The application of squeeze casting in the automotive industry has been limited. The major application is for the infiltration of reinforced-ceramic fiber pistons for diesel applications. Porsche also uses the squeeze casting process to create the cylinder block banks for a horizontally opposed V6 engine block. In terms of structural materials, squeeze-cast A356-T6 aluminum has been used in a number of steering knuckles for low-volume vehicle lines. In the 1990s, Delphi undertook a study that determined that squeeze casting would be a viable production method for high-volume (>1.5M) vehicles (Gerken and Neal, 1999).

Semi-solid casting systems

Semi-solid casting systems are a relatively new set of casting technologies that have developed over the last three decades. In general, these systems do not fully melt the alloy to be cast; instead, very tightly controlled heating processes are used. These processes are adjusted so that up to 60% of the metal is melted; the remaining 40% remains solid. This alters the microstructure and gives the material a very low resistance to shear deformation, which can especially be seen in aluminum alloys, where the standard dendritic structure is replaced by a more rounded, globular microstructure. This semi-solid metal (SSM) is then quickly placed into a molding machine and forced, injected, or extruded into a metal mold. The resulting component has very high mechanical properties due to a very fine microstructure and an almost complete absence of porosity.

The key to this casting system is the controlled heating of the alloy. One method, thixocasting, begins by heating solid bars of the alloy to be cast into the temperature range between the solidus and liquidus. A second method, rheocasting, involves cooling a liquid alloy into the temperature range between the solidus and liquidus. A third method, thixomolding, involves feeding pellets or chips of material into a heated screw-driven injection molding machine. The first two methods have been used with aluminum alloys such as A356/357; the third has been more commonly used with magnesium alloys such as AM50/60 and AE42.

The structural applications for these casting systems have been limited to a number of short production-run components for specialty vehicles due to the high costs associated with the material and equipment. One example was the control arms for the Ford GT, which were originally rheocast from an A356-T6 alloy.

7.7 Enablers

To significantly increase the use of light alloys in the production of automotive parts, the cost of manufacturing of these parts needs to be competitive with the existing products. The following have been identified as enablers for the use of light alloys:

- Reduce the cost of material
- Reduce weight of body structures and closures
- Reduce manufacturing costs
- Select process and design to match production volume

Presently, automotive researchers are working to develop technologies and practices on reducing manufacturing cost and material cost. Topics under consideration are listed below.

Areas of research: reduce manufacturing costs

- Formability
    - Alloy development
    - Optimization of conventional stamping (lubricants, CAE, etc.)
    - Tailored blanks
    - Hydroforming
    - Superplastic forming
    - Warm forming

  - Electromagnetic forming
- Joining
  - Adhesive bonding
  - Surface treatments
  - Self-piercing rivets
  - Spot welding
  - Laser welding
  - Friction-stir welding
  - Non-destructive evaluation

Areas of research: reduce material costs

- Sheet production cost
  - Primary aluminum smelting
  - Continuous casting
  - Economy of scale
  - Process optimization
- Reduce component weight
  - Design optimization
  - Improved alloys
- Minimize scrap
  - Stamping process improvement
- Increase scrap value
  - Scrap sorting (plant and post-consumer)

7.8 Promising metal forming processes for automotive applications

7.8.1 Warm forming

Warm forming refers to sheet forming in the temperature range of 150 °C to 350 °C using heated matched die sets similar to those used in conventional stamping. The advantage of forming aluminum alloys at warm temperatures to achieve higher levels of ductility has been known for over 30 years. The main advantage of using warm forming over room temperature stamping is the fact that the increased formability of aluminum at warm forming temperatures allows for part consolidation and the forming of more complex geometries; thus there is a potential for cost reduction. Additionally, the use of warm forming opens up the range of aluminum alloys that may be utilised in automotive manufacturing to include higher strength alloys that can be used in structural components, such as pillars and intrusion beams. There are also several advantages of warm forming aluminum alloys over high temperature forming processes (e.g. superplastic forming) including the amount of energy required to heat the sheet and the tooling, and the ability to form traditional automotive quality aluminum alloys such as AA5182, AA5754, AA5454, and AA6111. The potential for part consolidation offered by the process can lead to weight savings through improved part design, and can enable manufacture of parts that are traditionally produced by casting, which poses joining challenges and increased manufacturing cost. The limitations of the process include low cycle times compared with stamping and the potential for die distortion due to thermal cycling.

Typically, the response of aluminum alloys to warm forming (i.e. increased formability) improves with increasing Mg content for binary alloys (Shehata *et al.*, 1978); however, Krajewski (2005) noted that magnesium content as well as alloy composition, for complex alloys, affect the ductility. Additional improvements in formability may also be achieved by using non-isothermal warm forming. Preliminary studies indicate that using a punch that is cooler than the die and the blankholder leads to an improved depth of draw (Li and Ghosh, 2004).

Research on implementing the warm forming process (isothermal and non-isothermal) for automotive production is being performed by the major automotive manufacturers; however, more work is still needed to make the process commercially viable. Early trials on the process of Mg have also shown that a significant improvement in formability can be achieved (Krajewski *et al.*, 2007).

7.8.2 Tube hydroforming

The tube hydroforming process is a near-net shape metal forming process in which a straight or a pre-bent tubular blank is placed in a closed die cavity and its cross-sectional shape is changed using internal hydraulic pressure that forces the tube to conform to the shape of the die cavity. Punches may be mounted in the die to pierce holes in the tube wall during the forming operation. The desired final shape is achieved either by the effect of internal pressure, or by the combined effects of internal pressure and die displacement. In either case, axial feeding of the tube blank from the ends, called end feeding, is sometimes incorporated in the process to achieve larger circumferential expansion. High pressure levels (over 350 MPa) may be required to expand the tubular blank to the desired geometry of the closed die. A schematic of the process is shown in Fig. 7.10.

Warm tube hydroforming, also referred to as thermo hydroforming, is a promising technology for improving the formability of aluminum and magnesium tubes (Aue-u-lan *et al.*, 2006). The process uses tooling that is similar to the room temperature process with the addition of heaters to achieve the desired forming temperature. In the process, the tube is heated to the required temperature, and a fluid (heated liquid and heated gas are being investigated) is used to apply the forming pressure.

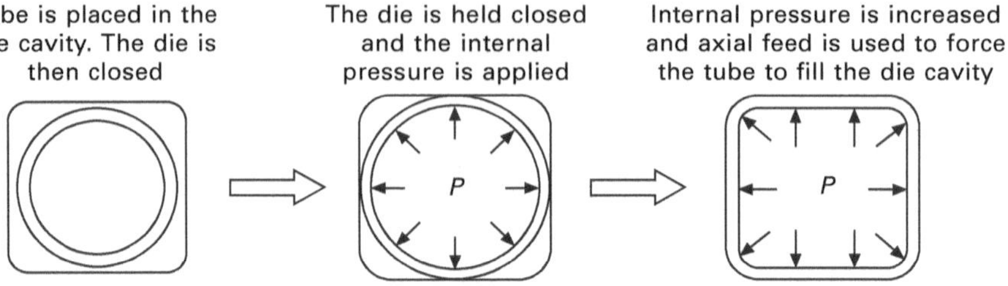

7.10 Schematic of the high-pressure tube hydroforming process.

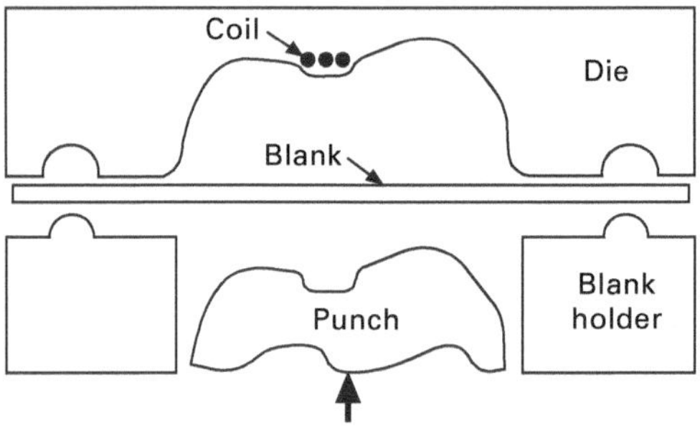

7.11 Schematic of electromagnetic-assist forming tool showing coils near part detail that cannot be conventionally stamped.

7.8.3 Electromagnetic forming

The electromagnetic forming process (EMF) is a high strain rate forming process with the potential to significantly increase the formability of aluminum alloys. The process uses electromagnetic pressure to form the workpiece. The forming pressure is generated when capacitors are discharged through a coil. Upon discharging, the generated magnetic field produces eddy currents in the workpiece, which result in an opposite magnetic field. The presence of these two opposing magnetic fields leads to a force that repels the sheet metal blank toward the walls of the forming tool at speeds of over 100 m/s. The process is limited by the coil size, capacitor life and cost. The process has been used in crimping applications, and is being investigated for applications in sheet metal forming as well as forming of tubular products. EMF is a candidate for use as a secondary forming operation, coupled with conventional stamping, to form part details for regions that cannot be formed by stamping alone, as shown in Fig. 7.11.

7.9 References

Agnew S R (2004), 'Wrought magnesium: A 21st century outlook,' *Journal of Metals*, 56(5), 20–21.

Aue-u-lan Y, Esnaloa J A, Guza D and Altan T (2006), *Warm forming magnesium, aluminum tubes – A high-temperature process for lightweight materials*. The Fabricator.com. Available from: http://www.thefabricator.com/Hydroforming/Hydroforming_Article.cfm?ID=146e [Accessed 7 April 2008].

Behrens B-A, Kurz G and Hübner S (2004), 'Heated hydro-mechanical deep drawing of magnesium sheet metal,' in K Siegert, *New Developments in Sheet Metal Forming*, MAT-INFO, Werkstoff-informationsges, Stuttgart, Germany.

Campbell J (1988), 'The ten casting rules: Guidelines for the production of reliable castings – a draft process specification', in *Materials Solutions Conference 1998*, ASM International, Materials park, OH.

Dröder K and Janssen St (1999), 'Forming of magnesium alloys – A solution for light weight construction,' *Technical Paper No. 1999–01–3172*, SAE International, Warrendale, Pennsylvania.

Gerken D and Neal R (1999), 'Squeeze cast (SCPM) light weight aluminum front knuckle case study', *Technical Paper No. 1999–01–0344*, SAE International, Warrendale, Pennsylvania.

Hadley S W, Das S and Miller J W (1999), *Aluminum R&D for Automotive Uses and the Department of Energy's Role: Technical Report No. ORNL/TM–1999/157*, Oak Ridge National Laboratories, Oak Ridge, TN.

Hatch J E (1984), *Aluminum: Properties and physical metallurgy*, ASM International, Materials Park, OH.

Hilditch T B and Hodgson P D (2005), 'Development of the sheared edge in the trimming of steel and light metal sheet. Part 1 – Experimental observations', *J. of Materials Processing Technology*, 169, 184–191.

Hunt W H and Herling D R (2005), Cost assessment of emerging magnesium sheet production methods. Available from: http://www1.eere.energy.gov/vehiclesandfuels/pdfs/alm_05/2j_herling.pdf [Accessed 5 May 2008].

Koganti R and Weishaar J (2008), 'Aluminum Vehicle Body Construction and Enabling Manufacturing Technologies,' *Technical Paper No. 2008–01–1089*, SAE International, Warrendale, Pennsylvania.

Krajewski P E (2005), 'The warm ductility of commercial aluminum sheet alloys,' *Technical Paper No. 2005–01–1388*, SAE International, Warrendale, Pennsylvania.

Krajewski P E, Friedman P A, Oikarinen K and Cedar D (2007), 'SAMP project MD307: Pan forming of aluminum and magnesium,' Presented at the *2007 MS&T Conference and Exhibition* in Detroit, Michigan.

Kridli G T, Friedman P A and Sherman A M (2000), 'Formability of aluminum tailor-welded blanks,' *Technical Paper No. 2000–01–0772*, SAE International, Warrendale, Pennsylvania.

Li D and Ghosh A K (2004), 'Baxial warm forming behavior of aluminum sheet alloys,' *J. of Materials Processing Technology*, 145(3), 281–293.

Meyer T N, Kinosz M J, Bradac E M, Mbaye M, Burg J T, Klingensmith M A (1999), 'Ultra large castings to produce low cost aluminum vehicle structures', *Technical Paper No. 1999–01–2252*, SAE International, Warrendale, Pennsylvania.

Powers W F (2001), 'Automotive materials in the 21st century', *Advanced Materials & Processes*, 157(5), 38.

Shehata F, Painter M J and Pearce R (1978) 'Warm forming of aluminum/magnesium alloy sheet,' *J. of Mechanical Working Technology*, 2, 279–290.

The Aluminum Association, Inc. (1998a), *Aluminum for automotive body sheet panels (AT3)*.

The Aluminum Association, Inc. (1998b), *Aluminum automotive extrusion manual (AT6).*
The D Chemical Company (1984), *Fabricating Magnesium.*
Ullmann's Encyclopedia of Industrial Chemistry (1990), 'Magnesium alloys', VCH Verlagsgesellschaft mbH, Weinheim, A15, 559–593.

8 Joining for lightweight vehicles

P. K. MALLICK, University of Michigan-Dearborn, USA

Abstract: This chapter gives an overview of the joining technologies applicable in the design and manufacturing of lightweight automobile bodies. It covers liquid phase welding, solid phase welding, mechanical joining and adhesive bonding for steels, aluminum alloys and magnesium alloys. The process parameters and their effects on joint performance for each process are described. Benefits, limitations and applicability of each process for automotive body construction are also discussed. A brief review of the joining techniques for polymer matrix composites is also provided.

Key words: resistance spot welding, arc welding, laser welding, friction stir welding, self piercing riveting, clinching, adhesive bonding.

8.1 Introduction

With the increasing trend toward lightweight vehicles, the body materials for automobiles are changing from mostly low carbon steels to a mix of materials that includes both low carbon steels and high strength steels, light non-ferrous alloys, such as aluminum and magnesium alloys, and fiber reinforced polymers. The acceptance of these materials will depend not only on their structural performance and manufacturability, but also on the joining methods that can be used to assemble them in a rapid, robust and reliable manner. The structural performance of a joint depends on the joining method, joint geometry and joint quality. The joining method often depends on the materials being joined, but under mass production conditions it must also be quick, less dependent on fit and dimensional variations of the parts being joined and as defect free as possible. Joining is an important consideration in the design of an automobile, since joints are usually the weakest areas in the structure and they are often the failure initiation locations in service. Joining becomes even more important as the body material changes from a single material to a mix of materials, since the compatibility between different materials in terms of their mutual joinability, surface characteristics, corrosion, assembly stresses, etc. may pose many technical challenges and needs to be considered in the design process and selection of materials.

Since high strength steels and aluminum alloys are emerging as the principal material options for body construction in the near future, this chapter considers the joining methods that are most applicable for these two types of materials. The joining methods considered can be divided into four categories: (i) liquid phase welding or fusion welding, (ii) solid phase welding, (iii) mechanical joining and (iv) adhesive bonding. There are several different techniques available under each of these categories (Table 8.1). For example, liquid phase welding or fusion welding includes resistance spot welding, arc welding and laser welding. The two most common methods of arc welding are gas metal arc welding (GMAW) and tungsten gas arc welding (TGAW); but there are many other forms of arc welding methods that are used either in special cases or in limited applications. Solid phase welding, which is accomplished without melting the materials, includes friction stir welding and a few other emerging technologies, such as magnetic pulse welding and ultrasonic welding. Mechanical joining includes the conventional mechanical fastening methods using bolts and nuts, but also the newer techniques, such as self-piercing riveting and clinching. Adhesive bonding is also an established method of joining, but a combination of adhesive joining and welding or riveting is relatively new and is finding wider application, especially in space frame designs. These joining methods are discussed in this chapter. In addition, the joining methods currently used with polymer matrix composites are briefly reviewed.

8.2 Liquid phase welding

Liquid phase welding is performed by locally heating the materials to be joined to their liquid phases, applying pressure to create fusion bonding between them and allowing the fusion bonded liquid phase to cool under pressure to form the joining. The principal liquid phase welding processes used in the automotive industry are resistance spot welding and arc welding. They are described in this section.

8.2.1 Resistance spot welding

Resistance spot welding (RSW) is the principal joining method used for low carbon steel body construction. The welding practice for resistance spot welding uncoated steel sheets is well established. A typical steel body-in-white (B-I-W) contains nearly 5000 resistance spot welds. The reasons for its widespread use are its low cost, fast operation, and robustness to part dimensional variations. Nearly all of the resistance spot welded joints today are created by robots and therefore they are not subject to human error. Other advantages of RSW are that the nugget at the joining of the two sheets is completely internal and is not visible from the outside and, unlike arc welding, no filler material is necessary to form the weld.

In the resistance spot welding operation, electric current is passed through two electrodes and the sheets to be joined to generate intense localized heating that causes melting and coalescence of a small volume of material

Table 8.1 Various joining methods used with sheet metals in the automotive industry

Liquid phase welding	Solid phase welding	Mechanical joining	Adhesive joining
• Resistance spot welding	• Friction stir welding	• Bolting	• Adhesive bonding
• Arc welding	• Friction stir spot welding	• Screwing	• Weld-bonding
Gas metal arc welding	• Magnetic pulse welding	• Self-piercing riveting	• Rivet-bonding
Gas tungsten arc welding	• Ultrasonic welding	• Clinching	
Flux cored arc welding			
Plasma arc welding			
• Laser welding			

at the interface between the two sheets. The welding nugget is formed as the coalesced material cools down (Fig. 8.1). The pressure from the electrode tips holds the sheets together during the entire welding operation.

The strength of a spot weld depends on the size and strength of the nugget. The welding parameters that control the nugget size include weld current, welding time, hold time and electrode force. The surface conditions, fit-up between the sheets and electrode tip design also have significant influence on the nugget.

The nugget diameter increases with increasing weld current (Fig. 8.2). For each combination of materials, there is a range of currents over which acceptable welds are produced. If the current is low, the nugget formation may not be complete or the nugget may fail at the interface. On the other

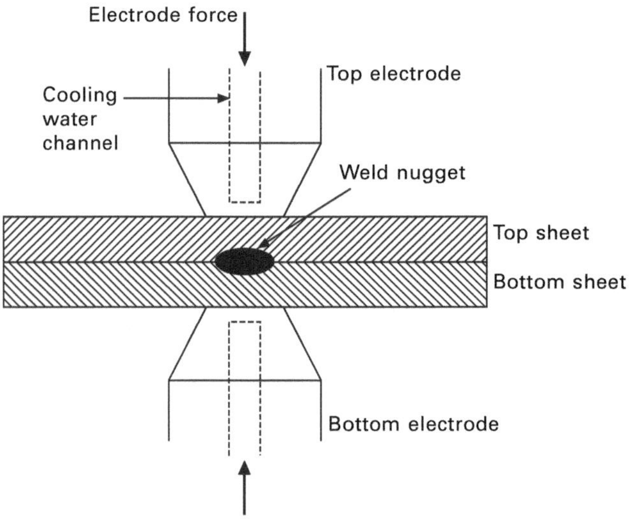

8.1 Resistance spot welding (RSW) process.

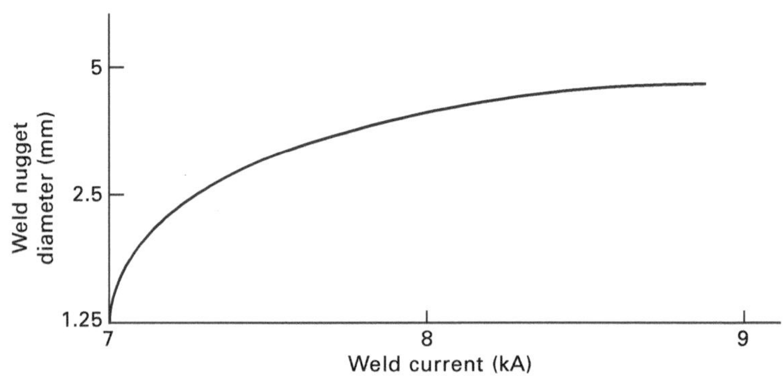

8.2 Weld nugget diameter vs. weld current.

hand, with very high currents, there may be expulsion of materials, which also produces an unacceptable nugget. The welding time is also important, since the range of weld currents that produces acceptable nuggets depends on the welding time. The minimum acceptable nugget diameter for automotive steels is between $4t^{1/2}$ and $5t^{1/2}$, where t is the nominal sheet thickness in mm.

The welding time–weld current relationship is shown by weldability lobes (Fig. 8.3). The wider the lobe, the greater is the current range and the higher is the tolerance to variations in production conditions. The location and the width of the weldability lobes are influenced by the electrical resistivity of the sheet materials being joined and sheet thickness as well as sheet surface conditions. For example, compared with uncoated sheet steels, longer welding time and/or higher weld currents are needed to form acceptable spot welds in zinc-coated sheet steels (Table 8.2), which are used in body applications for their improved corrosion resistance. Thus, the weldability lobes for zinc-coated sheet steels are shifted toward higher current levels.

The electrode material used in RSW is usually a copper based alloy, such as a chromium–copper alloy. With repeated use, the electrode tip tends to mushroom out and break down due to pitting and wear. The increase in tip diameter reduces the current density in the weld region, which slows down

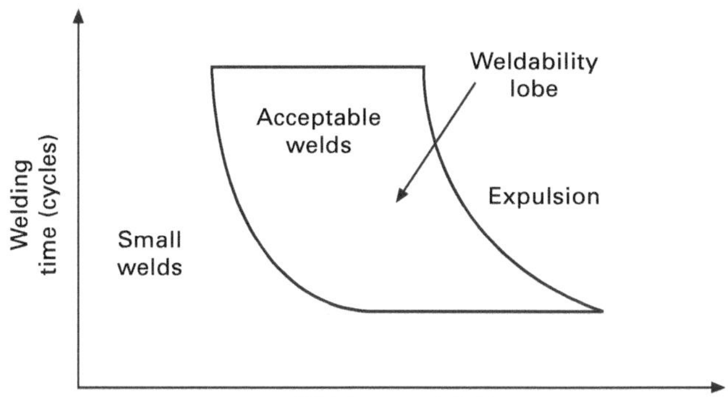

8.3 Weldability lobe for resistance spot welding process.

Table 8.2 Spot welding parameters for steel and aluminum sheets (sheet thickness = 0.9 mm)

Parameter	Steel (uncoated)	Zn-coated steel (hot-dipped)	Aluminum (uncoated)
Welding time (cycles) at 50 Hz cycles	7–10	9–12	3
Current (kA)	7–10	9–11	18–23
Force (kN)	1.9–2.6	2.2–2.9	4.1–5

Source: Barnes and Pashby, 2000a.

the nugget growth and decreases the nugget size. The electrode life, defined as the number of welds that can be made before the nugget size decreases below a specified minimum value, is another parameter used in evaluating the spot weldability of a material. In the automotive industry, the desired electrode life is between 2000 to 3000 welds. The electrode life with zinc-coated steels is lower than that with uncoated steels. The zinc from the coating combines with copper in the electrode to form brass at the electrode tip, which greatly accelerates the electrode tip wear.

Resistance spot welding parameters for high strength steels are slightly different from those used for low carbon steels. Some adjustments are required in welding current, since the bulk resistivity of high strength steels is higher. Higher electrode force may also be required to overcome the greater springback of high strength steels. For 2-mm thick DP steel sheets, the welding current is in the range of 8.3 to 11 kA, a welding time of 63 cycles and an electrode force of 4.2–5.3 kN were found to produce good quality spot welds (Tumuluru, 2006). (1 cycle is equivalent to $\frac{1}{60}$th of a second if 60 Hz AC current is used.) For some advanced high strength steels containing high carbon content, such as TRIP steels, longer welding times, controlled cooling rates and post-weld tempering may be necessary to reduce weld brittleness.

Resistance spot welding can also be used with aluminum alloys and has been applied to join aluminum sheets. However, there are a number of differences between steels and aluminum alloys that make resistance spot welding more difficult for aluminum alloys:

- Due to aluminum's significantly lower electrical resistance and higher thermal conductivity (Table 8.3), approximately two to three times higher weld current and only one-third to one-half the welding time are needed compared with steel. The weld current for aluminum is in the range of 15 to 30 kA compared with only 8–10 kA for steels, which means larger welding machines are required for aluminum. The energy consumption and the cost of resistance spot welding are also higher for aluminum alloys.
- The majority of weld development in aluminum occurs in 40 to 65

Table 8.3 Thermal and electrical properties of steel and aluminum alloys

Property	Steel	Aluminum alloys
Melting point (°C)	1460	660
Specific electrical resistance (Ohm. $mm^2.m^{-1}$)	0.14–0.16	0.0286
Thermal conductivity ($W.m^{-1}.K^{-1}$)	48–58	204
Coefficient of thermal expansion (10^{-6} K^{-1})	11.5	23.9
Affinity for oxygen	Medium	High

milliseconds compared with approximately 200 milliseconds required for steel. This is because the temperature difference between softening and melting of aluminum is relatively small. Thus, accurate control of electrode alignment and weld pressure prior to applying the weld current is more important for aluminum than for steel.

- The primary difficulty in spot welding aluminum is due to the presence of an aluminum oxide layer that forms spontaneously as bare aluminum surface is exposed to oxygen in the atmosphere. The melting point of the oxide layer is much higher than that of aluminum, and breaking down the oxide layer, which is required for forming an acceptable nugget, requires high heat input. This is another reason for higher energy consumption for resistance spot welding aluminum sheets compared with steel sheets.
- The surface condition of the aluminum sheet is extremely important, since it controls the number of welds that can be made before the electrode must be replaced or its tip must be dressed. With repeated welding, the electrode tip picks up aluminum, begins to either pit or erode, loses its shape and may even start sticking to the sheet surface. The aluminum oxide layer that spontaneously forms on an uncoated aluminum surface when exposed to air, also reduces the tip life. Removal of the oxide layer, either by mechanical means or by chemical etching, may become necessary to improve the tip life.
- Lubricants are often applied on aluminum sheet surface for improved formability. If the lubricants are not removed prior to resistance spot welding, the electrical resistivity of the surface is altered, which may lead to weld defects, including porosity and cracks.
- Higher current is required for 6000-series alloys than for 5000-series alloys, which is attributed to the higher bulk resistivity of the 5000-series alloys (Auhl and Patrick, 1994). However, 5000-series alloys tend to exhibit more weld expulsion.

Dissimilar materials with widely differing melting temperatures, such as aluminum and steel, cannot be joined directly using resistance spot welding. Furthermore, in the case of aluminum joining with steel, brittle intermetallic compounds are formed at the interface of the two materials, which makes the spot weld between them very weak. To resistance spot weld aluminum with steel, a sheet of aluminum clad steel can be used as the transition material between the two (Haynes and Jha, 1999). As the weld current is applied, the higher resistance of the steel–steel interface causes the steel side to heat up rapidly while the aluminum–aluminum side acts as the heat sink. The weld current required to form an acceptable nugget in such a joint is lower than that required for an aluminum–aluminum joint, but higher than that required for a steel–steel joint.

8.2.2 Arc welding

In arc welding, the heat needed to melt the metals at the joint is produced by an electric arc between an electrode and the parts to be joined. Two types of electrodes are used: (i) a consumable rod or wire electrode that not only conducts current, but also melts and supplies filler material at the joint, and (ii) a non-consumable rod electrode that simply conducts current to the weld area. The arc produces a temperature of about 3500 °C at the electrode tip and creates a pool of liquid metal at the weld area. When the pool solidifies behind the electrode as it is moved away from the joint, a metallurgical bond is created between the adjoining parts. To prevent chemical reaction between the liquid metal and the oxygen or nitrogen in the surrounding air, the weld area is shielded by a supply of inert gas or slag.

In the automotive industry, arc welding is used with both steel and aluminum. However, the arc welding practices for steel and aluminum are different due to the differences in their melting points, thermal conductivities and coefficients of thermal expansion (see Table 8.3). Arc welding of aluminum is also affected by the presence of the aluminum oxide layer on its surface. The melting point of the oxide layer is approximately 2035 °C, which is three times higher than that of aluminum. This oxide layer tends to absorb moisture from the air and since moisture is a source of hydrogen, it causes porosity in aluminum welds. Hydrogen may also come from oil, lubricants, paint and various surface contaminants. Since hydrogen is soluble in liquid aluminum, it is dissolved in the liquid weld pool. However, as the temperature decreases during cooling, the solubility of hydrogen in aluminum decreases and the dissolved hydrogen is rejected during solidification. At rapid cooling rates, free hydrogen is trapped in the weld and causes porosity. The aluminum oxide layer must therefore be removed from the aluminum surface prior to arc welding. In addition to causing hydrogen porosity, small oxide particles dislodged from the oxide layer may be entrapped in the weld and cause a reduction in ductility, incomplete fusion, and cracking.

Weld metal integrity is not usually a problem for low carbon steels. However, care is required when arc welding zinc-coated steels, since zinc vapor may cause porosity in the welds with high speed welding processes. In general, an arc weld in a low carbon steel is as strong as the base steel; but in most cases, the arc weld in an aluminum alloy is weaker, often to a significant degree, than the base aluminum alloy. For the non-heat treatable 5000-series alloys, the weld area will have zero-temper, annealed properties, regardless of the initial cold work. For the heat treatable 6000-series alloys, the weld area properties will be significantly lower than the properties of the T6 temper. Post-weld heat treatment can help restore the properties of the weld area in heat treatable alloys.

Among the aluminum alloys used for automotive body applications, the

5000-series alloys have higher weldability than the 6000-series alloys. The 5000-series alloys can be welded with or without any filler material, while the 6000-series alloys need filler material to prevent shrinkage cracking, generated during the solidification of the liquid weld pool. The filler material commonly used with aluminum alloys is a high Mg content aluminum alloy, such as the 5356 alloy (Al-5% Mg). A second filler material used with the 6000-series alloys is a high Si-content aluminum alloy, such as the 4043 alloy (Al-5% Si). Another problem associated with arc welding of aluminum alloys is thermally induced distortion, which can create considerable problems in maintaining dimensional fit-up.

With the increasing use of advanced high strength steels, it has also become necessary to consider their arc weldability. Table 8.4 gives the weld strength values determined by single lap shear tests on HSLA, which is a conventional high strength steel, and four advanced high strength steels, namely two dual phase (DP) steels and two martensite (M) steels. The joint efficiency, defined as the ratio of the weld strength and the base metal strength, is very high for both HSLA and dual phase steels, but is significantly lower for the martensite steels. The low joint efficiency for the martensite steels is attributed to softening of the heat affected zone (HAZ) due to tempering in the cooling stage. Interestingly, the fatigue strength of these steels is not affected by the softened HAZ, and is found to be insensitive to the static strength of the base material (Yan *et al.*, 2005).

Gas metal arc welding

Gas metal arc welding (GMAW), also called metal inert gas (MIG) welding, is an arc welding process in which the heat for melting the metal is generated by an electric arc between a consumable electrode and the metal (Fig. 8.4). The electrode is a solid wire, fed continuously through the arc into the weld pool, which ultimately becomes the filler metal at the weld. The wire type is selected to match the weld metal strength with the base metal

Table 8.4 Weld strength and joint efficiency for gas metal arc welding (GMAW) of high strength steels

Steel grade	Surface coating	Base metal strength		Weld strength (MPa)	Joint efficiency (%)
		Yield strength (MPa)	UTS (MPa)		
HSLA350	Bare	350	512	508	99
DP600	Hot dip galvanized	379	617	586	96
DP695	Bare	645	980	726	74
M900	Electro-galvanized	848	965	468	49
M1300	Bare	1157	1353	610	45

Source: Yan *et al.*, 2005.

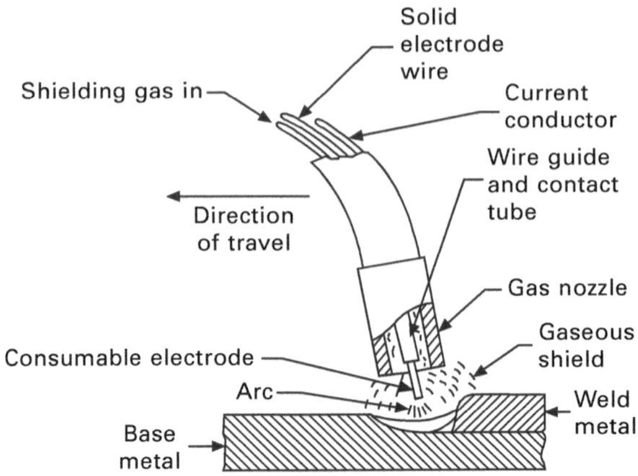

8.4 Gas metal arc welding (GMAW) process.

strength. A mixture of inert gases, such as helium and argon, is streamed into the weld area to shield and protect the arc, weld pool, electrode and the base metal adjacent to the weld from interacting with the atmosphere. The welding parameters that are controlled to produce acceptable welds are arc current, arc voltage, wire feed speed, electrode travel speed, current density and preheat temperature. Preheating involves heating the base metal in the region surrounding the joint prior to welding and is often used to reduce the residual shrinkage stresses and increase the resistance to cracking in the weld area.

GMAW can be used for joining dissimilar metals with close melting points and metallurgical compatibility. Close melting point is required to produce controlled melting on both sides of the joint. Metallurgical compatibility is required to prevent cracking in the heat affected zone (HAZ) or in the base metals, and to produce a microstructure in the weld zone that can provide adequate joint performance and corrosion resistance. For example, in welding a low carbon steel with a high alloy steel, the fusion boundary may contain unacceptable levels of very hard, brittle martensite phase, which will lower the joint strength. For some metallurgically incompatible metals, it may be possible to make a satisfactory weld using a suitable filler material.

Steel and aluminum alloys are not compatible for arc welding, since (i) there is a large difference between their melting points (see Table 8.3), (ii) iron has nearly zero solubility in aluminum and (iii) brittle intermetallic compounds, such as Fe_2Al_5 and $FeAl_3$, are formed at the weld. Furthermore, the large differences in their thermal properties, such as coefficient of thermal expansion and thermal conductivity, lead to internal shrinkage stresses after welding. For these reasons, fusion welds between steel and aluminum experience cracking and brittle failure in service.

Gas tungsten arc welding

In gas tungsten arc welding (GTAW), also called tungsten inert gas (TIG) welding, the electric arc is created between a non-consumable tungsten electrode and the parts to be joined. As in GMAW, inert gas shielding is used around the arc to protect the weld pool. Tungsten is a good electrode material because of its high melting point of 3410 °C. In the case of aluminum, the electrode itself is used to break up the oxide layer on the aluminum sheet surface. A filler material may be needed in some cases. When a filler material is used, it is fed into the weld site from a separate rod or wire instead of feeding it through the electrode. The filler material is melted by the arc and is added to the weld pool. GTAW is slower than GMAW, but the welds made by GTAW have much better surface appearance and require little or no finishing operation, since no spatter is created in GTAW.

8.2.3 Laser welding

In laser welding, the heat source is a focused high energy laser beam that impinges on the sheet surface. The primary laser welders are based on CO_2 lasers that emit a laser beam at 10.6 μm, i.e. at a near infrared wavelength; and Nd:YAG (Neodymium: Yttrium–Aluminum–Garnet) lasers that emit a laser beam at 1.064 μm, i.e. at a far infrared wavelength. The energy produced by the laser is focused over a 0.13 to 1 mm diameter spot to produce an energy density of 100–110 watts/cm^2.

Some advantages of laser welding are its deep, narrow heat-penetration with a narrow heat affected zone, and low heat input to the sheet material. There is very little heat distortion and the welds are highly formable. Laser welding does not require physical contact with the welded area, since the laser beam source can be located away from the weld area and is easily manipulated using a series of mirrors. The laser beam can be transmitted through air without any power loss and no vacuum chamber is needed. With Nd:YAG lasers, the laser beam can be transmitted using fiber optic cables, which provides great flexibility of maneuvering the laser beam to difficult-to-reach areas in complex geometries. The welding speed for laser welding is 5 to 10 cm/s compared to 2 to 3 cm/s for GMAW, which means higher productivity can be obtained with laser welding. As with arc welding, inert gas shielding may also be necessary with laser welding to prevent oxidation or degradation of the weld.

Laser welding requires precise joint fit up with very little gap and mismatch between the sheets to be joined, so that the narrow laser beam can be targeted accurately to the desired welding spot. Another disadvantage of laser welding is the high investment and running costs, the two main reasons for its low usage at the present time.

Laser welding has been applied to both steels and aluminum alloys to make spot welds as well as continuous welds. One of the current applications of laser welding with sheet steel involves fabrication of tailored blanks for inner door panels, floor panels, shock towers, etc., where sheets of different thicknesses or grades are welded prior to stamping for improved performance at reduced weight. One problem in the laser welding of zinc-coated steel is that the intense heat associated with the process may vaporize the zinc coating, and if the zinc vapor is trapped in the weld pool, it creates porosity at the joint. The power requirement for laser welding aluminum is higher for several reasons. Aluminum has high reflectivity, which causes loss of energy due to the reflection of the laser beam by the weld pool and reduces the absorptivity of the laser energy. Added to this is aluminum's high thermal conductivity, which causes rapid heat dissipation away from the weld area. The vaporization and subsequent loss of the alloying elements (e.g. Mg and Zn) from the weld pool is another problem observed in the laser welding of aluminum alloys, since this can cause degradation of the mechanical properties of the weld (Zhao *et al.*, 1999). In spite of these problems, using proper process parameters, it is possible to produce high quality laser welded joints in aluminum alloys that are stronger than the base material; however, the elongation at fracture is generally much lower than that of the base material (Martukanitz *et al.*, 1996). Research has also been done on the use of laser welding with magnesium alloys, such as AZ91D and AM 50 (Cao *et al.*, 2006). Much like the aluminum alloys, they also have low laser energy absorptivity, high thermal conductivity, strong affinity for oxygen and high solubility for hydrogen in the liquid state, and therefore the problems encountered in laser welding magnesium alloys are similar to those for aluminum alloys. Even then, the laser welded joints in magnesium alloys have been found to be as strong as the base material, but their elongation at fracture is low.

8.3 Solid phase welding

In solid phase welding, two sheets of the same material or different materials are joined by creating metallurgical bond without melting the materials. The weld quality is usually excellent, since porosity, grain boundary cracking and other problems commonly associated with liquid phase welding do not occur in solid phase welding. The mechanical properties, in general, are equal to or better than those obtained by liquid phase welding. Another advantage is that by not creating a molten pool of material which shrinks significantly on solidification, distortion of the part and residual stresses after welding are relatively low. The process is environmentally friendly, since no fumes or spatter are generated, and there are no arc glares or reflected laser beams to contend with.

The two common solid phase welding methods used with steel, aluminum and a variety of other metals are friction stir welding, known as FSW, and friction stir spot welding, known as FSSW. Both processes were initially developed by The Welding Institute (TWI) of UK. FSW was developed in 1991 and since then it has been gaining popularity for welding aluminum alloys in both automotive and aircraft industries. FSSW was developed later and is seeing increasing applications in the automotive industry. In both processes, the materials at the joint are brought to the softened states by the frictional heat generated between a rotating welding tool and the sheets or plates to be joined. They have the ability to join materials that are difficult to fusion weld; for example, the 2000- and 7000-series aluminum alloys. Another major benefit of FSW and FSSW is the reduced energy consumption. The electricity consumption for friction stir welding aluminum is only 5% of that of RSW. Since FSW does not need the use of large current, coolant, and compressed air that are required for RSW, the total energy consumption in FSW is significantly low.

There are two other solid phase welding methods that are worth mentioning, since considerable research is being done on both of these processes for joining dissimilar materials, such as steel and aluminum. They are magnetic pulse welding and ultrasonic welding. Magnetic pulse welding (MPW) is a cold solid state process in which the weld is created without heating up the parts to be joined and without any filler material (Shribman *et al.*, 2001). MPW involves discharging a high current over a very short time period through a cylindrical coil surrounding the parts to be joined. This creates a high density magnetic flux around the coil and induces an eddy current in the parts. Since the eddy current opposes the magnetic flux, a repulsive force is created, which drives the two parts toward each other at extremely high speed (in the range of 200 to 500 m/s) and join them together with an explosive or impact-type weld (Aizawa *et al.*, 2007).

Ultrasonic welding involves high frequency (typically 20 kHz) mechanical vibration between two contacting parts held under pressure. The frictional heat generated at the interface increases the temperature and creates a bond between the two surfaces without melting the material. The important process parameters in ultrasonic welding are the welding power, vibration frequency, welding time and pressure. Depending on the processing conditions selected, the temperature at the interface is in the range of 40 to 80 percent of the melting point of the material (Elangovan *et al.*, 2009).

8.3.1 Friction stir welding

In friction stir welding (FSW), a non-consumable rotating cylindrical tool with a specially profiled shoulder and a pin is inserted into the abutting edges of the two parts to be joined and is traversed slowly along the joint line (Fig. 8.5). The parts are clamped onto a backing bar to prevent the joint faces from being forced apart. Frictional heat generated between the rotating tool and the parts causes the materials at the joint to soften and reach the plasticized state. The rotary motion of the tool creates the mixing necessary to transport material from one side of the joint to the other. The heat-softened plasticized material is transferred from the leading edge of the pin to the trailing edge of the pin and is forged by the contact pressure from the tool. As a result, a solid state bond is created between the two parts being joined.

The tool geometry in FSW serves a critical role in the complex material flow that occurs as the rotating tool traverses along the weld line. Tools with a concave shoulder and a threaded cylindrical pin are commonly used;

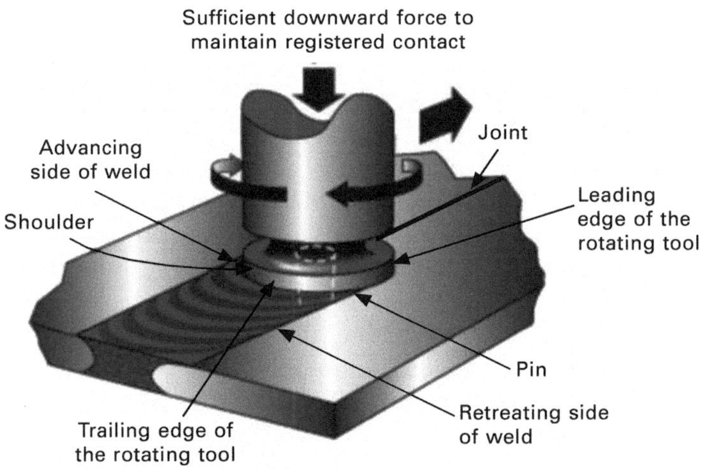

Schematic of the friction stir welding process

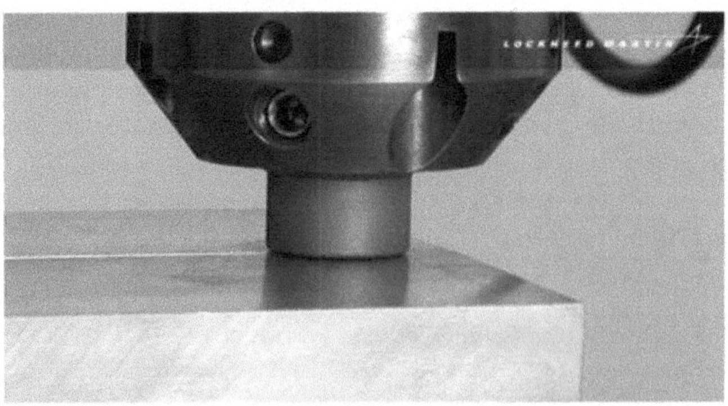

Photograph of a friction stir welding operation

8.5 Friction stir welding (FSW) process.

but a variety of pin and tool shoulder designs has evolved to improve the mixing and flow of material, increase the heat generation and reduce the welding force (Mishra and Ma, 2005). For soft metals such as aluminum and magnesium, the FSW tool can be made of hardenable tool steel, such as H13 steel. One tool can typically be used for up to 1000 m of weld length in 6000-series aluminum alloys. For steels, the FSW tool material is typically a hard ceramic, such as polycrystalline cubic boron nitride (PCBN), silicon nitride (Si_3N_4) and silicon carbide (SiC).

During FSW, the material at the joint undergoes intense plastic deformation at elevated temperature, resulting in microstructure that can be divided into three distinct zones, as shown in Fig. 8.6:

(i) *Thermo-mechanically affected zone (TMAZ)*: In this region, the material has been subjected to high plastic deformation and thermal effect. In the case of aluminum, the TMAZ contains a dynamically recrysatllized area near the center, which is surrounded by material that has undergone significant plastic strain without recrystallization. The recrystallized area in the interior of the TMAZ is called the nugget (Fig. 8.6), which contains fine and nearly equiaxed grains, whereas the grains in the non-recrystallized area outside the nugget are larger and highly distorted.
(ii) *Heat affected zone (HAZ)*: In this zone, the material experiences a thermal cycle that modifies the microstructure as well as the mechanical properties. However, there is no plastic deformation occurring in this area.
(iii) *Unaffected material or the base metal*: This material is outside the heat affected zone and is remote from the weld. Even though it may have experienced a thermal cycle, its microstructure and mechanical properties are not affected by the heat.

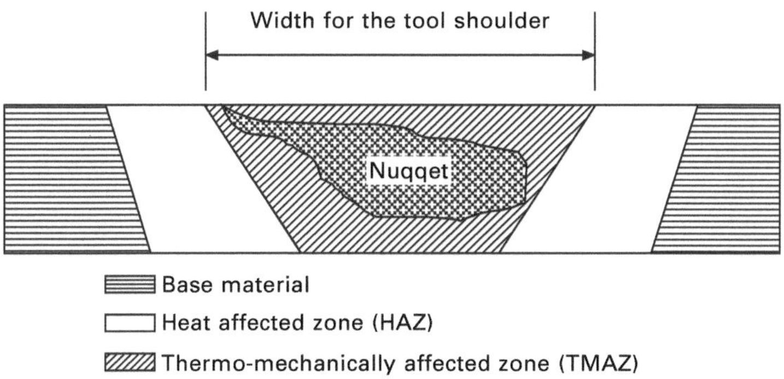

8.6 Schematic representation of the microstructure observed in friction stir welded joints.

FSW can be used to create both butt welds and lap welds between similar as well as dissimilar materials. The tensile strengths of friction stir welded joints in low carbon steels and high strength low alloy (HSLA) steels are very similar to those of the base material. The friction stir welded joints in aluminum alloys have shown up to 35% lower tensile strength than the base material (Mishra and Ma, 2005). Table 8.5 shows a comparison of the tensile properties of friction stir welded and gas tungsten arc welded joints in three aluminum alloys considered for automotive body applications. As can be seen in this table, the welded joints obtained by both FSW and GTAW in the two 5000-series alloys have very similar tensile properties, but for the 6000-series alloy, the welded joint obtained by FSW is better than the welded joint obtained by GTAW. This is attributed to the tempering effect on the heat-treatable 6000-series alloys in GTAW, which causes a greater amount of softening in the joint area.

The principal advantages of FSW result from the fact that joining by FSW takes place in the solid phase and the materials at the joint do not experience the melting–solidification cycle. The advantages that follow from solid phase welding are

- Low distortion, even in long welds
- No porosity
- Low shrinkage
- No arc, fume and spatter
- No consumables
- No shielding gas needed
- Low energy consumption

Another benefit of FSW is that it has only three process variables to control: tool rotational speed, tool traverse speed and pressure. Depending on the materials being joined and their thicknesses, the tool rotational speed and traverse speed can vary in the range of 100–2000 rpm and 10–1000 mm/min, respectively.

However, there are also a few limitations of the FSW process.

- The parts to be joined must be rigidly clamped, which means this process is most suitable for making long, straight flat joints.
- The welding speed is relatively slow, typically 0.3–0.6 m/min for 1.6–3.2 mm thick aluminum sheets
- A keyhole (tool exit hole) is created at the start and end points of the weld, which must be filled after welding.

The application of FSW in the automotive industry is limited because in the present body-in-white constructions there are not too many straight flat joints for which FSW is usually more suitable. Figure 8.7 shows an application where FSW was used to join a stamped 5754 aluminum part and

Table 8.5 Tensile properties of base material, friction stir welded and gas tungsten arc welded aluminum alloys

	AA5754-O			AA5182-O			AA6022-T4		
	YS (MPa)	UTS (MPa)	Elong. (%)	YS (MPa)	UTS (MPa)	Elong. (%)	YS (MPa)	UTS (MPa)	Elong. (%)
Base material	94.4	236.3	26.5	131	285.4	27.3	163.6	263.3	29.6
FSW specimen	98.1	235.6	24.9	138.2	285.7	25.7	140.1	235	12.9
GTAW specimen	96.6	235.5	24.1	127.4	283.2	27	119.7	207.4	8.2

Source: Miles *et al.*, 2005.

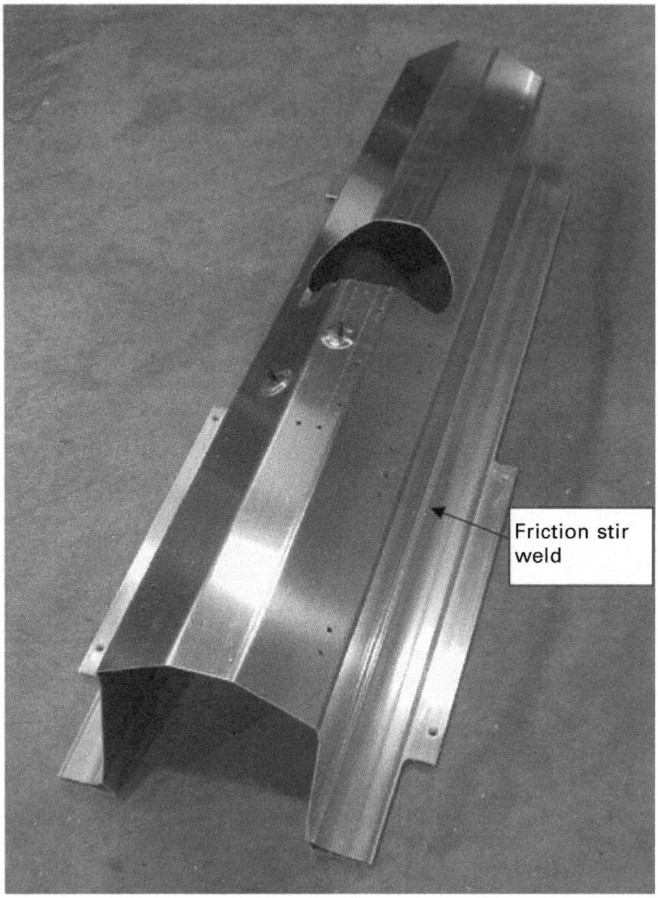

8.7 Example of a friction stir welded part.

a 6061-T6 extrusion to produce the center tunnel of a sports car. Among other applications of FSW are wheel rims, upper suspension links and rear seat back frames.

8.3.2 Friction stir spot welding

The friction stir spot welding (FSSW) process is similar to the friction stir welding (FSW) process, except that in FSSW the tool does not traverse laterally. FSSW is used for making spot welds in overlapping sheets, whereas FSW is used mostly for making linear butt joints, although it can also be used for making linear lap joints.

The fundamental mechanism of the FSSW process is shown schematically in Fig. 8.8. The rotating tool, usually with a concave-shaped shoulder and a

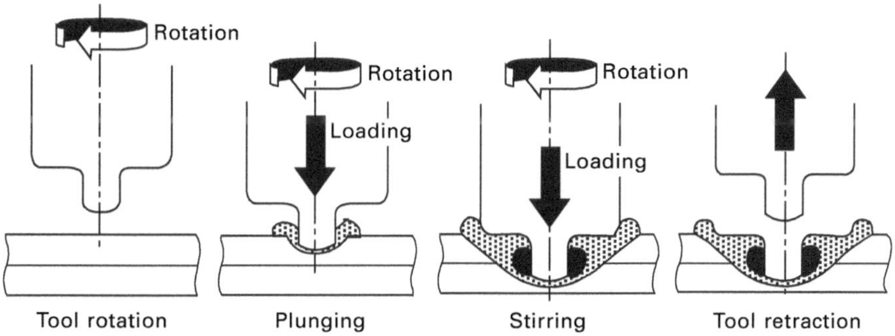

8.8 Friction stir spot welding (FSSW) process.

threaded pin at its end, is plunged into the upper sheet of two overlapping sheets with an axial force. A backing plate or an anvil is used below the lower sheet to counteract the axial tool force. The rotating tool's contact with the top sheet generates frictional heat, which softens both the top and the bottom sheets. The heat-softened material is stirred and starts to flow in the vertical direction along the threads of the pin as the tool is rotated at a constant speed and simultaneously displaced downward. The tool displacement is continued until either a predetermined level of displacement (in the displacement control mode of operation) or a predetermined level of axial tool force (in the force control mode of operation) is reached. After the maximum tool displacement is reached, the tool rotation is continued for a specified dwell time and then the tool is retracted from the sheets. A metallurgical bond is formed at the interface between the two sheets (Fig. 8.9), as the material at the joint cools down.

The tool's rotational speed in FSSW is usually between 1000 and 3000 rpm. The other process control parameters in FSSW are plunge rate, plunge depth (i.e. maximum displacement) and dwell time. The process parameters that produce the highest joint strength depend on the materials being joined, sheet thickness and tool geometry. The FSSW process can be automated using robots. There are also several variations of the FSSW process, such as the Refill FSSW process (Allen and Arbegast, 2005), which can fill the key hole left on the top sheet as the tool exits from the joint at the end of the dwell time.

FSSW has many more potential applications in automotive body applications than FSW, since in a typical body-in-white construction, there are numerous spot welds and very few long, linear joints. FSSW is more suitable than resistance spot welding for joining aluminum and has been used in assembling aluminum body panels since 2003 (Pan, 2007). Some of the early applications included the aluminum rear doors and hoods on the Mazda RX-8 and the decklids and hoods on Prius hybrid vehicles. Another example of FSSW is the joining of aluminum trunk lids to steel bolt retainers in 2005 Mazda

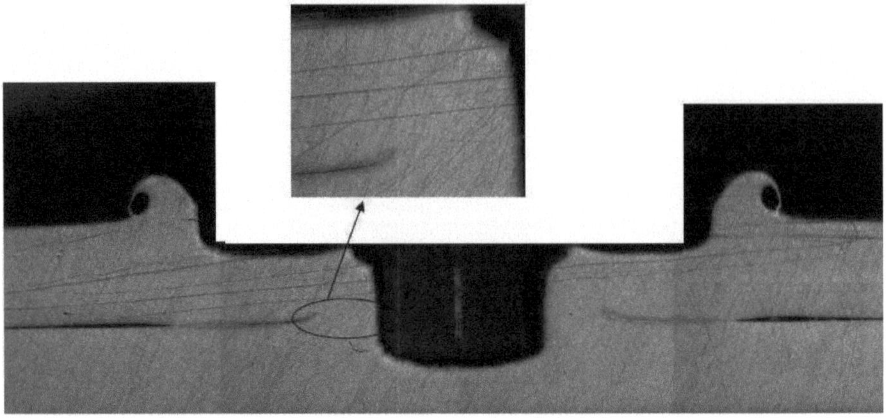

8.9 Joint configuration of friction stir spot welded joint.

MX-5 sports cars. FSSW has also been explored to join high strength steels (Feng *et al.*, 2005) and magnesium alloys (Mallick and Agarwal, 2009).

8.4 Mechanical joining

Conventional mechanical joining methods using bolts and nuts are not very common in automotive body applications. Although bolted joints have the advantage that they can be disassembled easily if replacement or repair of parts is needed, they require drilling of matching holes in the parts to be joined and the drilled holes must align accurately during the assembly process. Although they are used in assembling many chassis and suspension components, they are considered too slow for assembling body components compared to the available welding processes. In recent years, two relatively new mechanical joining methods are finding greater use in assembling sheet metals for body applications. These two methods, called self-piercing riveting and clinching, are described in this section.

8.4.1 Self-piercing riveting

Self-piercing riveting (SPR) is essentially a cold forming operation by which two sheets of materials are joined by piercing the top sheet with a semi-tubular rivet and then flaring the rivet shank into the bottom sheet to form a mechanical interlock between the two sheets (Fig. 8.10). No pre-drilled holes are needed in the SPR process, which eliminates the alignment problems often encountered in bolted joints. Even though SPR is used mainly to join aluminum sheets, it can be used to join steel sheets and sheets of dissimilar materials, such as aluminum to steel (Abe *et al.*, 2006). It can also be used for joining multiple sheets. For aluminum alloys, a total joint thickness up

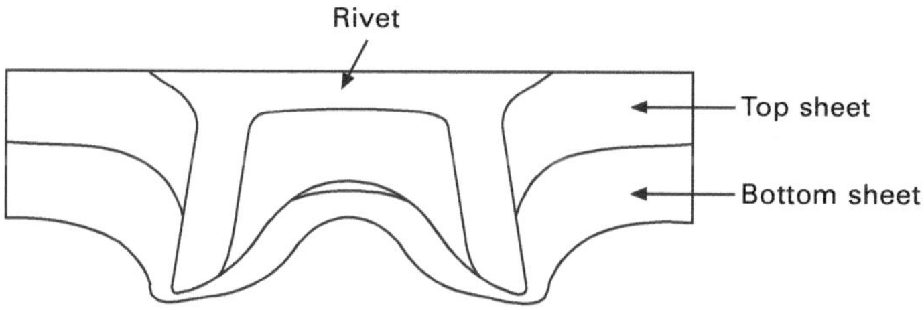

Joint configuration in self-piercing riveted joints

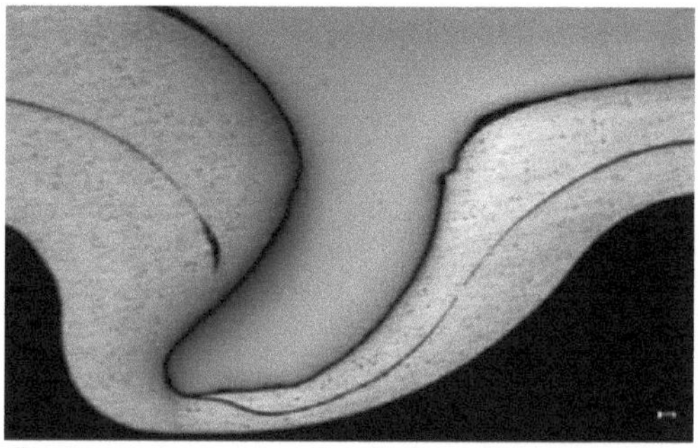

Photograph showing the flaring of the rivet in a self-piercing riveted joint

8.10 Self-piercing riveted joint.

to 10 mm can be riveted, whereas for steels, the total joint thickness may be limited up to 6 mm (He *et al.*, 2008). For optimum performance, the riveting direction should be from the thinner to the thicker sheets and from the harder material to the softer material. Since cold forming is involved in self-piercing riveting, the sheet materials must have sufficient ductility to deform and flow without fracture or crack formation during the riveting operation. A large difference in the flow stresses of dissimilar materials can cause a problem in self-piercing riveting.

The process of self-piercing riveting is divided into the following four steps (Fig. 8.11).

(i) *Clamping*: The sheets are clamped between a pre-shaped die and a rivet gun, which includes a rivet and a punch.
(ii) *Piercing*: The rivet is pushed by the punch to pierce through the top sheet and partially through the bottom sheet.
(iii) *Flaring*: The material in the bottom sheet flows into the die and the rivet shank begins to flare outward without perforating the bottom sheet.
(iv) *Releasing*: The punch is stopped and retracted when a predetermined value of either riveting force or stroke is reached.

Both rivet design and die geometry (Fig. 8.12) play critical roles in the

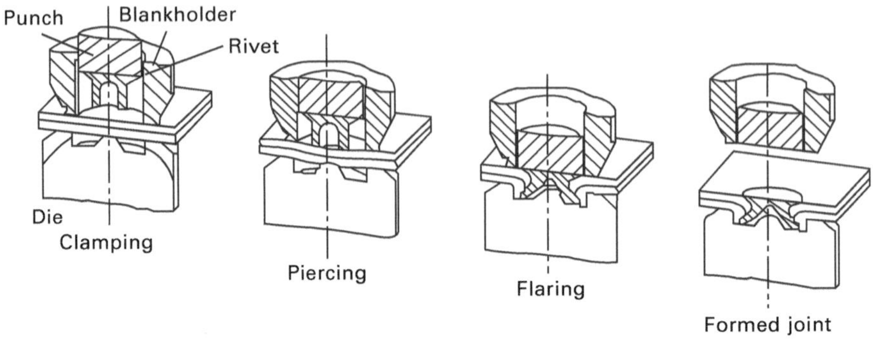

8.11 Self-piercing riveting (SPR) process.

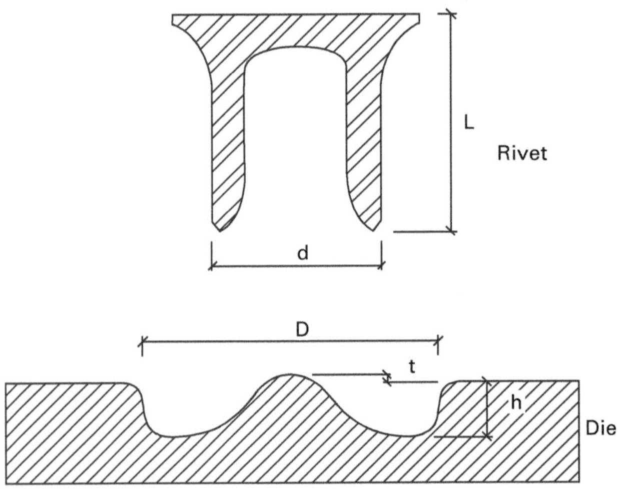

d = Rivet diameter, L = Rivet length
D = Die diameter, h = Die height, t = Die tip height

8.12 Rivet and die shapes used in self-piercing riveting.

formation of self-piercing riveted joints (Fu and Mallick, 2001). The important parameters related to rivet design are the rivet diameter, rivet length and rivet head style (e.g. flat head vs. countersunk head). The die depth, die diameter and die tip height are the three most important parameters for the die. The rivets are made of high strength steel and are surface coated for corrosion protection. The hardness of the rivet material is also important, since it controls the rivet deformation as it is pushed into the sheets. Coating on the rivet is particularly important for aluminum sheets, since otherwise, galvanic corrosion can take place between the steel rivet and the aluminum sheets.

Studies have shown that the static strength of self-piercing riveted joints is usually lower than that of resistance spot welded joints in both steel and aluminum. However, the fatigue strength of self piercing riveted joints is superior to that of resistance spot welded joints (Booth *et al.*, 2000). At 10^6 cycles, the reported fatigue strength of self-piercing riveted joints in aluminum alloys is 25 to 100 percent higher than that of resistance spot welded joints.

Several other advantages of self-piercing riveting compared to welding are:

- joining is possible between dissimilar materials and multiple stacks,
- cycle times are fast, usually between 0.25 to 1 second per rivet,
- energy requirements are low, and
- no fumes or spatters are produced and therefore, it is a much safer operation.

There are also several limitations:

- Self-piercing riveting cannot be used with outer body panels unless a finishing operation is used to cover the rivet head.
- The joints are not flush on the tail side, since the bottom sheet bulges out in the form of a button as it deforms into the die to accommodate the flaring rivet.
- Self-piercing riveting action requires access from both sides of the joint.
- Relatively high forces are required for the forming process involved in self-piercing riveting and therefore, the tooling can be heavy.
- High forming forces can also cause joint distortion, which may negatively affect the appearance of the sheet surfaces.

Self-piercing riveting is already an accepted joining method in the automotive industry. It has been employed for assembling aluminum body components in both body-in-white (e.g. in Jaguar XJ) and space frame (e.g. in Audi A8 and A2) constructions. It works well with aluminum, since it provides better joint strength than spot welding and it is simpler to use than adhesive bonding. It has also been used with high strength steels, such as in Volvo FH12 truck cabs where traditional resistance spot welding did not provide the required fatigue performance (He *et al.*, 2008).

Self-piercing riveting has also been tried with magnesium alloys for which localized heating is required at the riveting area. This is because magnesium has low ductility and poor formability at room temperature and cracks are formed in the magnesium sheets if attempts are made to join them by self-piercing riveting without raising the temperature to increase formability. The methods used for localized heating include resistance heating, induction heating and laser-assisted heating (Easton *et al.*, 2008).

8.4.2 Clinching

Clinching is also a mechanical interlocking process, like self-piercing riveting, except that rivets are not used in clinched joints. In the clinching process, shown schematically in Fig. 8.13, joining occurs by the action of a punch which forces both sheets into a die and forms an interlocking button in the shape of the die. The materials in the button area undergo plastic deformation, first due to drawing and then due to cold upsetting.

Clinched joints are used with aluminum for low to medium load applications. The failure load of clinched joints in aluminum alloys is typically 50–60% that of self-piercing riveted joints (Saathoff and Mallick, 1998). Even though clinched joints have lower static failure load than spot welded joints, their fatigue strength at 10^6 cycles is approximately 25 percent higher (Krause and Chernenkoff, 1995).

8.5 Adhesive joining

8.5.1 Adhesive bonding

Adhesive bonding offers improved joint stiffness compared to spot welding and mechanical joining, since it produces continuous joining instead of localized, discrete joining. This also results in more uniform stress distribution over a larger surface area. A properly designed adhesive joint produces high joint strength and is capable of high energy absorption. It also provides good noise and vibration damping. Another advantage of adhesive joints in automotive applications is that they prevent ingress of water and debris into the joint area, thus acting like a seal. However, adhesive bonding may require surface preparation, which may include surface cleaning and surface pretreatment. Furthermore, the adhesive may need heat curing, which is not only time-consuming, but also may affect the properties of the base material.

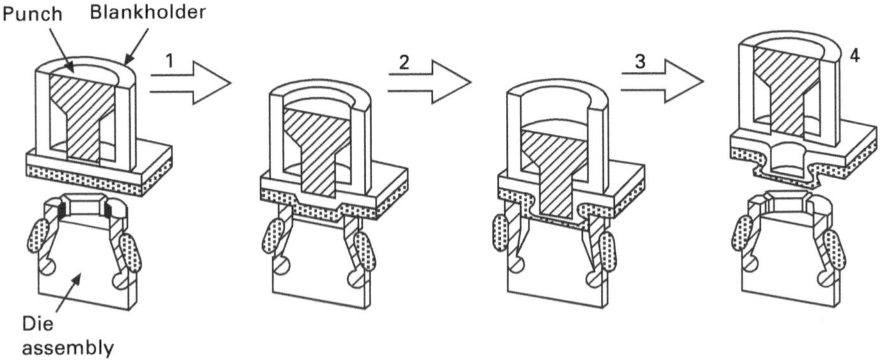

8.13 Clinching process.

Long-term durability of adhesive joints and the possibility of adverse effects of humidity, temperature, salt water and other environmental conditions on the properties of the adhesive need to be considered before using adhesive bonding in automotive applications. Adhesive joints also have low peel strength, which can pose problems in certain bonding situations and must be taken into consideration at the design stages, particularly with respect to crashworthiness.

Structural adhesives are classified as either one part or two part adhesives. One part adhesives are easier to dispense, but they cure slowly, often requiring heat. The heating required for curing one part adhesives can be provided in a separate curing oven or during the paint baking cycle for the body-in-white. Two part adhesives require mixing and they are usually harder to dispense, but they cure rapidly. Four types of structural adhesives are used in the automotive industry: epoxies, urethanes, acrylics and cyanoacrylates. Among these, the first three are available as both one part and two part adhesives. One part epoxies generally have superior environmental durability and resistance to oil, which makes them useful for bonding coated or lubricated surfaces after stamping. Urethanes require clean or primed surfaces for bonding and they are used mostly for bonding glass to metal. Acrylics are suitable for metal surfaces and do not require much surface preparation. Cyanoacrylates react with moisture on the sheet surface to induce curing. They cure very rapidly, which makes it difficult to use them for bonding large areas. They also require surface cleaning, since they cannot bond to oily surfaces.

Currently, adhesive bonding has limited use in automotive body construction. Closure body panels, such as doors, hoods and lift gates, are often produced by adhesive bonding a stamped inner panel or reinforcement to a stamped outer body panel. Although adhesive bonding is not very common in the current steel body-in-white, it will experience greater use as other materials, such as aluminum, magnesium and polymer matrix composites, become more accepted as body construction materials in the automotive industry. Adhesive bonding has the ability of joining dissimilar materials that cannot be welded, such as steel and polymer matrix composites, or steel and magnesium. For aluminum or magnesium, which are inherently more difficult to weld than steel, or for thermoset matrix composites that cannot be welded at all, adhesive bonding may be a better option than mechanical fastening. In general, adhesive bonds exhibit higher fatigue performance than mechanically fastened joints. One reason for this is that mechanical fastening creates stress concentration in the material due to the presence of drilled or punched holes, which is completely avoided in adhesive bonding.

The performance of an adhesive joint depends on a number of joint design parameters, such as bond thickness, overlap length and sheet thickness. The selection of adhesive is also important, since it controls the bond strength, failure mode (adhesive failure vs. cohesive failure) and energy absorption.

The adhesive properties to be considered in adhesive selection are its shear modulus, shear strength and toughness. The surface energy of the adhesive relative to the surface energy of the sheets is another important factor, since it determines the wetting of the surface by the adhesive.

Comparison of adhesive joints and resistance spot welded joints shows that adhesive joints are much superior in fatigue performance. For example, in the case of two 0.9 mm thick overlapping steel sheets, the fatigue failure load at 10^6 cycles of a bonded joint was nearly four times greater than that of a spot welded joint (Jones and Williams, 1986). The adhesive bonding in this case was made with a toughened epoxy and the overlap length was 25 mm.

Adhesive bonding is practiced more commonly with aluminum than with steel, since aluminum is much more difficult to weld than steel. Several aluminum-intensive concept cars, as well as low volume niche cars, have been produced using adhesive bonding as the primary joining method (Barnes and Pashby, 2000b). The adhesive for joining aluminum sheets used for body applications must be compatible with the pretreatment and lubricant applied to their surfaces at the end of the production mill (Schroeder, 1996). The chemical pre-treatment is applied not only to prevent surface oxide formation, but also to improve paint adhesion and corrosion protection. The lubricant is added on the pre-coated surface to facilitate the stamping and other press forming operations needed to manufacture the parts. Other important characteristics of the adhesive are listed below.

- Proper viscosity for uniform and easy dispensing without running or dripping
- Curing temperature in the range of 170–210 °C so that curing can take place in the paint baking oven
- Long shelf-life
- Ability to maintain high strength after exposure to elevated temperature and high humidity (e.g. 50 °C and 90% relative humidity) under stressed condition.
- High fatigue resistance, impact resistance at sub-zero temperatures (e.g. at –40 °C) and creep resistance at elevated temperatures

8.5.2 Weld bonding and rivet bonding

Weld bonding combines resistance spot welding and adhesive joining in the same joint. The adhesive carries the main structural load and the spot welds help reduce the peel stresses and improve the crashworthiness by carrying the out-of-plane loads. In making weld-bonded joints, the adhesive (usually a one-part heat-curable epoxy) is applied first and spot welds are made afterward using the resistance spot welding method. An adhesive-free zone between the two surfaces is created by removing the adhesive locally at the desired weld site to make an electrically conducting path between the two joining metal surfaces. This usually occurs during the squeeze time when the electrodes are pressed against the sheets prior to applying the weld current. Control of the adhesive viscosity is important so that it fills the joint without excessive overflow. The electrode temperature, which is controlled by cooling water surrounding the electrodes, may affect the adhesive viscosity. Resistance spot welding through the adhesive often causes irregularly shaped nuggets, which may not be desirable.

Rivet-bonding (or simply riv-bonding) is a combination of self-piercing riveting and adhesive bonding. In this process, the two sheets to be joined are first coated on one side each with an adhesive, typically a one-part heat-curable epoxy, and then joined together using self-piercing rivets. The adhesive is later cured at an elevated temperature.

Depending on the adhesive type and the joint design, the strengths of both weld-bonded and rivet-bonded joints are in general higher than the respective simple spot welded or riveted joint (Table 8.6). The fatigue strength is also higher with both weld-bonded and rivet-bonded joints. For example, a study conducted on joints in 0.8 mm thick zinc-coated low carbon steels showed that the fatigue strength at 10^6 cycles improved from 1.1 kN to 3 kN when adhesive was used with resistance spot welding and from 1.5 kN to 2.2 kN when adhesive was used with self-piercing riveting (Booth *et al.*, 2000).

8.6 Joining of polymer matrix composites

Adhesive bonding and mechanical fastening are the common joining methods for polymer matrix composites. In general, adhesive joining is preferred over mechanical fastening for its advantages listed in Table 8.7. Thermoset matrix composites, such as sheet molding compounds (SMC) and carbon fiber reinforced epoxy, cannot be welded. Thermoplastic matrix composites can be welded; but the welding processes used for them are different from the ones used for metals. The welding processes for thermoplastic matrix composites are described in Chapter 5.

In joining polymer matrix composites, special attention should be given to the behavior of these materials under mechanical and thermal loads, which are vastly different from that of metals. Unlike steels and aluminum alloys, most polymer matrix composites do not exhibit yielding and plastic deformations that are considered the common stress redistribution mechanisms in ductile metals. Failure modes of joints in these composites are also different from the failure modes observed in metals. Furthermore, polymer matrix composites containing continuous fibers are not isotropic materials, so that their modulus and strength vary in different directions of the material. This non-isotropic behavior creates a level of design complexity in both bolted and adhesive

Table 8.6 Strength of weld-bonded and rivet-bonded joints compared to resistance spot welded and self-piercing riveted joints

Substrate material	Ratio of weld-bonded joint strength and resistance spot welded joint strength				Ratio of rivet-bonded joint strength and self-piercing riveted joint strength			
	Lap shear strength		T-Peel strength		Lap shear strength		T-Peel strength	
	Adhesive 1	Adhesive 2	Adhesive 1	Adhesive 2	Adhesive 1	Adhesive 2	Adhesive 1	Adhesive 2
Zinc-coated low carbon steel (0.8 mm thickness)	1.60	1.38	1.94	1.25	1.74	1	1.44	1.02
AA6016 aluminum alloy (1.2 mm thick)	2.67	2.08	3.42	3	1.57	1	1.02	1.29

Note: Adhesive 1 is a one-part rubber-based heat curing adhesive and Adhesive 2 is a two-part acrylic adhesive.
Source: Booth *et al.*, 2000.

Table 8.7 Adhesive bonded joints vs. mechanically fastened joints for polymer matrix composites

Adhesive bonded joints	Mechanically fastened joints
• Distribute the load over a larger area • Require no pre-drilled holes • Higher fatigue resistance due to lower stress concentration • Are susceptible to failure due to out-of-plane loading • Add very little weight to the structure • Increase the stiffness of the structure • Are difficult to disassemble without either damaging or destroying the substrates • Are difficult to inspect for joint quality (e.g. unbonded or disbonded sections and voids) • May require surface preparation, including grit blasting and priming, to improve bonding • May require curing at elevated temperatures • May be affected by service temperature, humidity and other environmental conditions	• Permit quick and repeated disassembly for repairs or replacements without destroying the substrates • Greater resistance to out-of-plane loading • Require pre-drilled holes that cause stress concentration, interrupt fiber continuity and introduce local damages (such as delamination), which may reduce the strength of the composite • Add weight to the structure • Are easy to inspect • No special surface preparation needed • May create a potential galvanic corrosion problem, for example, in an aluminum fastener if it is used for joining carbon fiber reinforced epoxy composite

joints that does not exist in metals. The design methodologies and failure prediction in joints of polymer matrix composites are also different from the ones practiced with metals.

Currently, random glass fiber reinforced composites, such as sheet molding compounds (SMC) and structural reaction injection molded composites (SRIM), are the most common thermoset matrix composites used in the automotive industry. Adhesive bonding, mechanical fastening and a combination of the two are used with these composites. Bolts are the most common mechanical fasteners, although in some cases, rivets are also used. Large diameter flat washers with smooth faying surface are recommended for distributing the clamping load evenly around the bolt hole. Design of bolted joints in these composites requires the use of proper spacing between the bolt holes, bolt hole to part edge distance, washer edge to part edge distance and bolt tightening torque (Mallick and Little, 1994).

Future lightweight vehicle structures are expected to contain significant amounts of carbon fiber reinforced composites, since they have a much higher weight saving potential than the currently used materials, including advanced high strength steels and aluminum alloys. Carbon fiber reinforced epoxy laminates are used in significant amounts in the aerospace industry and the joining methods used with them are mechanical fastening, adhesive bonding and the combination of mechanical fastening and adhesive bonding. Co-curing of two composite parts at the joint, with or without an adhesive layer between them, is also used.

Mechanical fasteners used in the aerospace industry are rivets, pins, two-piece bolts and blind fasteners made of titanium, stainless steel and aluminum. Since these fasteners are very expensive, low-cost fasteners and rapid fastening techniques need to be developed if carbon fiber reinforced composites are to be used for automotive applications. The important fastener design considerations for carbon fiber reinforced composite laminates (Thoppul *et al.*, 2009) are as follows:

- Differences in thermal expansions and contractions of the fastener material and the composite, which can influence the clamping load
- The effect of drilling holes on the integrity of the hole as well as the material surrounding the hole, since drilling introduces damages in the form of delamination, fiber breaks, etc. in the composites at the edge and in the vicinity of the hole
- Water intrusion between the fastener and the composite, which can lead to water absorption in the composite and reduction in its properties
- Possibility of galvanic corrosion in the fastener

Adhesive joining in polymer matrix composites requires special attention to the joint design, adhesive selection and surface preparation. Even though single lap joints are the most common joints used in the automotive industry, they may not produce the highest joint strength in composites. Other joint designs, such as strap joints, stepped lap joints and scarf joints, may need to be considered (Fig. 8.14). Fiber orientation in the surface layers of the composite adjacent to the bond line is also an important design consideration. Three different types of adhesives are used with composites: epoxies, acrylics and urethanes. Epoxies are preferred if the composite is based on an epoxy matrix. For adhesive joining of composites in the automotive industry, a one-component adhesive and short curing time are preferred. The effects of the automotive environment on both short-term and long-term performance of the adhesive, as well as the adhesive joints, are important in selecting an adhesive for a polymer matrix composite (Erdman *et al.*, 2000).

8.7 Conclusion

Joining is a critical design and manufacturing issue in the development of lightweight vehicles. This chapter has broadly reviewed the most promising joining methods for high strength steels and aluminum alloys, some of which require further studies on their process capabilities and performance. Each of the joining methods has its advantages and limitations. A study (Briskham

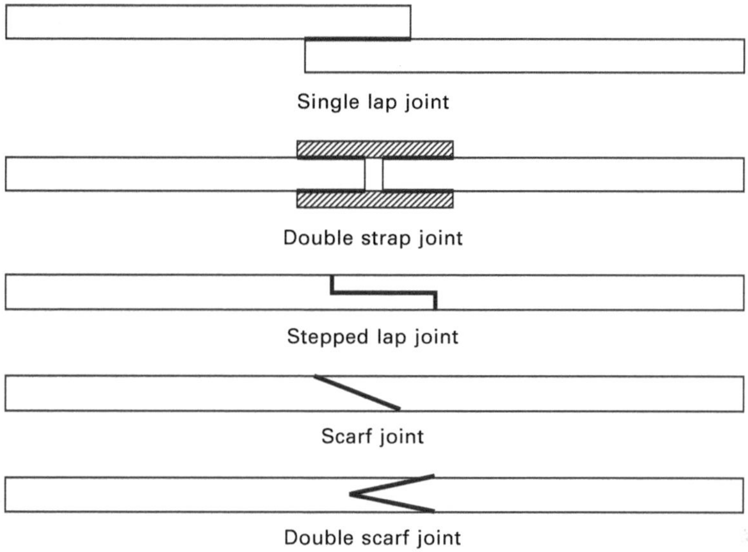

8.14 Various adhesive joint configurations used with polymer matrix composites.

et al., 2006) comparing the resistance spot welding (RSW), friction stir spot welding (FSSW) and self-piercing riveting (SPR) showed that, among these three spot joining methods between similar as well as dissimilar aluminum alloys, SPR produces the highest joint strength and energy absorption capability. It also has the best potential for joining dissimilar materials. Its main limitations are the running piece-cost of the rivets and the requirement for different rivet/die configurations as the sheet thickness is changed. Since SPR uses steel rivets, the end-of-life recycling of aluminum parts becomes more complex. RSW has the highest flexibility of making joints with varying sheet thickness. It also has the lowest consumables and investment costs; however, the energy consumption per joint is almost 10 times higher with RSW than with either SPR or FSSW. The weld consistency and electrode life are the main limitations of RSW, particularly with aluminum alloys. RSW is also more sensitive to the surface condition than either SPR or FSSW. The principal advantage of FSSW is its low energy consumption and low running costs. FSSW has similar process time to SPR and RSW if the total stack up thickness is 3.8 mm or lower, but it requires much longer process times for greater stack up thickness. It also tends to produce lower strength than either SPR or RSW. Out of the three types of joining methods, RSW, even with frequent electrode tip cleaning and maintenance schedule, is a more economically favorable option than either SPR or FSSW for aluminum body construction.

The assembly of lightweight vehicles will be accomplished with a combination of joining methods. No single joining method will be suited for all the joints, especially since the body construction, whether it is a body-in-white or a space frame, will contain not only mixed materials, such as aluminum and high strength steel, but also a variety of manufactured configurations, such as sheets, castings and extrusions. The suitability and efficiency of the joining methods will depend not only on an understanding of the optimum joining conditions and performance measures, but also on the design methodologies and analytical tools that can be used for determining the durability, crashworthiness and life-cycle costing of the produced joints. Significant research is needed in all of these areas for successful development of lightweight vehicles.

8.8 Acknowledgment

The author wishes to acknowledge the help of Dr. T.Y. Pan for providing him with some of the figures used in this chapter.

8.9 References

Abe Y, Kato T and Mori K (2006), 'Joinability of aluminum alloy with mild steel sheets by self-piercing rivet', *Journal of Materials Processing Technology*, 177, 417–421.

Aizawa T, Kashani M and Okigawa K (2007), 'Application of magnetic pulse welding for aluminum alloys and SPCC steel sheet joints', *Welding Journal*, 86, 119s–124s.

Allen C D and Arbegast W J (2005), 'Evaluation of friction spot welds in aluminum alloys', *2005 SAE World Congress*, Paper No. 2005-01-1252, Warrendale, PA, Soc. of Automotive Engineers.

Auhl J R and Patrick E P (1994), 'A fresh look at resistance spot welding of aluminum automotive components', *1994 SAE World Congress*, Paper No. 940160, Warrendale, PA, Soc. of Automotive Engineers.

Barnes T A and Pashby I R (2000a), 'Joining techniques for aluminum spaceframes used in automobiles. Part I – solid and liquid phase welding', *Journal of Materials Processing Technology*, 99, 62–71.

Barnes T A and Pashby I R (2000b), 'Joining techniques for aluminum spaceframes used in automobiles. Part II – adhesive bonding and mechanical fasteners', *Journal of Materials Processing Technology*, 99, 72–79.

Booth G S, Olivier C R, Westgate S A, Liebrecht F and Braunling S (2000), 'Self-piercing riveted joints and resistance spot welded joints in steel and aluminum', *2000 SAE World Congress*, Paper No. 2000-01-2681, Warrendale, PA, Soc. of Automotive Engineers.

Briskham P, Blundell N, Han L, Hewitt R, Young K and Boomer D (2006), 'Comparison of self-piercing riveting, resistance spot welding and spot friction joining for aluminum automotive sheet', *2006 SAE World Congress*, Paper No. 2006-01-0774, Warrendale, PA, Soc. of Automotive Engineers.

Cao X, Jahazi M, Immarigeon J P and Wallace W (2006), 'A review of laser welding techniques for magnesium alloys', *Journal of Materials Processing Technology*, 171, 188–204

Easton M, Beer A, Barnett M, Davies C, Dunlop G, Durandet Y, Blacket S, Hilditch T and Beggs P (2008), 'Magnesium alloy applications in automotive structures', *JOM*, 60, 57–62.

Elangovan, S, Semer S and Prakasan K (2009), 'Temperature and stress distribution in ultrasonic metal welding – an FEA-based study', *Journal of Materials Processing Technology*, 209, 1143–1150.

Erdman D, Battiste R, Boeman R and Klett L (2000), 'Characterization of a structural adhesive in automotive environments', *2000 SAE World Congress*, Paper No. 2000-01-1559, Warrendale, PA, Soc. of Automotive Engineers.

Feng Z, Santella M L, David S A, Steel R J, Packer S M, Pan T, Kuo M and Bhatnagar R S (2005), 'Friction stir spot welding of advanced high-strength steels – a feasibility study', *2005 SAE World Congress*, Paper No. 2005-01-1248, Warrendale, PA, Soc. of Automotive Engineers.

Fu M and Mallick P K (2001), 'Effect of process variables on the static and fatigue properties of self-piercing riveted joints in aluminum alloy 5754', *2001 SAE World Congress*, Paper No. 2001-01-0825, Warrendale, PA, Soc. of Automotive Engineers.

Haynes G and Jha B (1999), 'Joining aluminum to steel with transition material', *1999 SAE World Congress*, Paper No. 1999-01-0660, Warrendale, PA, Soc. of Automotive Engineers.

He X, Pearson I and Young K (2008), 'Self-pierce riveting for sheet materials: State of the art', *Journal of Materials Processing Technology*, 199, 27–36.

Jones T B and Williams N T (1986), 'The fatigue properties of spot welded, adhesive bonded and weldbonded joints in high strength steels', *1986 SAE World Congress*, Paper No. 860583, Warrendale, PA, Soc. of Automotive Engineers.

Krause A R and Chernenkoff R A (1995), ' A comparative study of the fatigue behavior of spot welded and mechanically fastened aluminum joints', *1995 SAE World Congress*, Paper No. 950710, Warrendale, PA, Soc. of Automotive Engineers.

Mallick P K and Agarwal L (2009), 'Fatigue of spot friction welded joints in Mg-Mg, Al-Al and Al-Mg alloys', *2009 SAE World Congress*, Paper No. 2009-01-0024, Warrendale, PA, Soc. of Automotive Engineers.

Mallick P K and Little R E (1994), 'Design of mechanically fastened (bolted) joints in automotive composites', *1994 SAE World Congress*, Paper No. 940623, Warrendale, PA, Soc. of Automotive Engineers.

Martukanitz R P, Altshuller B, Armao F G and Pickering E R (1996), 'Properties and characteristics of laser beam welds of automotive aluminum alloys', *1996 SAE World Congress*, Paper No. 960168, Warrendale, PA, Soc. of Automotive Engineers.

Miles M P, Decker B J and Nelson T W (2005), 'Formability and strength of friction-stir-welded aluminum sheets', *Metallurgical and Materials Transactions A*, 35A, 3461–3468.

Mishra R S and Ma Z Y (2005), 'Friction stir welding and processing', *Materials Science and Engineering*, R 50, 1–78.

Pan, T (2007), 'Friction Stir Spot Welding (FSSW) – A Literature Review', *2007 SAE World Congress*, Paper No.2007-01-1702, Warrendale, PA, Soc. of Automotive Engineers.

Patrick E P and Sharp M L (1992), 'Joining aluminum auto body structure', *1992 SAE World Congress*, Paper No. 920282, Warrendale, PA, Soc. of Automotive Engineers.

Saathoff D G and Mallick P K (1998), 'Static and fatigue strength evaluation of clinched joints in an aluminum alloy', *1998 SAE World Congress*, Paper No. 980693, Warrendale, PA, Soc. of Automotive Engineers.

Schroeder K J (1996), 'Structural adhesives for aluminum vehicles' *1996 SAE World Congress*, Paper No. 960166, Warrendale, PA, Soc. of Automotive Engineers.

Shribman V, Gafri O and Livshitz Y (2001), 'Magnetic pulse welding and joining – a new tool for the automotive industry', *2001 SAE World Congress*, Paper No. 2001-01-3408, Warrendale, PA, Soc. of Automotive Engineers.

Thoppul S D, Finegan J and Gibson R F (2009), 'Mechanics of mechanically fastened joints in polymer matrix composite structures – a review', *Composites Science and Technology*, 69, 301–329.

Tumuluru M D (2006), 'Resistance spot welding of coated high-strength dual-phase steels', *Welding Journal*, 85, 31–37.

Yan B, Lalam S H and Zhu H (2005), 'Performance evaluation of GMAW welds for four advanced high strength steels', *2005 SAE World Congress*, Paper No. 2005-01-0904, Warrendale, PA, Soc. of Automotive Engineers.

Zhao H, White D R and DebRoy T (1999), 'Current issues and problems in laser welding of automotive aluminium alloys', *International Materials Reviews*, 44, 238–266.

9
Recycling and life cycle issues for lightweight vehicles

S. DAS, Oak Ridge National Laboratory, USA

Abstract: This chapter addresses recycling and life cycle considerations related to the growing use of lightweight materials in vehicles. The chapter first addresses the benefit of a life cycle perspective in materials choice, and the role that recycling plays in reducing energy inputs and environmental impacts in a vehicle's life cycle. Some limitations of life cycle analysis and results of several vehicle- and fleet-level assessments are drawn from published studies. With emphasis on lightweight materials such as aluminum, magnesium, and polymer composites, the status of the existing recycling infrastructure and technological challenges being faced by the industry also are discussed.

Key words: recycling; life cycle; automotive lightweight materials; energy savings.

9.1 Introduction

The average weight of a North American-made light vehicle has steadily increased from 3694 lb in 1995 to 4017 lb in 2005 as demand for more vehicle power continued (WAG, 2007). Although the amount of conventional steel in vehicles remained constant at about 2100 lb, lightweight materials use has increased. For example, between 2000–2005, high and medium strength steel and aluminum content of a North American light-duty vehicle has increased by 83 lb (20%) and 48 lb (18%), respectively (WAG 2007). Fuel economy has remained almost the same, while vehicle weight and horsepower have increased by 11% and 32%, respectively (EPA, 2007). The use of major lightweight materials such as aluminum, magnesium, and plastics and plastic composites has increased in terms of weight by 37%, 150%, and 38%, respectively, during this period. Magnesium use is limited to about 10 lb/vehicle today, whereas the other two lightweight materials each account for about 330 lb of the vehicle. Each year, more than 50 million vehicles reach the end of their service life worldwide, 15 million of them in the United States. Vehicle composition will continue to evolve as more and more lightweight materials will be used in vehicles to meet the proposed 2007 Corporate Average Fuel Economy standards and because of pressure to lower vehicle fuel consumption due to higher fuel prices as seen lately in the market worldwide.

The North American post-consumer recycling industry has an inverted pyramid structure, consisting of less than 6000 scrap collecting and dismantling yards, about 200 scrap shredders, close to ten sink–float plants, and one metal sorter. Approximately 95% of vehicles retired from use each year are processed for recycling, in comparison to 99% of auto batteries, 35% of tires, and 45% of aluminum beverage cans; and at least 84% of a car's material content is recycled today. With the exception of plastic, rubber, glass, and other non-metallic materials that are discarded as automotive shredder residue in landfills, most other automotive materials are recycled. Specifically, more than 75% of automotive materials are profitably recycled via parts re-use, parts and components re-manufacturing, and ultimately by the scrap processing industry. However, only conventional steel has a recycling rate of 99% with closed loop recycling. As use of lightweight materials increases, there is concern that the percent of material recycled from each car will decline. This is in contrast to recycling mandates elsewhere. The European 2000 End-of-Life Vehicles Directive requires that, beginning in 2006, a minimum of 85% of vehicles are reused or recovered (including energy recovery) and at least 80% must be reused or recycled and, by 2015, 95% must be reused or recovered (including energy recovery) and 85% reused or recycled. Japan's Automotive Recycling Law has automotive shredder residue (ASR) targets requiring a 30%, 50%, and 70% decrease in landfilled ASR by 2005, 2010, and 2015, respectively.

Life cycle analysis (LCA) tools provide a holistic view of decisions related to the selection of automotive materials by considering energy and emissions impacts at each stage from cradle-to-grave of a product system. Recycling has the potential to reduce materials production energy consumption by 70–95%. The durability of metals — they are repeatedly recycled and maintain their properties in comparison to many hydrocarbon based materials — enhances their life cycle performance. When considering life cycle effects, recycling is critical to a sustainable future for materials. LCA allows the consideration of both the end-of-life material/vehicle life cycle stage and the reduced energy consumption at the vehicle use phase that in many cases offsets their higher energy use at the first material production stage. A discussion of recycling and life cycle issues related to lightweight materials starts with a brief introduction of LCA methodology, followed by a presentation of the existing automobile recycling infrastructure. A summary of LCA studies is provided both at the specific component level and at complete vehicle level, to illustrate the importance of recycling in the context of a life cycle analysis. Finally, trends and issues in lightweight materials, particularly aluminum, magnesium, and polymer composites, that the industry currently faces are discussed.

9.2 Life cycle analysis

The methodology most commonly used in life cycle assessment is based on ISO 14040/44 (ISO, 2006a and 2006b). LCA is a technique compiling a quantitative inventory of relevant inputs and outputs of a product system; evaluating the potential environmental impacts associated with those inputs and outputs; and interpreting the results of the inventory and impact phases in relation to the goal and scope of the study, as illustrated in Fig. 9.1 (ISO, 2006a). The goal and scope phase of LCA is acknowledged to be the most crucial part of the complete LCA for receiving valuable results, as is the selection of the functional unit under study consideration. Parametric modeling of the vehicle life cycle can also be done using this methodology and therefore it has similarities with previous studies such as Sullivan and Hu (1995).

Life cycle inventory assessment (LCI) is an emerging method for evaluating environmental sustainability. It quantifies all of the resources consumed and all of the emissions to our natural environment associated with an activity such as recycling or with a product. LCI provides a quantitative summary of energy, water, and resource consumption, all of the major wastes, water contamination, and air pollution associated with a product from its 'cradle' to its disposal, recovery, or recycling. Note that scrap collection and secondary smelting (i.e. metal recycling) is an essential part of any life cycle assessment.

Three broad considerations of particular importance in LCI analyses are the selection of data categories, attention to data quality, and definition of allocation procedures. Data categories need to be consistent with the goal

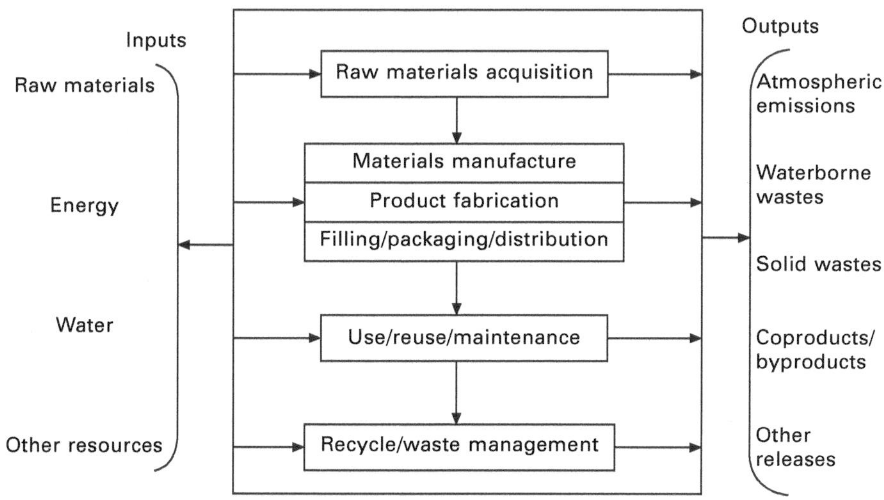

9.1 Lifecycle stages of a product life.

and scope of the LCI study. Precision, completeness, representativeness, and consistency are some of the data quality indicators generally used for data quality. ISO 14040 series of Standards on LCA has set out allocation procedures for reuse and recycling. Within these standards a distinction is also made between open and closed loop recycling. Open loop recycling is used to describe open loop product systems where the material is recycled into a new product or where the inherent material properties change. Closed loop recycling applies to products which are recycled to produce the same product or where the material properties do not change. Of the two different allocation procedures used for metal recycling, the 'attributional approach' considers the history of the product, the origin of the material (primary or secondary), and applies appropriate discounts at the end-of-life stage of the product under study (Geyer, 2006). It recommends avoiding allocation wherever possible through further division of unit processes or expansion of the product system. The end-of-life would then be limited mainly to the flow directly associated with the operation of shredders. With this approach, there is only a marginal benefit for the original user to recycle the metal at the end of life and also a high 'recycled content' of the product leads to better results in an LCA study.

The other approach, 'consequential approach,' supports the 'end-of-life' recycling by inclusion of process changes instead of inclusion of additional processes in a given state, and has been strongly supported by the metals industry (Atherton, 2007). It needs to account for any significant change in elementary flows, by studied change in the product system. This approach supports the consideration of end-of-life at the design stage and incentives are given to ensure that metal will be recycled at the end-of-life of the product after the use phase. The recycled metal in this case substitutes or displaces the necessity to mine, i.e. primary production. However, it is fairly usual to find LCA studies (for example, a study that focused on greenhouse gas impacts of competing automotive materials) that contain elements of both methodologies, such as attributional studies using consequential system expansion (Geyer, 2007). It is also important that the chosen allocation method employed in the sensitivity analysis of the results be explicitly assessed and communicated.

According to ISO, LCAs investigate product systems with regard to a broad variety of environmental impact categories, such as resource depletion, ozone depletion, acidification, eutrophication, photochemical oxidation, toxicity, and also climate change. This raises the possibility that changes in the product system decrease some environmental impacts but increase others, e.g. reducing ozone depletion but increasing climate change. Since there is no scientific method of reducing LCA results to a single overall score, such trade-offs cannot be assessed scientifically but need to be evaluated based on the relative societal importance of environmental impact categories, i.e.

value judgments. A modified life cycle analysis (MLCA) approach has been suggested, which provides a manageable set of metrics after reduction from the original LCI output, thereby providing a fairly broad perspective of the environmental performance of the product systems studied (Sullivan *et al.*, 1998a). Lack of a uniformly accepted method for weighting environmental impacts has caused a decline in LCA studies of environmental impacts.

Recent studies argue that since new product designs are introduced progressively in a fleet, only a fleet level analysis can account for the temporal effects of use and the distribution of energy over a vehicle lifetime (Field *et al.*, 2000). Since the product manufacture and use are distributed over long periods of time that are not simple linear combinations of single product life cycles, it is all the products in use over a period of time, rather than a single product, that are more appropriate for the life cycle analysis. This approach allows one to predict the time when benefits will be realized. A past study looking at the rolled products for body-in-white applications (in which aluminum provides the greatest weight saving potential compared to steel and ultra light steel autobody) found that it would take about four and ten years, respectively, for an aluminum vehicle to achieve a life cycle energy equivalence with a steel and an ultra light steel autobody on a single vehicle basis (Das, 2000). It would take twice as long to realize the benefits of aluminum at the fleet level.

9.3 Recycling

Recycling of materials is strongly associated with the economics of producing raw materials. Dalmijn and De Jong (2007) provide an excellent overview of the existing processing technologies in the recycling industry, although their emphasis is on Europe. For a system to be financially self-sustaining, the value of the recycled materials must exceed the costs of producing them. If closing a loop is preferred in view of sustainability, recycling will take place only if it can be sustained financially. The costs of recycling can be roughly divided into logistical costs and processing costs, the latter depending heavily on the complexity of the recycled object. The recycling cost of end-of-life vehicles has been estimated to be $199–$241/tonne in Europe compared to $108–$128/tonne in the United States (Dalmijin and De Jong, 2007). The higher cost in Europe is mainly due to higher landfilling costs, lesser economies of scale, and more processing to reach the 85% recycling target.

The new waste-reduction legislation in Europe has forced the recycling industry to develop new and economical recycling processes and the manufacturing industry to integrate recycling aspects into the design of their future products. The extensive use of polymers combined with the increasing complexity of passenger cars has led to a substantial decrease in the maximum recycling rate achievable with current separation technology. The recyclability of end-of-life vehicles has dropped from roughly 90% in 1955 to about 75% in 2005.

Figure 9.2 is a schematic representation of the current automobile recycling infrastructure. Recycling of end-of-life vehicles begins at dismantling, where spare parts and hazardous components (e.g. batteries, fuel, and oil) are removed. Subsequently, size reduction is carried out in car shredders to obtain appropriate liberation of the materials to be separated. The shredded material is first pre-concentrated by air classification to remove dust and light non-metallic materials, such as foam, textile, foil, and wood. This light fraction, known as automotive shredder residue (ASR), amounts to approximately 25 weight% of the shredder input and is mostly landfilled today. Further processing of ASR is being encouraged by ever increasing landfilling costs and stricter regulations such as by the European Commission (EC). In Europe, one of the potential recycling/recovery options is energy recovery. Stricter specifications for ASR used as a fuel require a process which combines an eddy current separator, metal sensor, and color and material sorters.

Steel, comprising 65% to 70% of the weight of the car, is the main constituent of the remaining material after air classification and is recovered by magnetic separation. Until recently, separated steel has been sufficient to produce saleable grades of secondary steel, but increasing copper content in the steel scrap has led to a costly hand sorting operation after magnetic separation. The non-magnetic fraction – 10% to 12% – of the shredded scrap, consists of various nonferrous metals and stainless steel as well as non-metals such as glass, stone/dirt, plastics, and rubber. This fraction is subsequently screened into several size fractions using rotary screens for separation since it is heterogeneous with respect to material composition, particle size, and shape. Separation of the non-ferrous mix starts with the removal of light non-metallic components, such as plastic, textile, and wood using the common and efficient eddy-current method. Subsequently, the resulting nonferrous pre-concentrate is subjected to a two-step sink–float separation in media having firstly a high density and then a lower density. Different densities of liquid media are used for preferentially separating out the lighter and heavier material constituents of the concentrate. At the lower liquid density, magnesium together with plastics and rubber are removed. At the higher liquid density, the aluminum fraction is separated from the heavy non-ferrous metals. The aluminum fraction contains 10–15% of similar density materials, including glass, stone, and copper wires, which can be further concentrated by eddy current separation to achieve a purity of 99% and a recovery rate of over 98%. The float fraction obtained after these separations is sometimes subjected to a further sink–float step to separate unliberated aluminum pieces. The sink product of this fraction is a hard-to-separate heavy non-ferrous mix, the significant portion of which is shipped

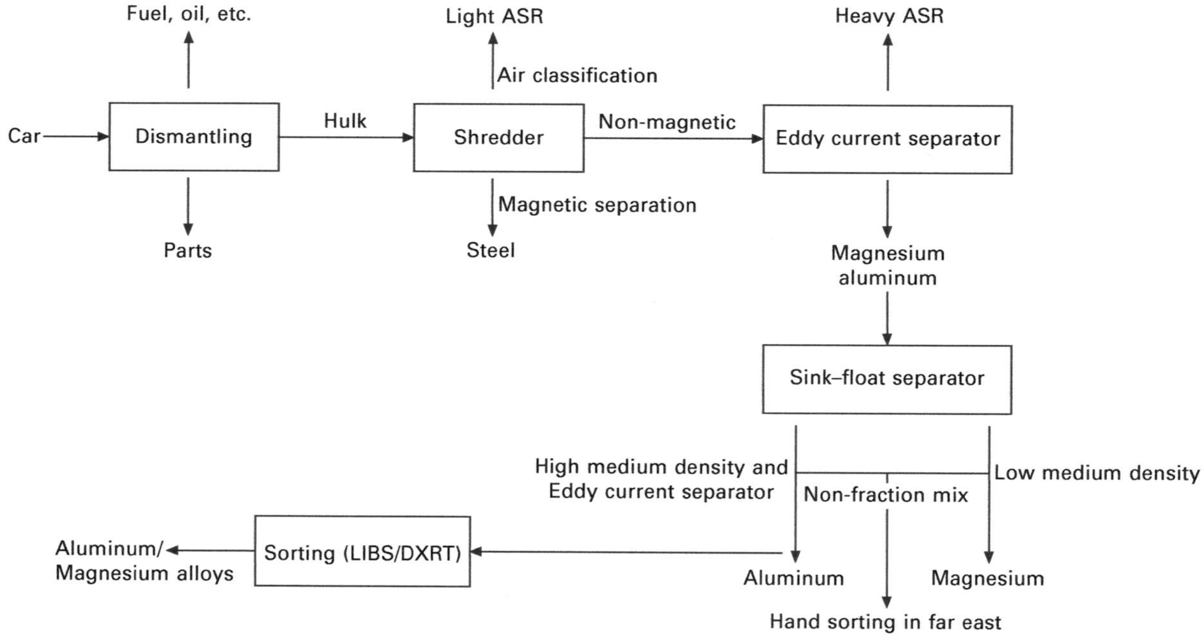

9.2 Schematic representation of a recycling infrastructure.

to China and India where it is hand sorted. Low labor costs make hand sorting cost-effective and have slowed the introduction there of sophisticated arrangements of eddy current separators and automated sorting devices.

The recycling and recovery of materials from automobiles will increase particularly in Europe, with the European Commission legislation (EC) for product responsibility. The EC recycling target for end-of-life vehicles in 2015 is 95% by weight, and 10% maximum is allowed for energy recovery. There is a need to develop optimized sorting methods to improve recycling. A combination of bulk processing, dry sorting, data processing, intermediate products, and new sensors would lead to more complete recovery of materials. An example where processes have been combined for optimum recovery is dry sorting for aluminum alloys. The process starts with an eddy current separator, which concentrates the oversize fraction. The rejects of the eddy current separator can be processed by a metal sensor to separate the metals from the nonmetals to improve recovery. The ~95% metal concentrate from the eddy current separator is processed with a differential X-ray transmission (DXRT) sorter to make an aluminum concentrate of about 99%. This aluminum concentrate from the DXRT sorter can be further processed with laser-induced breakdown spectroscopy (LIBS), developed by Huron Valley Steel Corporation, to recover different aluminum alloys. Most technologies available for upgrading of the light-metal scrap are characterized by high productivity, low processing costs, and negligible energy requirements.

The scrap value of a vehicle at end-of-life has increased as its light-metal content has increased. Light metals offer other cost advantages: lower recycling processing costs when compared to primary metal reductions (e.g. about 5% for aluminum) and lower melting temperature that make remelting costs lower than for conventional steel. Recycling, therefore, significantly favors the substitution of light metals for steel in lightweighting applications. However, the price of light-metals scrap is dictated by the price of the prime metals it displaces and composition requirements in specific applications. Recyclability of light metals plays an important role in the economic viability of current integrated material recycling systems. If newer, lighter vehicles require more aluminum, aluminum's higher value will be a positive impact on the recycling industry, given that cost-effective technologies are developed to sort different aluminum alloys (Das, 1999).

A sustainable light-metals recycling system requires (i) relatively pure prime alloys to act as diluents for the recycled scrap; (ii) a few large-volume secondary alloys with high recycled content to provide a sufficient market for all the recycled scrap; (iii) a low-cost technology system for managing the alloying elements in scrap and using these elements to alloy new metals (e.g. hand-sorting in Asia); and (iv) unrestricted flow of metal scrap between market segments and geographical locations (Gesing, 2004). Besides low wages and lack of environmental regulations, high custom duties on prime metals, elimination of value added tax for recyclers in China, and fixed exchange rates allow scrap to flow to Asia at higher prices. The export of scrap to Asia has caused a serious shortage of scrap needed for the domestic US and Europe light-metals industries, thereby raising metal prices. Recycled metal has been growing in importance in satisfying the global demand for aluminum metal. Although transportation demand has contributed to the growth in aluminum consumption, automobiles' share of in-use aluminum represents only a small fraction of total use today. It is anticipated with the increased trend of aluminum use in the transportation sector, that a significant amount of aluminum (e.g. an additional 1.5 million tonnes contained in recovered scrap in 2009) would be available for other applications, for which scrap sorting would be important (Gesing, 2004).

9.4 Importance of recycling in the context of life cycle analysis

The major stages of 'life' included in the life cycle analysis of automotive lightweight materials are material extraction, part production, use (or operation), and end-of-life stages. Life cycle assessments with the consideration of 'cradle to grave' aspects of a material used in automobiles have been instrumental in providing a more holistic view of energy consumption, considering energy consumption beyond the materials production stage and demonstrating a significant potential for energy savings. A detailed and complete examination of the life cycle energy consumption of a generic family US sedan, weighing 1500 kg and having 120 000 lifetime miles, estimates total energy use to be about 974 GJ. This corresponds to the consumption of 46 000 lb of hydrocarbons (coal, oil, and natural gas) and the generation of 130 000 lb of CO_2 (Sullivan *et al.*, 1998b). Figure 9.3 shows life cycle energy impacts of an automotive liftgate inner, where comparison has been made between cast aluminum and conventional stamped steel materials (Das, 2005).

The use phase has the biggest impact on energy consumption, contributing up to 84% of the total energy consumption. Material production accounts for the second largest share of energy. The part production and end-of-life life cycle stages consume considerably less energy than the first and third stages and have only sparse data about them available in the literature today. Reducing a vehicle's weight by using lightweight materials is shown to significantly reduce the life cycle energy of a vehicle because most energy is used during the vehicle's use phase.

Life cycle analysis has shown that high energy consumption during production of virgin aluminum and magnesium is offset by the reduced energy consumption in the use phase, bringing the 'break-even point' to within the lifetime of the car (Sivertsen *et al.*, 2003). When the break-even point is reached depends on the amount of lightweight materials included,

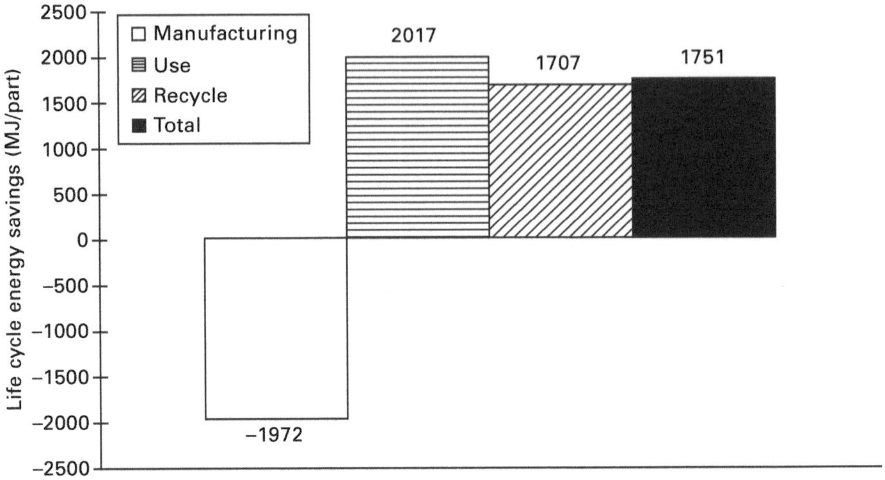

9.3 Life cycle energy savings of a cast aluminum liftgate inner.

the amount of recycled material used for production of the part, and the system and conditions assumed in the study.

The transportation necessary within each of the phases has often been included in past studies, but additional work is needed to address the vehicle assembly stage and secondary weight reduction due to primary lightweight material substitution effects. The life cycle energy benefit in switching to a lighter weight material is not just a function of weight saved; it is also dependent on material production energies and substitution factors and vehicle operational efficiency. For example, a radical use of carbon fiber composites and aluminum (i.e. replacement of all steel and iron with carbon fiber composites and aluminum, respectively, resulting in overall 40% vehicle weight savings) is predicted to yield a life cycle energy saving of 16% for a conventional mid-size vehicle (Sullivan and Hu, 1995). Similarly, the life-cycle energy impacts of different body-in-white materials indicate that aluminum substitution would result in a total life cycle energy saving of 52 GJ/vehicle as compared to steel, and 22 GJ/vehicle as compared to the advanced high strength steel (Das, 2000).

Table 9.1 shows the primary- and secondary-production energy of various materials used in automobiles today. There is considerable variation – as much as 40% – in reported values of material production energies, depending on the estimation methodology used and assumptions about production technologies and their efficiency. The primary production energy is the sum of all energies required to extract raw materials from the earth (e.g. mine ore or pump oil) and to process (wash, concentrate, or refine) it into a usable form (ingot or rolled sheet). Much of this material-production energy is avoided by recycling, although getting a recycled material of a comparable purity to virgin stock

Table 9.1 Primary and secondary production energy for automotive materials

	Energy (MJ/kg)	
Material	Primary	Secondary
Steel	30	9
Cast iron	34	24
Wrought aluminum	201	29
Cast aluminum	189	26
Cast magnesium	284	27
Reinforced plastic	56	37
Unreinforced plastic	79	14
Copper	140	35
Zinc	53	16

Source: Das (2004)

is often not readily achieved due to the difficulty in removal of impurities. The recycled content is an important factor in the overall life cycle analysis because less than 15% of the energy used for production of virgin material is needed for remelting aluminum and magnesium, compared to 30% in the case of steel, as shown in Table 9.1. Energy used in recycling operations is minimal: 35–50 kWh per tonne for segregating ferrous and non-ferrous metals and about 5 kWh per tonne of shredder residue for separating the polyolefins using the wet separation system (Pomykala *et al.*, 2007). These authors also estimate a recovery rate of more than 75 kg of ferrous and non-ferrous metals and about 60 kg of polyolefins per tonne of shredder residue. Assuming energy savings per kg of metal based on cast aluminum and unreinforced plastic from Table 9.1 as representative for ferrous and non-ferrous metals and polyolefins, respectively, total energy savings are then estimated to be about 16 GJ per tonne of shredder residue. The use of recycled materials improves life cycle energy consumption considerably, e.g. a 7.5% life cycle energy reduction for a particular aluminum substitution becomes 12% when the part is 50% recycled aluminum (Sullivan and Hu, 1995). Curlee *et al.* (1994) estimate that, with a 100% recycling rate of ferrous and non-ferrous metals, energy savings for the total potential number of vehicles recycled in 2000 would have been around 326 petajoule (PJ) – about 2% of total highway transportation energy consumption by light-duty vehicles during that year.

Life cycle energy comparisons on a single product basis are quite sensitive to the underlying assumptions made for major input parameters. For example, a comparative assessment of material substitution examining automatic transmission cases made from die-cast magnesium and aluminum found aluminum use more favorable, despite magnesium's significantly greater weight reduction potential. This result is assumed to be due to the different recycling rates for the two materials (Reppe *et al.*, 1998). On the other hand, magnesium substitution ranging from 10–30% in automotive parts such as gearbox housings and dashboards indicated a breakeven point around 40 000 miles (Sano *et al.*, 1999). The magnesium substitution can save more life cycle energy than the aluminum substitution, but recycling of magnesium parts is indispensable for efficient CO_2 reduction (Hakamada *et al.*, 2007). CO_2 emissions during new ingot production of magnesium are much higher than those of conventional steel and aluminum, resulting in part from the use of sulfur hexafluoride in cases when used as the cover gas, it having 22 800 kg equivalent CO_2 emissions per kg during melting.

Several life cycle analyses on a single product basis indicate that, although aluminum production is about ten times more energy intensive than steel production, aluminum in automotive applications is less energy intensive during its life cycle than conventional steel. Since vehicle use dominates overall life cycle energy use, 'lightweighting' with aluminum increases fuel efficiency enough to reduce overall energy use. If aluminum-intensive light-duty vehicles had been commercialized on a mass scale by 2005, the reported national petroleum energy savings would have been quite significant, about 4% and 5% by 2020 and 2030, respectively (Stodolsky *et al.*, 1995). A study on future post consumer aluminum scrap availability impacting the energy intensity of the industry and life cycle analyses of aluminum products indicates supply adequacy in terms of the industry as a whole for the foreseeable future, but the combination of transportation and container/packaging sector are adequate until around 2010, although the development of separation technologies for different alloys (discussed in Section 9.2.4) would be critical (Choate and Green, 2004). Curlee *et al.* (1994) estimated the life cycle energy implications per vehicle for the changing vehicle composition for the period 1976–2000. Energy requirements during vehicle manufacturing have increased, due primarily to the increased contents of plastics and aluminum, whereas substantial energy savings have occurred because of improved fuel economy, and to a lesser extent, energy savings at the recycle stage. The decline in vehicle life cycle energy requirements has been 66%, and the trend is anticipated to continue in the future if the greater use of lightweight materials continues. The impacts of lightweight materials on non-conventional vehicles such as electric and hybrid vehicles are comparatively less due to the use of highly energy efficient powerplants that reduce energy consumption in the vehicle use stage, but on the other hand lightweight materials help in improving the cost-effectiveness of advanced powerplants by lowering power requirements of a vehicle with equivalent performance (Aluminum Association, 2008).

For non-energy impact categories, the operational phase is similarly dominant; for example, generating 87% of the life cycle CO_2. Thus, reducing fuel consumption in vehicle operation has the greatest effect in producing sustainable transportation from a greenhouse gas point of view. The material production phase dominates solid waste impacts in the life cycle analysis. The relative contribution of the end-of-life phase is 5% or less for most environmental impact categories, if credits are not allocated to this phase (Schmidt *et al.*, 2004). Solid waste, the only burden category where the end of life phase shows a significant contribution to the life cycle, is about 7% of the total waste production. Of the LCA impact types, energy consumption and CO_2 emissions are the most sensitive to vehicle weight during the vehicle use phase; solid waste and SOx emissions are most sensitive during the material production and vehicle manufacturing stages. Maintenance and repair stages of the life cycle are least sensitive to vehicle weight (Sullivan and Cobas-Flores, 2001).

9.5 Trends and issues in lightweight materials recycling

9.5.1 Aluminum

Recycling of aluminum is important because it can save almost 90% of the energy to produce virgin aluminum, as shown earlier in Table 9.1. With the decrease in US production of primary aluminum, recycled aluminum has become an increasingly important component of metal supply for which effective and efficient technology is needed. Aluminum recycling in sectors such as transportation and construction is about 95% in North America, compared to only about 45% for recovered beverage cans. The reclamation of aluminum scrap is a complex interactive process involving collection centers, remelt facilities, metal processors, and consumers. There are three major scrap streams of aluminum, viz. used beverage cans (UBCs), automotive scrap, and municipal scrap. The alloy compatibility of the components of the beverage can makes it uniquely suitable for the closed-loop recycling concept and is responsible for the consistently high value of UBCs; nevertheless, its recycling rate is considerably lower than the transportation and construction sectors today. Unlike the can recyclers, who are totally dedicated to a single product of two compatible aluminum alloys, automotive recyclers must deal with a number of different alloys with different destinations and relatively low values. The volume of recycled aluminum coming from the transportation sector – of which 80% was derived from automotive components – exceeded the recycled metal coming from UBCs for the first time in 2005 (Choate and Green, 2004). As discussed in Section 9.2 and seen in Figs. 9.2 and 9.4 (to be discussed), aluminum recycling occurs at several sequential recycling steps, viz. eddy current separator, sink-float separator, and sorting, with the more efficient alloy separation taking place at the latter stages. The aluminum automotive recycling world would be ideal if recycling of all discarded vehicles were standard (happening in Europe today to an increasing extent

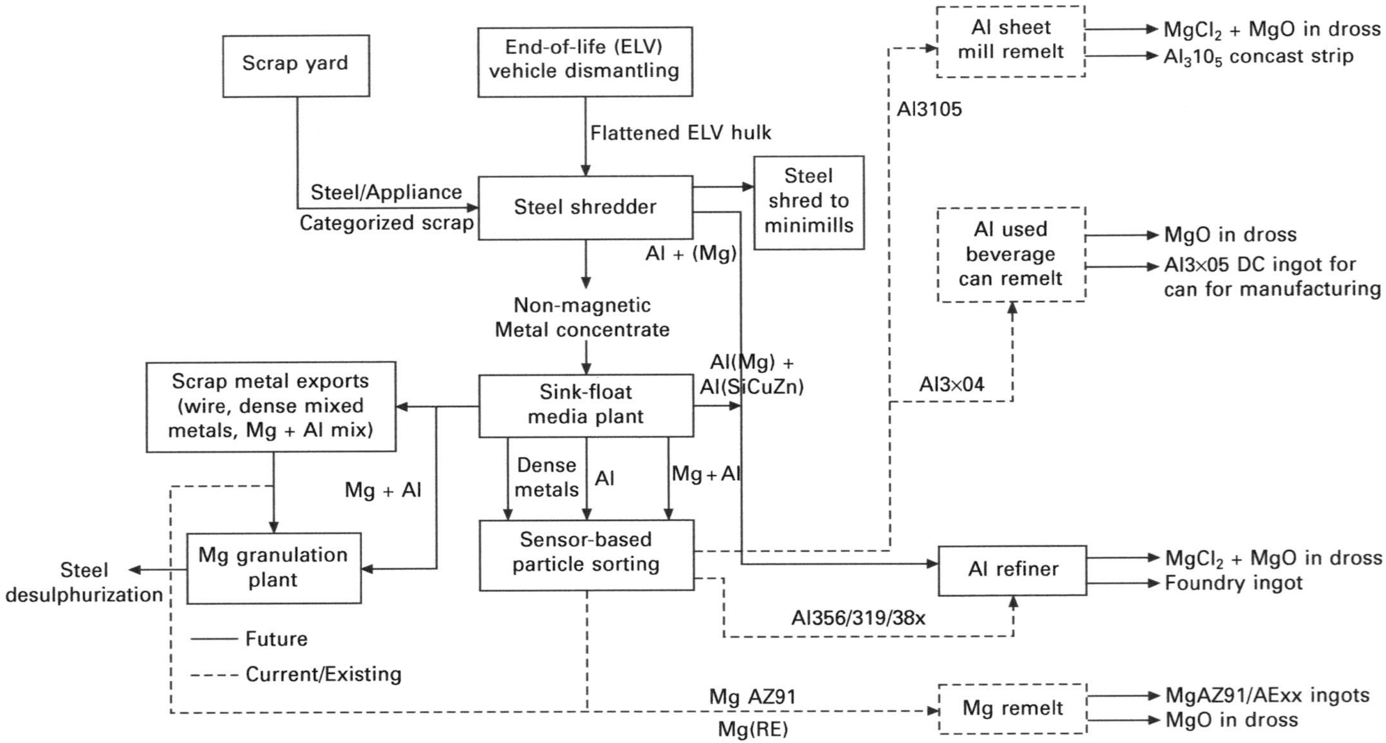

9.4 Detailed schematic representation of light metals recycling infrastructure.

through vehicle recycling centers and take back vehicle programs); obvious aluminum components, such as bumpers, wheels, and body panels, would be regularly disassembled and retained separately (by alloy class such as 5xxx, 6xxx, etc.) or remelted to the degree practical; the remainder of the vehicle is put through an automated shredding where sorting technology, such as laser-induced breakdown spectroscopy, would be subsequently applied. Under these circumstances, no extra purification steps would be required before the aluminum is recycled into high-performance products. This would allow development of new alloy options for potential direct use in vehicle components, either cast or wrought.

The two key challenges in optimizing the recycling of automotive aluminum alloys and products are controlling the dismantling and presorting processes to maximize opportunities to control metal stream composition, and developing new alloys or modifications of existing alloys that come directly from the metal stream. 7xxx alloys used for bumpers greatly complicate the reuse of a melt containing the wrought alloys 2xxx, 5xxx, or 6xxx used for body panels and structural elements. Cast alloy scrap differs significantly from wrought alloy scrap, notably with a higher total alloy content, higher silicon content, and, depending on which cast alloys are involved, higher copper or zinc content. The exceptionally high silicon content of the castings works against mixing castings with sheet or extrusion alloys. As a group, however, castings may be rather readily reused as castings. Limited applications suitable for direct production from the recycled metal exist today, particularly for wrought alloys, due to strict composition requirements for the alloying elements silicon and iron, as well as other impurities such as magnesium, nickel, and vanadium, which are difficult to control in recycled metal and tend to increase modestly the more often the metal is recycled. Even segregated wrought scrap can have relatively widely varying compositions due to the use of different alloys for wrought applications; for example, 7xxx and 2xxx alloys used for body panels and bumper applications contain higher copper and zinc, respectively. Compositions resulting from mixed wrought and cast scrap are the most difficult to use directly, except perhaps in some casting alloys.

Resource maximization in aluminum recycling would be improved with the development of new aluminum alloys and better recovery and sorting technologies. New aluminum alloys that fit recycled metal streams and do not require any further post-processing for reuse (similar to the recyclability of beverage cans today) would offer significant benefits. Directly reusing recycled secondary metal reduces post-remelting process costs and limits the need to 'sweeten' the metal with more costly and energy-intensive primary metal in order to meet the performance requirements of many alloys and product specifications before it is used to produce new products. Otherwise, recycled metal tends to be used primarily for lower-grade casting alloys and products that are typically the most tolerant to the presence of undesirable alloying elements. Presorting cast from wrought components is easier (because of the volume of cast components used today), but reuse of cast components perhaps is the largest challenge. It has been suggested that new recycle-friendly automotive alloys can be developed under four major categories, viz. body sheet alloys, structural elements, bumpers, and cast alloys, if proper presorting/pre-shred dismantling of components can be done (Das *et al.*, 2007).

Considerable attention has been devoted to improving the recovery of used aluminum components for recycling and the effectiveness of automated shredding and sorting technologies. These efforts have been led primarily by the Aluminum Association and metal recovery businesses such as Huron Valley Corporation. Development of one or two 'unialloys' (alloys that meet all of the requirements for a large number of automotive components) has proven to be difficult because of the diverging performance requirements of different automotive applications. There is much to be gained by adopting disassembly and presorting technique that provides separate alloy and/or metal pools. Elemental analysis sorting technologies such as laser beam sensors (LBS) or X-ray fluorescence (XRF) sensor-based particle sensors need to be developed. As shown in Fig. 9.4, sensor-based particle sorting would allow sorting of aluminum and magnesium alloys, whereby these could be cost-effectively used in three major aluminum markets, viz. continuous strip casting, direct casting of ingots for can manufacturing, and foundry ingots.

9.5.2 Magnesium

Magnesium recycling has become more important because of the increasing use of the metal in the automotive die casting market. There are eight main alloying elements used in the production of various magnesium alloys, aluminum being the most widely used because of its favorable effect on casting. The family of Mg–AZ alloys containing aluminum, zinc, and manganese dominate the die casting market, while other alloy families such as AM series (Mg–Al–Mn), AS series (Mg–Al–Si) and ZK series (Mg–Zn–Zr), are less commonly used. Today, only high-grade magnesium scrap is recycled from die casting operations, using either flux or fluxless refining. This is because of the inability of current refining technologies to adequately clean the variety of scrap produced during die-casting. It becomes uneconomical due to the extensive pretreatment, chemical adjustment, and refining procedures required to convert low-grade magnesium scrap into a useful product.

The challenge of enabling the recycling of post consumer magnesium alloys involves the need for a collection system, a recycling system, and a financial profit incentive. Until recently, there was not sufficient Mg alloy content in the metal concentrate shredder product to warrant the separation of the Mg as a separate scrap product, nor was there demand for such a product. Figure 9.4 shows the existing as well as future plausible options (denoted by solid and dotted lines, respectively) for the recycling of automotive magnesium based on the latest assessment of life cycle scenarios for magnesium automotive components (Gesing and Gesing, 2007). Post consumer magnesium separation begins when end-of-life vehicles enter as flattened hulks into a steel shredder plant. Using magnetic separation at the shredder facility, the pieces undergo several sorting stages to separate the magnetic from the nonmagnetic metal shredder fraction residue (NMSF) that contains primarily magnesium and aluminum. The NMSF is about 22% (by weight) of the input feed and is further refined by using several air-separation steps. The remaining NMSF is sent to the third and final sorting step at the shredder facility, eddy current separation, from which a product containing more than 95% aluminum is derived. The remainder is sent to sink–float plants. Since Mg has a response very similar to that of aluminum in the eddy current separator, a substantial portion of Mg input to the shredder ends up in this aluminum concentrate rather than as NMSF bound for sink–float plants. The aluminum concentrate obtained at this step is either exported for hand-sorting in Asia or sold to secondary aluminum smelters for foundry alloy production, where the Mg is refined out of the melt by chlorination and ends up in the aluminum dross as $MgCl_2$.

The non-magnetic, electrically conducting metal particles are sold to wet sink–float plants, where flotation and eddy current separation methods are used to produce clean aluminum shred products as well as light and dense mixed-metal products. The light-metal mix contains virtually all the magnesium that comes to the sink–float plant and is recovered mostly in the magnetite-float fraction. Both light and dense metal mixes consisting of ~75% aluminum beverage cans and extrusions and ~25% magnesium castings, are almost exclusively sold today for export to Asia for hand-sorting by parent metal category, where the sorted Mg alloys are generally used for steel desulfurization. Using different specific gravities for liquid media in a multi-stage process, different density fractions of non-magnetic materials are separated at each stage of the sink–float process. The last fraction obtained from the sink–float separation process can also be processed through an eddy current separator to separate the magnesium and hollow form aluminum from the other non-metals.

Today's limited recycling of post consumer magnesium either by hand-sorting in Asia or lost as a dross during recycling of the aluminum concentrate as foundry alloys, can be expanded to three other potential markets, with the separation of magnesium alloy particles from the aluminum scrap bound for refining into foundry alloys. These markets are in aluminum can alloys or as a master alloy for magnesium alloying element addition to other aluminum based alloys, additives for steel and nodular iron production, and magnesium die casting alloys in non-critical applications such as housings for portable electronics, or automotive interior components such as instrument panels. With the anticipated growth in these markets, there is very little additional life cycle advantage to further tightening the recycling loop and directing the post-consumer magnesium scrap back into production of magnesium alloys or sorting these alloys to closed-loop alloys or product cycles. Both Mg–AM and Mg–AZ alloys are compatible in composition with aluminum can alloys; thereby magnesium scrap obtained by dense media sink–float separation can be used for satisfying magnesium alloying requirements of the aluminum can sheet. As long as expansion in the combination of these three markets outpaces the growth of supply of old, post consumer magnesium alloy scrap, there will be no magnesium metal recycling system need to include old scrap content in property-critical applications in automotive, marine, aerospace, and defense applications. In addition, AZ91 diecasting alloy most commonly used today in automotive applications has the required combination of alloying elements to become a consumer of magnesium mixed alloy scrap for non-critical applications. It appears that the optimum near-term strategy would be to direct as much mixed alloy magnesium scrap as can be accommodated to secondary versions of magnesium diecasting alloys and to granulate the rest for steel desulfurization, both having ready markets in China and India where hand sorting of exported magnesium scrap will likely continue to occur.

There will be little incentive in the near-term to sort alloy and upgrade scrap for property-demanding applications until there is a considerable proportion of high-performance alloys of magnesium (Mg–Re) containing elements such as calcium, strontium, and rare earths in the old scrap mix. At present, the value of Mg–Re alloys is more in the aerospace and defense industries than in the automotive industry. The incentive for the separation of these expensive exotic metals would be economic, rather than due to impacts on the energy consumption or greenhouse gas emissions. For the auto industry before a Mg–Re scrap recycle loop is established, an alloy that is composition-tolerant to a mix of Mg–Re alloying elements needs to be developed which could be used at first for an improved replacement for an Mg–AM or Mg–AZ diecasting alloy component, and perhaps later as a casting for a powertrain housing. For this elemental analysis, sorting technology such as laser-induced breakdown spectroscopy (LBS) or X-ray fluorescence (XRF) sensor based particle sensors need to be developed (Gesing and Gesing 2007).

9.5.3 Polymer composites

Polymers are generally more difficult to separate from waste streams than metals because differences in their properties are not as significant as for metals.

Addition of plasticizers, fire retardants, antioxidants, modifiers, and fillers to polymers for actual use causes overlap in their properties, although they have significantly different properties when in their pure states. The difficulty is further compounded by the fact that most polymers are not compatible with each other and, therefore, must be separated from each other.

Pomykala *et al.* (2007) provide an excellent overview of today's status of the recycling of polymer composites. The energy saved by recycling polymers depends on several factors, including the type and number of materials in the waste stream, separation process used, degree of incompatibility among the different polymers, and the presence of abrasive materials such as metals or glass that must be removed prior to recycling the plastics because they increase the wear and tear on the processing equipment. Variations in shapes, sizes, and surface roughness caused during the shredding and granulation processes affect the settling velocities of the different particles thereby causing difficulty during the separation processes used.

Most polymer composites are not recycled today; they are usually discarded as automotive shredder residue (ASR) in landfills (Fig. 9.2). Treated ASR is one of the items approved by EPA as a municipal landfill cover throughout the United States, this use being proven to be more cost effective than energy recovery (Gesing, 2004). Certain types of plastics are more difficult to recycle than others. For example, the class of plastics known as thermosets cannot be melted; therefore, these plastics cannot be remolded into new products. Although thermoplastics, the other main class of polymers used in automobiles today, can be melted down and reformed into new products, they cannot, as a general rule, be mixed across type. One common method for separating plastics is a sink–float technique based on density differences. It is mainly effective for separating simple mixtures that contain a few components, as well as for washing the plastics while they are being separated. Numerous other techniques are used that include manual dismantling, surface modification/froth flotation, selective dissolution, electrostatic separation, selective stickiness (thermal, infrared, or solvent), softening/centrifugation, color sorting, and infrared/ultraviolet/X-ray, laser identification/separation. Each technique has its own advantages and limitations. For example, manual dismantling with pre-identification of the polymers can result in purer products and very few by-products. Automotive parts marking with a universal code that identifies the specific resin contained in the part has been suggested to plastics parts suppliers. However, it is labor intensive and costly. Argonne National Laboratory has developed a process for separating plastics from end-of-life durable products (Pomykala *et al.*, 2007). The process consists of two sections. The first section is a dry mechanical separation process for separating the polymers from the highly dense non-polymers as well as large bulky materials such as flexible polyurethane foam pieces in the waste stream. It also includes shredders and granulators to reduce per-piece size, trommels or screens to separate materials by size or shape, magnets to remove ferrous metals, an eddy current separator to separate the nonferrous metals, and air classifiers to remove light fabrics and fibers from polymer concentrate. This process has achieved over 95% recovery of the ferrous and nonferrous metals (mainly aluminum) in the waste stream and over 90% of the polymers as concentrates from shredder residue.

The second section is a wet froth flotation section to separate the different plastics from the mixed polymer concentrate. It consists of two parts, the first employing a conventional gravity separation system based on density differences. It is used to fractionate the polymer concentrate. The second part is a froth flotation process, in which materials having similar or overlapping densities are separated in a series of tanks having the appropriate surface tension, specific gravity, and pH. By this method, the targeted plastic particles selectively float or sink.

Another technical approach to plastics recycling has been the retrieval of ASR's energy content through combustion. This approach, which raises environmental questions, would retrieve energy from the high-Btu stream and also reduce the weight and volume of residue that would require disposal. However, incineration of waste with energy recovery is important since up to 10% of end-of-life vehicles may be subject to energy recovery under the European End-of-Life Vehicles directive.

Most thermoset composite materials used today are recycled by two categories of process: those that involve mechanical communition techniques to reduce the size of the scrap to produce recyclates and those that use thermal processes to break the scrap down into materials and energy. Mixtures of polymer, fiber, and filler obtained after mechanical recycling can be used as a substitute for calcium carbonate filler in new sheet molding compound or bulk molding compound. The thermal recycling processes include fluidized bed processing, pyrolysis, and combustion with energy recovery. These processes have the advantage of being able to tolerate more contaminated scrap materials and produces fiber products, although more development work is needed before these fiber products, can be reused cost-effectively.

Jody *et al.* (2004) have developed a one-step thermal treatment method under different environments by which carbon fibers can be separated from polymer matrix composite materials made with thermoset substrates at residence times on the order of minutes, depending on the treatment temperature and on the substrate material. Initial results indicate that the recovered carbon fibers had properties that compare favorably with those of virgin carbon fibers produced from polyacrylonitrile, and this self-sufficient energy process can have a potential payback of less than two years.

Curlee and Das (1996) have examined the viability of recycling plastics by tertiary processes (i.e. manufacturing of monomers, basic chemicals, or fuels) as an alternative to secondary recycling (i.e. mechanical recycling), quaternary recycling (i.e. incineration with heat recovery) and disposal by landfilling. Tertiary recycling processes can tolerate different levels of impurities depending on the technology used, and they allow closed-loop recycling in the sense of either reducing the polymer to a monomer from which new polymers can be produced or producing more basic chemicals from which new polymers can be manufactured. From an energy balance perspective, tertiary recycling appears to hold no particular advantage. In addition, the depolymerization process used under tertiary recycling holds no particular advantage over secondary recycling in terms of cost, conservation of materials, and emissions or damages, since both recycling processes displace virgin polymers.

Recovery and recycling technology for plastics from end-of-life vehicles thus continues to receive attention from industry, government, and academia. Although several technical solutions are in sight, an effective recycling infrastructure does not currently exist for collecting, identifying, sorting, and processing polymers and composites. With the greater plastics content in future automobiles, recapturing the energy and material content of polymers for use in re-manufacturing will necessitate significant improvements in mechanical, chemical, and thermal processing technologies. Public perception and regulations remain a barrier to the implementation of clean energy and fuel recovery technologies. Instead of focusing on recycling alone, life cycle assessment of resource use and conservation provides a truer picture and the means to achieve environmental sustainability.

9.6 Conclusions

Life cycle analysis offers valuable contributions to a holistic view on decisions for the selection of an automotive materials tool. Using the LCA tools effectively can be challenging due to the complexity of parameters and assumptions that must be made. LCA does show that increased recycling of lightweight materials is an area that can reduce energy, emissions, and waste impacts. Open-loop recycling seems to be the most appropriate model for end-of-life scenarios, considering today's availability of technology. Trying to set up a system that monitors closed-loop recycling within any portion of the current integrated material recycling system is unlikely to add value or result in any more recycling. However, increasing dismantling, reducing the materials used, and separating alloys before re-melting will facilitate closed-loop recycling. Ultimately, the economic viability of recycling will be determined by the balance between its cost and the value dictated by the price volatility of virgin materials and energy inputs that recycled materials compete with.

9.7 References

Aluminum Association (2008), 'Aluminum vehicle structure: Manufacturing and lifecycle cost analysis: Hybrid drive and diesel fuel vehicles,' *Research Report No. 2008-05*, written by Ibis Associates, Inc. for the Aluminum Association.

Atherton J (2007), 'Declaration by the metals industry on recycling principles,' *Int. J. Life Cycle Assessment*, 12 (1), 59–60.

Choate W T, and Green J A S (2004), 'Modeling the impact of secondary recovery (recycling) on U.S. aluminum supply and nominal energy requirements,' *Light Metals*, Proceedings of the 133rd TMS Annual Meeting, held in Charlotte, North Carolina, Mar. 14–18, 913–918.

Curlee T R, Das S, Rizy C G and Schexnayder S M (1994), 'Recent trends in automobile recycling: An energy and economic assessment,' *ORNL/TM-12628*, Oak Ridge National Laboratory, Oak Ridge, TN.

Curlee T R and Das S (1996), *Back to Basic? The Viability of Recycling Plastics by Tertiary Processes*, Working Papers – Program on Solid Policy, School of Forestry and Environmental Studies, Yale University, New Haven, CT.

Dalmijn W L and De Jong T P R (2007), 'The development of vehicle recycling in Europe: Sorting, shredding, and separation,' *Journal of Metals*, 59(11), 52–56, Nov.

Das S (1999), 'Recycling of new generation of vehicles,' *SAE Paper No. 1999-01-0673*, Society of Automobile Engineers, Warrendale, PA.

Das S (2000), The life cycle impacts of aluminum body-in-white automotive material, *Journal of Metals*, 52 (8), 41–44.

Das S (2004), 'Material use in automobiles', *Encyclopedia of Energy*, Vol. 3, Elsevier Publishing Company, Inc., pp 859–869.

Das S (2005), 'Life cycle energy impacts of automotive liftgate inner,' *Resources, Conservation and Recycling*, 43, 375–390.

Das S, Green J A S and Kaufman J G (2007), 'The development of recycle-friendly automotive aluminum alloys,' *Journal of Metals*, 59 (11), 47–51.

Environmental Protection Agency (EPA) (2007), 'Light-duty Automotive Technology and Fuel Economy Trends: 1975 through 2007,' *Report No. EPA420-S-07-001*, September.

Field F, Kirchain R and Clark J (2000), 'Life cycle assessment and temporal distributions of emissions: Developing a fleet-based analysis,' *Journal of Industrial Ecology*, 4(2), 71–91.

Gesing A. (2004), 'Assuring the continued recycling of light metal in end-of-life vehicles: A global perspective,' *Journal of Metals,* 56 (7), 18–27.

Gesing A J and Gesing R (2007), 'Life Cycle Scenarios for Mg Automotive Components,' *Consulting Report# GC70916*, prepared for Natural Resources Canada, Ottawa, Ontario by Gesing Consultants *Inc., Tecumseh, Ontario, Canada*, Nov. 28.

Geyer R (2006), 'The impact of material choice in vehicle design on the life cycle greenhouse gas emissions,' presented at the Intl. LCA/LCM 2006 Conference, Oct. 4–6, held in Washington, DC.

Geyer R (2007), 'Life cycle greenhouse gas emission assessments of automotive materials – the example of mild steel, advanced high strength steel, and aluminum in body in white application: Methodology report,' report prepared for the *World Auto Steel,* International Iron and Steel Institute, Brussels, Belgium, Dec.

Hakamada M, Furuta T, Chino Y, Chen Y, Kusuda H and Mabuchi M (2007), 'Life cycle inventory study on magnesium alloy substitution in vehicles,' *Energy*, 32, 1352–1360.

ISO (2006a), ISO 14040: Environmental management – life cycle assessment – principles and framework, *ISO/FDIS 14040:2006(E)*, ISO, Geneva, Switzerland.

ISO (2006b), ISO 14044: Environmental management – life cycle assessment – requirements and guidelines, *ISO/FDIS 14044:2006(E)*, ISO, Geneva, Switzerland.

Jody B J, Pomykala Jr. J A, Daniels E J and Greminger J L (2004), 'A process to recover carbon fibers from polymer-matrix composites in end-of-life vehicles,' *Journal of Metals*, 56 (8) 43–47.

Pomykala Jr. J A, Jody BJ, Daniels E J and Spangenberger J S (2007), 'Automotive recycling in the United States: Energy conservation and environmental benefits,' *Journal of Metals*, 59 (11) 41–45.

Reppe, G, Keoleian G, Messick R and Costic M (1998), 'Life cycle assessment of a transmission case: Magnesium vs. aluminum,' *SAE Paper No. 980470*, Society of Automotive Engineers, Warrendale, PA.

Sano T, Winter A,. Saiki T De, Horiskoshi S, Fuchiazawa S and Sado S (1999), 'Reduction of environmental impact attained by magnesium alloys for automotive components,' *Proceedings of the 6th International Conference on Technology of Plasticity*, held in Nuremberg, Sept. 19–24, edited by M. Geiger.

Schmidt W P, Dahlqvist E, Finkbeiner M, Krinke S, Lazzari S, Oschmann D, Pichon S and Thiel C (2004), 'Life cycle assessment of lightweight and end-of-life scenarios for generic compact class passenger vehicles,' *Intl. Journal of LCA*, 9 (6) 405–416.

Sivertsen L K, Haagensen J O and Albright D (2003), 'A review of the life cycle environmental performance of automotive magnesium,' *SAE Paper No. 2003-01-0641*, Society of Automotive Engineers, Warrendale, PA, USA.

Stodolsky F, Vyas A, Cuenca R and Gaines L (1995), 'Life cycle energy savings potential from aluminum-intensive vehicles,' Proceedings of 1995 Total Life Cycle Conference, *SAE Paper No. 951837*, Society of Automotive Engineers, Warrendale, PA, USA, p. 47–57.

Sullivan J L and Cobas-Flores E (2001), 'Full vehicle LCAs: A review,' *SAE Paper No. 2001-01-3725*, Society of Automotive Engineers, Warrendale, PA.

Sullivan J L, Costic M M and Han W (1998a), 'Automotive life cycle assessment: overview, metrics, and examples,' *SAE Paper No. 980467*, Society of Automotive Engineers, Warrendale, PA.

Sullivan J L, and Hu J (1995), 'Life cycle energy analysis for automobiles,' *SAE paper No. 951829*, Society of Automotive Engineers, Warrendale, PA.

Sullivan J L, Williams R L, Yester S, Cobas-Flores E, Chubbs S T, Hentges S G and Pomper S D (1998b), 'Life cycle inventory of a generic U.S. family sedan: Overview of results USCAR AMP project,' Proceedings of 1998 Total Life Cycle Conference, *SAE Paper No. 982160*, Society of Automotive Engineers, Warrendale, PA, USA, p. 1–14.

Ward's Automotive Group (WAG) (2007), *Ward's Motor Vehicle Facts and Figures 2007*, Southfield, MI.

10
Crashworthiness design issues for lightweight vehicles

A. DEB, Indian Institute of Science, India

Abstract: The present chapter highlights considerations in vehicle crash safety design using lightweight materials when compared with conventional steel. Key concepts are first introduced in terms of general requirements of structural crashworthiness and occupant safety. Insight is provided into how steel can be substituted with aluminum as the material for a structural component at a substantial weight reduction without compromising functionality. Crashworthiness evaluation requires consideration of a full vehicle body assembly, and CAE (computer-aided engineering) is an indispensable tool for assessing and guiding design. Thus, salient features in the front impact safety design of an aluminum spaceframe vehicle are shown using LS-DYNA, a commercial finite element-based explicit nonlinear dynamic analysis package. A number of other relevant lightweight materials and design solutions for impact energy absorption are discussed.

Key words: crashworthiness; aluminum; fiber reinforced composites; foam; plastics.

10.1 Introduction

The content of lightweight materials such as plastics and aluminum in automobiles, when compared with extensively used steel alloys, has been steadily increasing. It is estimated that polymers in an average car comprise at least 10% of its total weight and applications of such materials in vehicles continue to rise.[1] Similarly, the usage of aluminum in vehicles has increased by 80% in recent times.[2] While vehicle designers are inclined to consider lighter-than-ferrous materials, they have to be mindful of the aggressive multi-attribute targets that today's vehicles need to meet. An area that is of particular concern is ensuring vehicle crashworthiness and occupant safety in design. As compared with many other vehicle attributes in which design standards are of a subjective nature and are often dictated by competition, vehicle crash safety standards are relatively well defined and enforced through regulations. In the present exposition, behaviors of a range of lightweight materials are considered under impact loading conditions at both component and system (i.e. vehicle body structure) levels. In modern vehicle design, CAE (computer-aided engineering) tools are extensively used. This is more so in the case of vehicle crashworthiness design, as enormous savings of time and cost in product development can result from the usage of finite element-based crash analysis tools such as LS-DYNA. As design is evolutionary, it is expected that proven concepts and safety countermeasures will play an important role in deciding an initial geometric design of a vehicle, also influenced by attributes such as styling, vehicle dynamics, NVH (noise, vibration and harshness), durability, etc. However, the assessment of a detailed design, especially in terms of occupant safety, can only be obtained through elaborate computational mechanics such as explicit nonlinear finite element analysis. Thus, emphasis is placed on illustrating how CAE techniques can be applied to the design of automobile bodies using lightweight materials such as aluminum alloys and fibre-reinforced composites.

10.2 Background of vehicle crash safety

Vehicle crash safety can be broadly divided in two areas, viz. active and passive safety. Active safety deals with countermeasures incorporated in vehicles such as ABS (Antilock Braking System), traction control, electronic stability control, rollover-detection sensors, active suspension systems, forward collision warning system, reverse backup sensors, emergency brake assist, lane departure warning system, etc., which can aid in avoiding crashes. Passive safety, on the other hand, is involved with designing a crashworthy vehicle structure along with restraint systems such as seat belt, airbag, etc. for protecting the lives of occupants in the event of a crash. The passive safety of vehicles is assessed through laboratory crash tests generally conforming to regulatory requirements such as various FMVSS (Federal Motor Vehicle Safety Standards) in the USA. The current trend is also to design vehicles for more demanding and publicized consumer information standards such as NCAP (New Car Assessment Program) in the USA or EuroNCAP in Europe, which are aimed at assigning ratings to new vehicles for crash safety performance. Additionally, risk to pedestrians due to collisions with motor vehicles is a major concern in densely populated areas.[3] Perhaps usage of lightweight materials combined with generous packaging space in the front structure of a vehicle can lead to improved performance in protecting pedestrians against severe injuries to head and lower limbs in low speed crashes. It may be noted that loss of vehicle occupant lives results primarily from front and side impact, and roll-over crashes.

The main objective of vehicle crash safety design is to manage impact kinetic energy in such a way that sufficient energy is absorbed by the vehicle structure, resulting in acceptable levels of deceleration and intrusion (i.e. encroachment of permanently deformed vehicle body structure into the passenger compartment). For a vehicle with two belted front dummies crashing head-on against a rigid barrier, as in the US regulatory front impact test (FMVSS 208) or the NCAP test shown in Fig. 10.1, a simple relation can be written expressing absorbed energy (E) in terms of mean impact

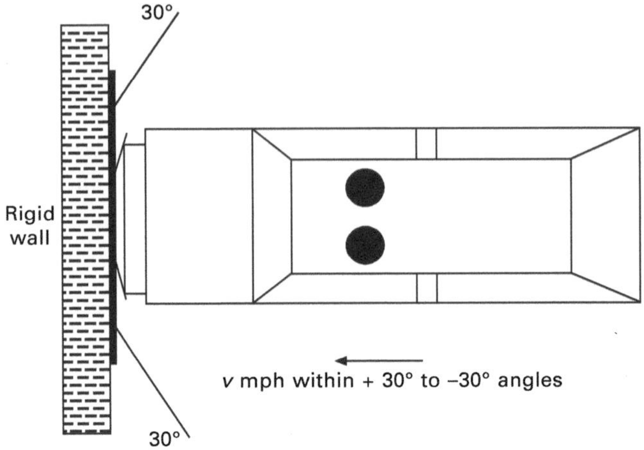

10.1 Schematic of vehicle front impact safety test according to FMVSS 208 (V = 30 mph) or NCAP (V = 35 mph) procedure.

force (*P*), vehicle mass (*M*), average vehicle deceleration (*a*), and maximum plastic deformation (*d*) as follows:

$$E = P \times d = (Ma) \times d \qquad [10.1]$$

In obtaining the mean force, it is ensured that the rectangular area in Fig. 10.2 is the same as the shaded area representing the energy absorbed by the impacting vehicle. For a given value of *E*, it is detrimental for passenger safety for either *a* or *d* to be unacceptably high. If *a* is high, the occupants are likely to be thrown inside a vehicle, sustaining severe to fatal injuries by striking against the interior surfaces of the vehicle; on the other hand, a high value of *d* can result in collapse of the passenger compartment of the vehicle and occupants can be subject to harmful impact by its intruding interior body parts. Thus, *a* should be maintained at a level such that chances of secondary impacts for occupants will be minimized; the safety of occupants can be further improved with passive restraint systems such as seat belts and airbags. While *a* is maintained at a desirable level, Eq. [10.1] implies that *d* should be of sufficient magnitude for the target for energy absorption (i.e. *E*) to be met. Thus, though the structural integrity of the passenger compartment should be maintained to avoid any catastrophic collapse, adequate crush zones should be provided so that desirable level of plastic deformation (i.e. *d*) can be reached by keeping the occupants out of harm's way. It needs to be mentioned here that there is a limit to which an ordinary vehicle can be designed for crash safety to make it viable for both the manufacturer and the buyer; in other words, if *E* goes beyond a range expected for common accidents (as is ensured by regulatory and new vehicle safety ranking tests), either *a* or *d* or both can reach levels leading to disastrous consequences for occupants.

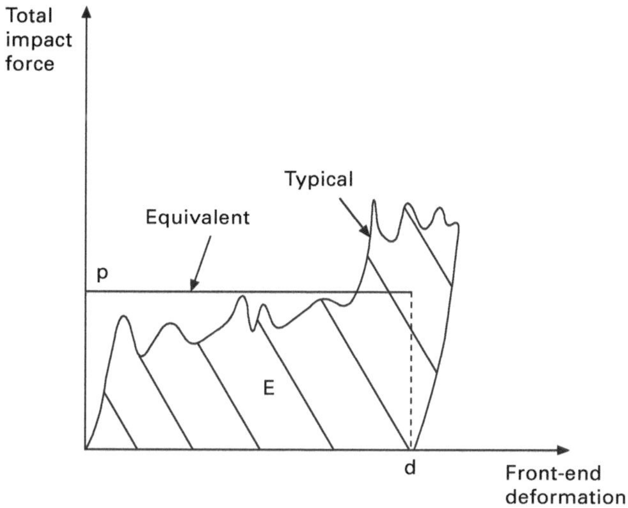

10.2 Typical and equivalent force–deformation variations in a vehicle front impact test.

Despite the overall goals of vehicle safety design being to absorb impact energy to the maximum extent while maintaining deceleration and intrusion within desirable levels, the indexes for vehicle safety assessment are primarily in terms of injury parameters such as HIC (head injury criterion), chest g (peak chest acceleration in g), chest compression, TTI (thoracic trauma index), VC (viscous criterion), pubic symphysis force, femur load, combined force and moment criterion for neck, etc., measured on anthropomorphic devices such as instrumented dummies. For front impact, it has been shown,[4] with the aid of linear regression analysis of actual crash test data, that major injury parameters such as chest g and HIC are directly dependent on the mean value of the structural crash pulse generated in a full-width frontal impact test against a rigid barrier. For an NCAP test speed of 35 mph (56 kph), a 5-star rating (based on combined values of HIC and chest g) was found to statistically correlate to an average vehicle deceleration of 23g; and the corresponding correlation for a 4-star rating was 27g.[4] In the case of 40% frontal offset EuroNCAP tests, the damage to the front structure of a vehicle (on the impacted side) is usually more severe, while vehicle peak deceleration is lower as compared to NCAP tests. Thus there is a strong likelihood of occupant injury parameters being affected largely by vehicle intrusion in front offset impact tests.

10.3 Designing for crashworthiness with lightweight materials

The discussion in the foregoing section establishes the broad objectives of vehicle safety design in terms of limiting mean deceleration and intrusion, and the implications of structural performance on occupant safety. A review of safety ratings of passenger vehicles based on NCAP tests reveals that a vast majority of them have achieved two 4-stars (one 4-star each for driver and front co-passenger), or above. The part of a vehicle that plays a significant role in absorbing impact energy is its body. Most of the current vehicles are made of steel. A lightweight metal that has been used primarily in selected low-volume passenger cars is aluminum. However, aluminum while being one-third the weight of steel for a given volume, is also three times more compliant than steel and possesses lower tensile strength. The main question then arises: how can an aluminum-intensive vehicle be equivalent to its steel counterpart in terms of structural performance under mechanical loads and yet retain its weight advantage?

An insight into the design strategy that can ensure the potential weight savings associated with aluminum without compromising its mechanical abilities when compared to steel can be obtained by considering the hollow cantilever beam of length l shown in Fig. 10.3. The deflection, δ, at the free end of the cantilever in Fig. 10.3 is given by the following familiar relation, based on classical beam theory:

$$\delta = \frac{Pl^3}{3EI} \qquad [10.2]$$

In Eq. [10.2], E is the elastic modulus and I is the moment of the hollow beam cross-section shown in Fig. 10.3 given by the following relation:

$$I = \frac{bd^3}{12} - \frac{(b-2t)(d-2t)^3}{12} \qquad [10.3]$$

The cantilever in Fig. 10.3 is considered to be made of either steel or aluminum with its length and applied load remaining unchanged; in the former case, it is assumed that $E = E_{St}$ and $I = I_{St}$, and in the latter case, $E = E_{Al}$ and

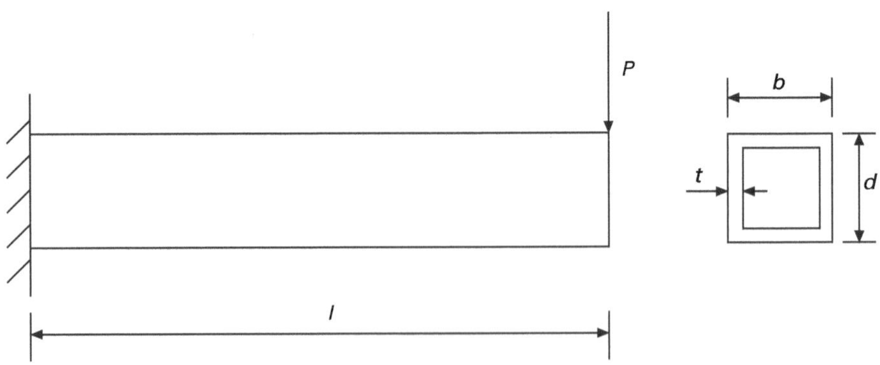

10.3 A cantilever beam subjected to a point load, P.

$I = I_{Al}$. Then for the steel and aluminum cantilever beams to undergo the same free-end deflection, it is required from Eq. [10.2] that

$$E_{St}\, I_{St} = E_{Al}\, I_{Al} \qquad [10.4]$$

Steel is roughly three times as stiff as aluminum, i.e. $E_{St} = 3E_{Al}$; it therefore follows from Eq. [10.4] that

$$I_{Al} = 3I_{St} \qquad [10.5]$$

Assuming the steel cantilever section dimensions as $b = d = 50$ mm and $t = 2$ mm, I_{St} from Eq. [10.3] is obtained as 147 712 mm^4. In order to have an aluminum beam with a sectional moment of inertia (about the trace of neutral plane) as three times this value according to Eq. [10.5], the dimensions of the original beam section need to be increased. However, the most straightforward design change would be to increase the thickness (t) of the section such that the following equality is satisfied:

$$\frac{bd^3}{12} - \frac{(b-2t)(d-2t)^3}{12} = \frac{50^4}{12} - \frac{(50-2t)(50-2t)^3}{12} = 3 \times 147712 \qquad [10.6]$$

Solution of the above equation yields $t = 9.463$ mm.

If the density of steel is denoted as ρ_{St}, the weight of the steel cantilever would be:

$$W_{St} = \rho_{St}[bd - (b-2t)(d-2t)]l = 384\ \rho_{St}l \text{ (in consistent units)} \qquad [10.7]$$

In a similar manner, assuming the density of aluminum as one-third of that of steel, and using the thickness of aluminum section calculated above (i.e. $t = 9.463$ mm), the weight of the aluminum cantilever would be:

$$W_{Al} = \frac{\rho_{St}}{3}[bd - (b-2t)(d-2t)]l = 511.5\ \rho_{St}l \text{ (in consistent units)} \qquad [10.8]$$

Comparing Eqs. [10.7] and [10.8], it can be concluded that, if only thickness is modified, an aluminum cantilever of equivalent bending stiffness as a steel cantilever will actually be heavier than the latter!

It may be noted that requiring the constancy of bending stiffness also implies that the fundamental natural frequency and Eulerian buckling load are the same for the steel and aluminum cantilevers, apart from the condition of equivalence of free-end deflection which initially led to Eq. [10.4]. An alternative strategy for modifying the section of the steel cantilever would be to increase its sectional depth, d, while maintaining its thickness, t, to its original value of 2 mm. In this case, the equivalence of Eq. [10.6] would be:

$$\frac{50d^3}{12} - \frac{46(d-4)^3}{12} = 3 \times 147712 \qquad [10.9]$$

The real solution of Eq. [10.9] corresponds to d = 79.45 mm. Using this value of sectional depth, the weight of the aluminum cantilever is recalculated as:

$$\begin{aligned} W_{Al} &= \frac{\rho_{st}}{3}[50 \times 79.45 - 46 \times 75.45]l \\ &= 167.27\ \rho_{st} l \text{ (in consistent units)} \end{aligned} \qquad [10.10]$$

Comparing Eqs. [10.7] and [10.10], the aluminum cantilever is seen to be now indeed lighter with a weight saving of as much as 56.4% with respect to the steel cantilever.

An important consideration in designing a structural component when it is subjected to a large bending moment due to a transverse load or an eccentric compressive load is its ultimate moment capacity, which corresponds to the formation of a fully plastic hinge at a section. The behavior of common metals, such as steel and aluminum, under tensile loading is often idealized with a bilinear true stress versus true strain curve, as shown in Fig. 10.4. In this figure, σ_y represents the yield point, i.e. the end of elastic behavior characterized by the Young's modulus, E, and the onset of plasticity; the plastic deformation is associated with strain hardening with a tangent modulus, E_l, culminating in failure at the tensile strength of σ_u and a breaking strain of σ_f. Assuming that the material of the cantilever in Fig. 10.1 behaves as in Fig. 10.4, its ultimate sectional moment capacity can be shown to be:

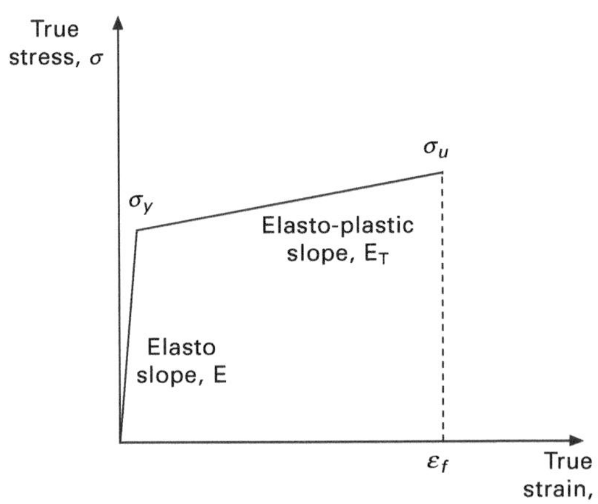

10.4 Idealized bilinear stress–strain behavior of common metals.

$$M_p = \sigma_u t\left[b(d-t) + 2\left(\frac{d}{2} - T\right)^2\right] \qquad [10.11]$$

It is noted that sometimes the stress–strain behavior under uniaxial tensile loading is assumed conservatively as elastic perfectly plastic. In such a situation, E_T in Fig. 10.4 is set to zero and σ_u in Eq. [10.11] needs to be replaced by σ_y.

Incorporating the sectional properties of the steel cantilever considered earlier on the right side of Eq. [10.11], its ultimate moment capacity can be obtained as:

$$M_p^{St} = 6916\sigma_u^{St} \qquad [10.12]$$

where σ_u^{St} represents the tensile strength of steel of a given grade. By using the revised depth (i.e. 79.45 mm) of an aluminum cantilever and other relevant sectional dimensions in Eq. [10.11], the ultimate moment capacity of the aluminum cantilever will be:

$$M_p^{Al} = 13\,438\sigma_u^{Al} \qquad [10.13]$$

where σ_u^{Al} represents the tensile strength of aluminum of a given grade. If mild steel is assumed, with a typical value of σ_u^{St} being 0.3 GPa, and σ_u^{Al} is assigned a value of 0.17 GPa for an aluminum extrusion of grade SAPA 6060 T6, the following values of ultimate moment capacity are obtained from Eqs. [10.12] and [10.13]:

$$M_p^{St} = 2074.8 \text{ kN mm} \qquad [10.14]$$

$$M_p^{Al} = 2284.5 \text{ kN mm} \qquad [10.15]$$

Clearly, from relations [10.14] and [10.15], the ultimate moment capacity of an aluminum cantilever can be comparable to that of a steel cantilever while being substantially lighter in weight.

The above results establish that merely replacing steel body parts with aluminum parts of the same geometry but increased thickness is not likely to give rise to a weight-efficient design. On the other hand, appropriate changes in geometric configuration offer a high potential for an aluminum-intensive design that retains the advantages of the lighter weight of aluminum and provides comparable structural performance to a steel-bodied vehicle. In the automotive industry, aluminum-intensive cars are relatively scarce owing primarily to the higher cost of the material and the joining of parts associated with aluminum; nevertheless some of the successful aluminum bodied vehicles, such as Audi A8 (Fig. 10.5[5]) and Audi A2 (Fig. 10.6[5]) can be seen to be based on a spaceframe architecture unlike the vast majority of steel bodied cars which are made of sheet metal panels with a unibody (i.e. unitized or integral body, or monocoque) design (as in Fig. 10.7). A spaceframe design

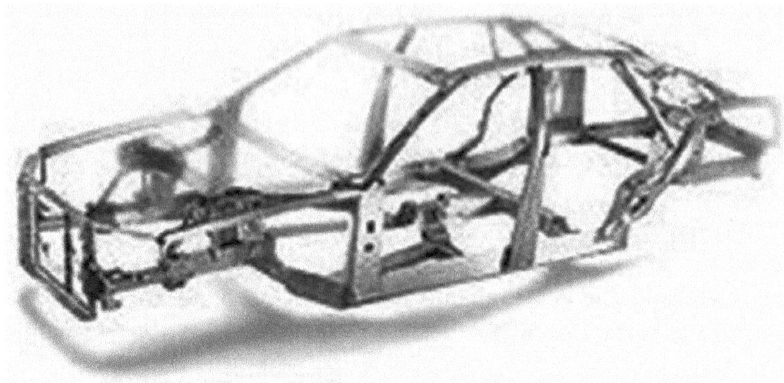

10.5 Audi A8 aluminum spaceframe body-in-white (BIW).

10.6 Audi A2 aluminum spaceframe BIW.

10.7 BIW of a unibody vehicle in which roof and floor panels are integral with the body.

consists of a three-dimensional grid of components (predominantly extrusions and castings) with welded joints and is an inherently rigid construction when compared with a unibody design in which thin sheet metal parts formed by stamping are joined along flanges with spot welds located at regular intervals. A spaceframe design does not have discrete spot welds (which are a source of discontinuity between mating parts) and the joints are structurally robust, contributing to the stiffness of the frame. However, the attachment of body panels to the frame requires extra effort and free-form styling can be difficult to achieve. Large volume production tends to be cheaper and efficient for unibody vehicles while spaceframe-based body-in-white (BIW) may be more suitable for low to medium volume production, with less initial investment on tooling as compared to stamping of a large number of body panels.

The Audi A8 spaceframe BIW consists of a combination of aluminum extrusions, stampings and castings which are joined using robotic laser welding. The Audi A2 has a predominance of extrusions, especially in the underbody construction. The Lotus Elise, a compact sporty convertible, relies heavily on aluminum extrusions in its spaceframe BIW and uses adhesively bonded joints with rivets. The Jaguar XJ is an exception in terms of aluminum-intensive body construction as it is based on a unibody design in which stamped aluminum sheets are joined along their flanges with adhesive bonding and self-piercing rivets. However, no data are available on the US-NCAP or EuroNCAP safety assessment of this vehicle. It may be noted that the 2000 model year Audi A8 received a double 5-star (i.e. the highest) US-NCAP safety rating.

Due to global stiffness requirements imposed by attributes such as NVH (noise, vibration and harshness) and vehicle dynamics, or due to constraints of packaging (such as an S-shaped bend in conventional unibody front rails), components of vehicle BIW that are important for crash energy absorption can become either too stiff or exhibit undesirable collapse modes such as bending. Such situations can sometimes be remedied by introducing deliberate weaknesses (such as holes, indents, etc.), also called 'triggers', in a component at desirable locations to initiate progressive folding of the component and maximize impact energy absorption.

10.4 Crash safety design using computer-aided engineering (CAE)

The crash safety design of a vehicle for modes such as front, side and rear impact, as well as roll-over, requires performance evaluation of the entire body assembly. Due to the complexity of BIW designs, nonlinear geometric and material behavior of body components with contact forces under impact loading can be best analyzed by explicit finite element codes such as LS-DYNA. Such numerical procedures, including data exchange and collaboration between various product development groups, are generally referred to as Computer-Aided Engineering (CAE). A CAD model of a vehicle BIW is usually the starting point for development of a detailed finite element model. The behavior of components such as extruded aluminum tubes are, however, important in gaining insight into failure mechanisms such as dynamic progressive buckling and global buckling, sizing of members for initial design of vehicle body structure, and development of reliable modeling and analysis procedures by comparing against experimental component crush data. A brief example is given here of the CAE-based design procedure of a small aluminum-intensive electric prototype vehicle with a spaceframe made exclusively from extruded members. A CAD model of the vehicle BIW is shown in Fig. 10.8.[5] The corresponding spaceframe finite element model is given in[6] and a more complete version of the same is shown in Fig. 10.9. The material properties given in Table 10.1 for SAPA 6060 T6 aluminum alloy extrusions were incorporated in an elasto-plastic constitutive model for analysis using LS-DYNA.

The BIW design in Fig. 10.8 assumes that the extruded members are connected at joints with the aid of MIG (metal inert gas) welding. This thermal welding procedure involves the use of an aluminum filler material that is heated to join two aluminum edges in the presence of an inert gas such as argon. The HAZ (heat affected zone) surrounding a weld line is generally weaker as compared to the parent parts that are joined. In the finite element analysis-based assessment of a spaceframe body, there may

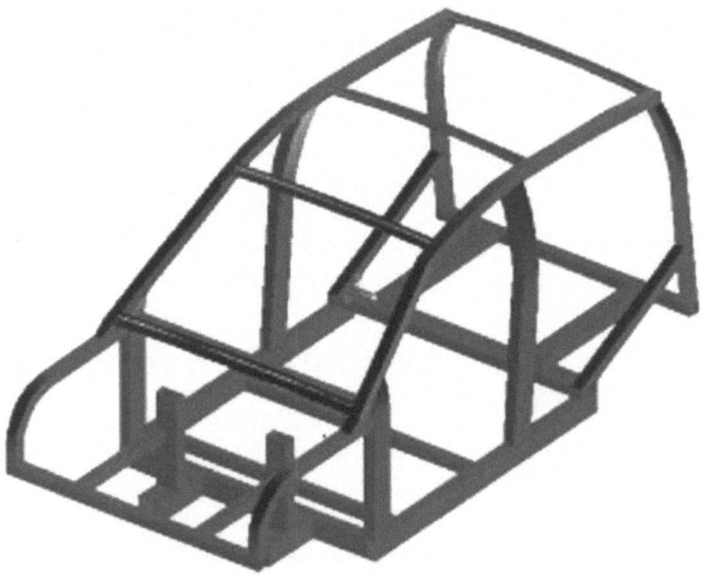

10.8 CAD model of aluminum extrusion-based spaceframe concept.

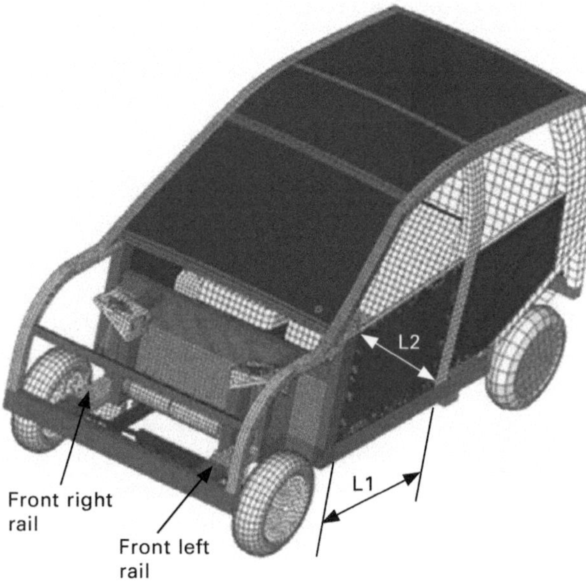

10.9 Finite element model of prototype space frame vehicle.

Table 10.1 Mechanical properties of aluminum extrusion used in LS-DYNA constitutive model

Property	Value
Young's modulus (E)	70 GPa
Poisson's ratio (ν)	0.33
Tensile strength (σ_u)	170 MPa
Yield strength (σ_y)	140 MPa
Failure strain (ε_f)	8%
Density (ρ)	2.7 E-06 kg/mm^3

be a need to use suitable properties for selected elements at a weld joint. In Ref.[6], these properties (as given in Table 10.2) were determined through a parametric study leading to optimal prediction of experimental behavior of a welded T-joint up to failure when one of its legs is subject to bending caused by a transverse load. It is observed from Table 10.2 that the average strength $\left(\frac{\sigma_y + \sigma_u}{2}\right)$ of the aluminum alloy described in Table 10.1 is 25% less in the weld zone, which is consistent with the findings reported by other researchers on reduction of strength in the HAZ around a fillet weld.

During a full frontal collision as mandated in FMVSS 208 regulation in the USA, the front body components that should participate substantially in impact energy management are the front rails indicated in Fig. 10.9. The

Table 10.2 Optimized weld properties for finite element analysis

σ_y (MPa)	σ_u (MPa)	ε_f
98	136	10%

desirable failure mode of these body components in front impact is dynamic progressive buckling as shown in Fig. 10.10 for a number of extruded aluminum tubes of circular, rectangular and square sections, which have been tested in a drop-weight impact rig. It is of interest to note that an undesirable collapse mode is global buckling without substantial fold formation in the walls of an axially impacted component. Under such a condition, minimal impact energy is absorbed. The rightmost component in Fig. 10.10 exhibits a combination of progressive buckling and bending, or transition to global buckling. The phenomenon of combined progressive buckling and bending may have been caused due to the centerline of the impactor being not well aligned with the center line of the specimen, or due to its lower slenderness ratio.

If a CAE-based approach is followed in predicting the performance of a vehicle in which the front rails are made of aluminum alloy extrusions, it would be prudent to ensure that the progressive dynamic buckling phenomenon described above can be simulated reliably. Semi-analytical relations are available for estimating the mean load developed in a section subject to progressive dynamic loading when inertial effects are negligible. For a square tube, one such relation for estimating the mean load P_m during progressive buckling is due to Abramowicz and Jones[7] as given below:

$$P_m = 13.06\ \sigma_0 h^{5/3} b_m^{1/3} \qquad [10.16]$$

where,

$$\sigma_0 = \frac{1}{2}(\sigma_y + \sigma_u),\ \text{i.e. mean strength} \qquad [10.16a]$$

h = thickness of tube wall [10.16b]

b_m = mean width of square tube (i.e. total width – h) [10.16c]

A finite element model of an AA 6063 T7 square tube of 1.88 mm wall thickness is shown in Fig. 10.11[6]. The walls of the tube are meshed with 4-node Belytschko-Lin-Tsay corotational shell elements. The deformed tube and load-displacement behavior obtained through analysis carried out with LS-DYNA are given in Figs. 10.12 and 10.13[6], respectively. The mean load in Fig. 10.13 was found to be 20 kN, which compared extremely well with the theoretical estimation of 20.6 kN yielded by the expression on the right side of Eq. [10.16].

Using the finite element model shown in Fig. 10.9, an analysis of full

10.10 Extruded aluminum tubes of different cross-sectional shapes crushed by axial impact loading.

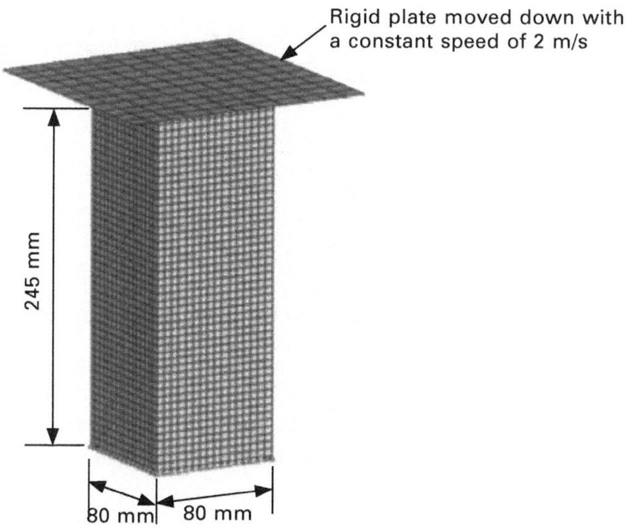

10.11 Finite element model for axial compression simulation.

frontal impact against a rigid barrier with a closing speed of 48 kph (30 mph or 13.33 m/s) conforming to the mandatory FMVSS 208 regulation in the USA was carried out in Ref.[8] The aluminum members of the vehicle frame were assigned the properties of SAPA 6060 T6 alloy given in Table 10.1. All members were of wall thickness 3.0 mm except the front fore-rails (comprising crush box and mid-rail), which had a wall thickness of 2.0 mm. The front fore-rails, which are designed to absorb substantial kinetic energy in front impact, had cross-sectional dimensions of 100 mm × 50 mm. The

10.12 Front view of simulated tube collapse after full compression.

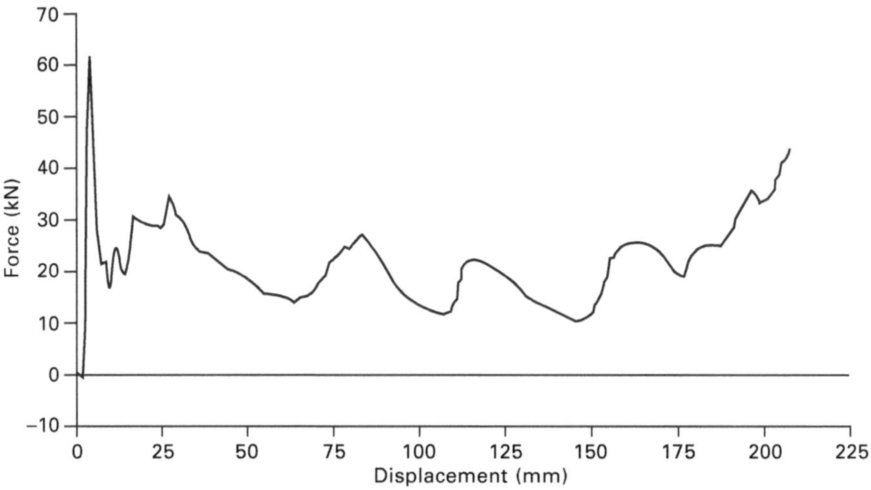

10.13 Axial load vs. displacement behavior for dynamic progressive buckling of square tube.

mass of the vehicle was 620 kg, including a power train comprising a 72 volt DC motor with front wheel drive transmission system and a battery pack with necessary ampere-hours for meeting the mileage target for a city car.

The full vehicle front impact simulation was carried out (a) with all frame joints being considered as monolithic, and (b) using weld elements at all body joints with optimized properties given in Table 10.2. Overall vehicle deceleration responses obtained for these two cases after analysis with the explicit LS-DYNA code, version 960, are shown in Fig. 10.14.[8] Each of these deceleration histories (sometimes referred to as a front 'crash pulse') can be looked upon as an average for the entire vehicle and was directly obtained in the LS-DYNA post-processor. It is seen in Fig. 10.14 that the peak deceleration is slightly lower when joints were represented with weld elements, perhaps because of increased overall compliance in the structure

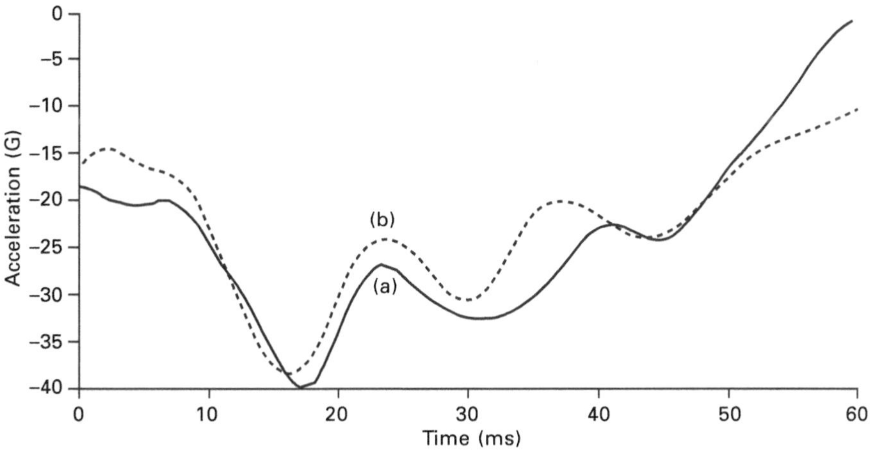

10.14 Acceleration histories for front impact simulation of (a) monolithic joint and (b) welded joint.

when compared with monolithic joint modeling. It needs to be mentioned that, in full frontal impact against a rigid barrier, components such as front rails, shotguns, floor pan, etc. are subject to compressive loading, and weld elements in joints may not be strained significantly, especially in tension or shear. This may explain the general agreement between welded and monolithic joint representation results in full frontal impact simulations as demonstrated in Ref.[8] The consistency of explicit nonlinear dynamic analysis can be verified by plotting total energy along with kinetic and internal energies, as shown in Fig. 10.15.[8] In a successful analysis, the total energy should remain nearly constant and the fall in kinetic energy and rise in internal energy with time should have monotonic trends as shown in Fig. 10.15.

The assessment of occupant injury cannot directly be made from structural crashworthiness as discussed above. Two belted 50th percentile Hybrid III dummies, along with deployable front airbags, are required for prediction of occupant injury parameters such as HIC and chest g. However, as pointed out previously, parameters such as mean deceleration can be correlated with injury parameters. The level of mean deceleration will, in turn, depend on the peak value of the crash pulse. The spaceframe vehicle being considered here has a relatively stiff BIW compared with unibody cars and can be considered as closer to SUVs, which are usually body-on-frame vehicles. On inspection of NCAP crash pulses[6] reported by NHTSA for a number of SUVs in the USA (such as Ford Escape and Explorer, Jeep Cherokee and Grand Cherokee, etc.), it was found that a peak deceleration of about 40g (see Fig. 10.14) tallied well with the deceleration peaks of those vehicles. As the SUVs mentioned recorded good front impact injury performance in NCAP tests, a preliminary conclusion would be that the present spaceframe

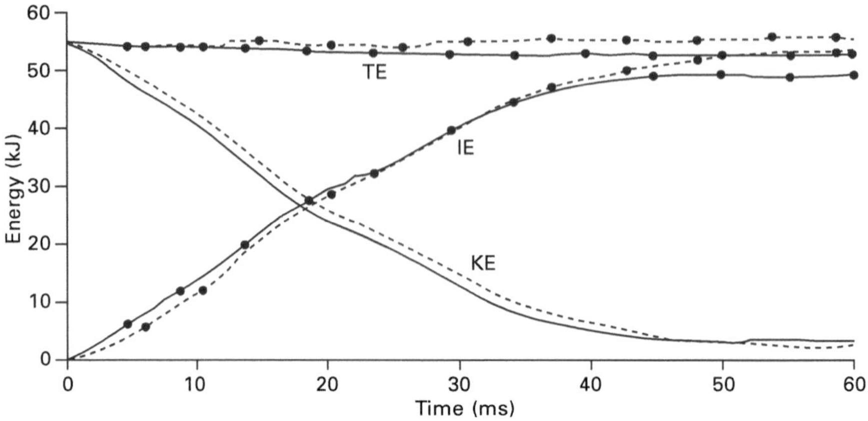

10.15 Energy plots for front impact simulation of monolithic joint (unbroken lines) and welded joint (broken lines). (note: KE = kinetic energy; IE = internal energy; TE = total energy).

vehicle would perform satisfactorily in terms of occupant safety when equipped with properly designed countermeasures such as collapsible steering column, seatbelts and front airbags.

As explained earlier, apart from deceleration, it is also necessary to ensure that the intrusions in the front occupants' cage are within reasonable limits. For case (a), a snap-shot of the analyzed vehicle is shown in Fig. 10.16, from which it can be inferred that the front passenger cabin remains largely intact following the full frontal impact simulation according to the FMVSS 208 procedure. An objective measure of front passenger compartment integrity can be obtained by examining changes in dimensions of the front doors following impact. Minimal changes in the geometry of the front doors will greatly increase their operability after a front collision event. Numerically computed temporal changes in door dimensions L1 and L2 are shown in Fig. 10.17.[8] It can be seen in this latter figure that the maximum predicted change in door apertures is quite low, being in the range of 7–8 mm, confirming the integrity of the front passenger compartment with a high probability that front doors can be opened after a full frontal crash against a rigid barrier at a test speed of 48 kph as stipulated in FMVSS 208.

Salient features of designing a vehicle, using CAE tools, for front impact safety according to the US regulatory standards have been given in this section. For the European market, vehicles have to comply with requirements for a 40% offset impact test against a fixed but deformable aluminum honeycomb barrier at a closing speed of 40 mph (64 kph). This safety test is demanding in terms of structural integrity of the front of a vehicle and actually complements the US-NCAP type tests in which vehicle deceleration needs to be within reasonable levels. The present aluminum spaceframe prototype was evaluated

10.16 Deformed shape of aluminum spaceframe vehicle after full frontal impact analysis.

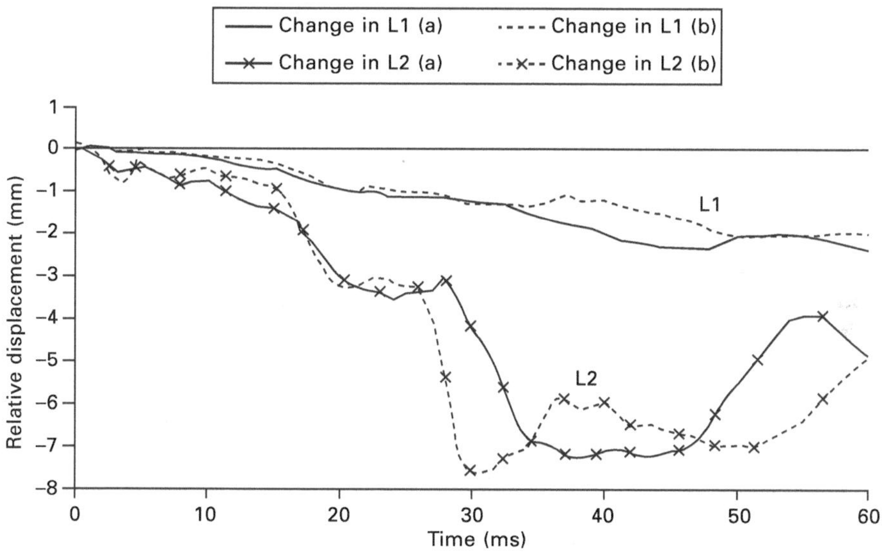

10.17 Changes in horizontal and diagonal front door apertures during impact.

for this condition using LS-DYNA. The finite element model of the vehicle positioned against the deformable barrier (represented with solid elements having similar overall barrier behavior as the actual honeycomb barrrier) at the start of analysis is shown in Fig. 10.18. In addition to front impact safety, vehicles have to be designed also to meet other compliance requirements pertaining to side impact, roof crush resistance, fuel system integrity, interior head impact, etc.

10.18 Finite element model of aluminum spaceframe vehicle for front offset impact safety evaluation.

10.5 Fiber reinforced composites for lightweight automotive body structures

Fiber reinforced composites (FRCs) would be naturally considered a strong choice for automotive body construction due to the high strength-to-weight ratio of such materials and the unique scope for adjusting material properties, especially with laminated composites, to suit a given requirement. Two common fiber materials are carbon and glass, with epoxy being a frequently used matrix. The elastic properties such as longitudinal and transverse moduli and Poisson's ratios of a composite can be estimated using the relevant elastic properties of the constituent materials, i.e. fiber and matrix.[9] It has been observed in Ref. 10 that the ultimate strength of a composite lamina in the longitudinal direction can also be assessed by using information on strengths of constituent materials in a suitable relation accounting for composite type (i.e. fiber or matrix dominated). However, the latter properties can be more reliably obtained by performing monotonic tests until failure on coupon specimens made from a fiber reinforced composite laminate. It appears that there is merit in designing and fabricating a composite-intensive vehicle body using a spaceframe architecture as in the case of aluminum-intensive vehicles. Other applications arise in the form of composite roof and body panels which can even be molded to a metallic body frame.

In designing a vehicle body structure with FRCs, it is important to understand the behavior of such materials under mechanical loads. Tensile stress–strain curves of two typical FRCs with glass and carbon fibers are given in Figs. 10.19 and 10.20.[10] It is immediately apparent from these figures that, although the behaviors of FRCs with different fibers can be similar,

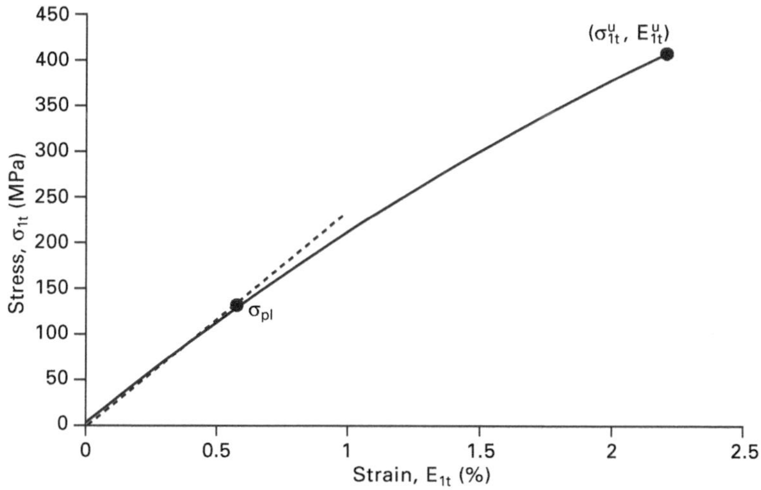

10.19 Stress–strain curve of woven glass/epoxy composite under uniaxial tensile loading in the warp direction (eight-harness satin weave; M10E E-glass/epoxy; E1 = 23.8 GPa, σ_{1t}^{u} = 408 MPa, e_{1t}^{u} = 0.022).

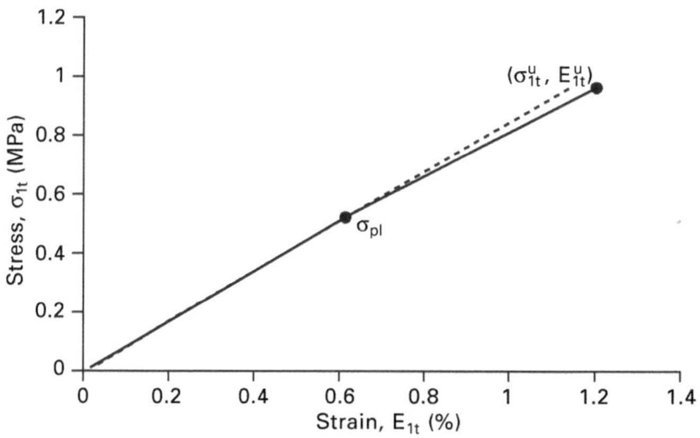

10.20 Stress–strain curve of woven carbon/epoxy composite under uniaxial tensile loading in the warp direction (five-harness satin weave; AGP370-5H/3501-6S; E1 = 77 GPa, σ_{1t}^{u} = 960 MPa, e_{1t}^{u} = 0.012).

these are in sharp contrast to metals due to the absence of a well-defined yield limit and plastic deformation. The failure strains of typical FRCs are also seen to be substantially lower than the corresponding values for metals. This latter characteristic (i.e. brittle behavior) poses a significant challenge to designing passenger cars with FRCs, as premature failure of body parts under impact loads can severely compromise occupant safety.

As in the case of aluminum-intensive vehicle design, studying the behaviors of laminated composite tubes under axial impact loading can provide valuable insight into their energy absorption capability and failure modes. A number of failure theories, such as those of Tsai-Wu[10] and Chang-Chang,[11] are available to model the failure of an FRC material. The Tsai-Wu criterion is a smooth quadratic failure criterion that does not distinguish between the various phases of failure of an anisotropic material; however, it has generally been known to fit well with the experimental data on failure points in stress space. The Chang-Chang failure criteria are a set of discrete failure conditions, such as matrix failure in shear and compression, fiber-matrix debonding, etc., and have been applied to the design of FRC-based structures.

A four-ply chopped strand mat (CSM)-based FRC tube with randomly oriented E-glass fibers is shown in Fig. 10.21 after being subjected to axial crush in a drop-weight impact testing rig. It can be seen in this picture that, unlike in the case of aluminum extrusions, failure takes place by progressive crushing leading to debonding of matrix and fibers. A typical force–displacement plot[12] for axial crushing of a composite tube under dynamic (impact) conditions is given in Fig. 10.22. It is observed that the crush force is maintained at nearly a uniform level throughout the history

10.21 Crushing of an FRC tube with glass fibers, caused by axial impact load.

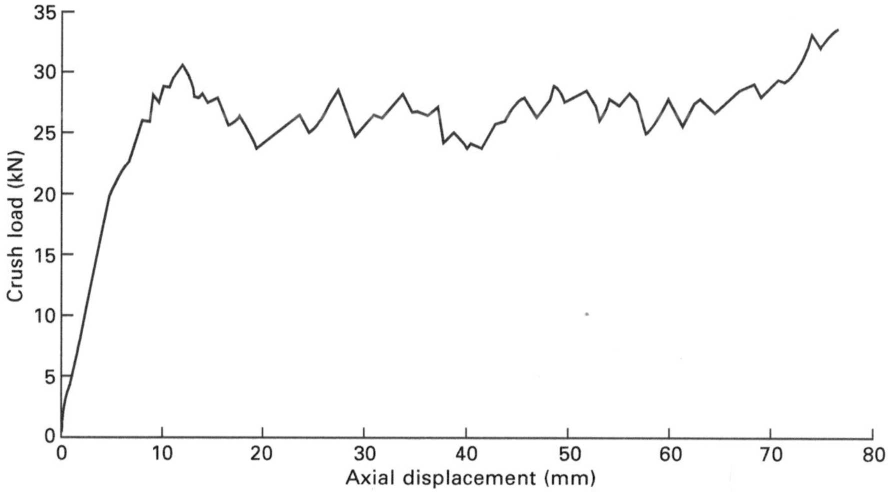

10.22 Typical load–displacement behavior of an FRC tube subjected to axial impact load.

of displacement (or time); thus an FRC component behaves as an efficient energy absorber when impacted axially. The chassis of high performance racing cars is often made of advanced FRCs using carbon fibers. Although such vehicles are also equipped with advanced passive restraint systems, the composite-dominated structure is expected to be remarkably competitive in crashworthiness due to the extremely high driving speed involved, exceeding 200 mph (320 kph). It can be thus said that FRCs can be a viable solution for automotive crash safety design provided the requirements of other attributes (including cost) are met, and catastrophic failure of structural components especially in the occupants' cage does not take place at sufficiently high impact loads.

10.6 Miscellaneous lightweight countermeasures

As pointed out in the previous section, a question mark on large-scale usage of FRCs in passenger car body construction is the low failure strain of such materials. Researchers[13, 14] working in the area of developing lightweight solutions for energy absorption have tried to overcome such inadequacies by combining metals with composites so that the reserve ductility of metals is available during impact with additional strength contributed by composite wrap-ups. Additionally, investigators[14–16] have also considered filling up the hollow inner cavity of thin-walled aluminum tubes with foam, thereby enhancing the load carrying capacity of such components at minimum increase of weight. Polymeric foams are inexpensive and are readily available in wide varieties and strengths as suitable filler material. Recently, aluminum foam has been studied for its high strength-to-weight ratio and specific energy absorption attributes, as a means for enhancing vehicle crashworthiness with minimum weight penalty. It is pointed out that such addition of foam fillers in components such as A- and B-pillars can also be advantageous for traditional steel-based vehicle BIW with potential to significantly increase roof crush resistance (included under FMVSS 216 in the USA) during roll-overs without increasing the sizes of those components (which are ordinarily hollow).

Crushable (i.e. rigid or semi-rigid) polymeric foams, generally in the form of padding under interior trim and instrument panel, have been used in vehicles to reduce occupant injury in front and side impact as well as interior head impact. Polyurethane (PU) and polystyrene foams are most common. The plateau strength of such foams is a function of density as shown in Fig. 10.23.[16] Foam-based countermeasures, when incorporated in the front bumper of a vehicle, can play an important role in pedestrian impact protection as well as in reducing damage to the front structure in a low speed collision against a fixed object or another vehicle. Foams have high specific energy absorption because of their cellular structure and are most effective under compressive loads.

Plastics of various types, such as polypropylene, ABS (acrylonitrile butadiene styrene), etc. are extensively used in the interior of a vehicle to provide aesthetic appearance and improve comfort. However, such materials can also act as measures for reducing occupant injury during secondary impacts of occupants with the interior surfaces of a vehicle during an accident. An important application of plastics for impact injury protection is in the form of internal energy-absorbing ribs[17] behind A-pillar and B-pillar trims

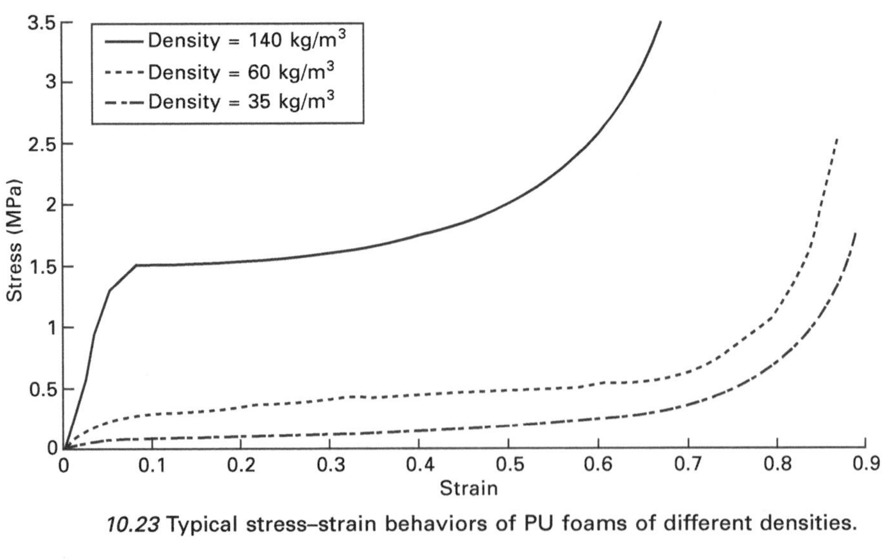

10.23 Typical stress–strain behaviors of PU foams of different densities.

in vehicles. These ribs (integrally molded with a trim panel or attached to it by heat staking) enable a vehicle to meet upper interior head impact safety requirements such as mandated by the FMVSS 201 regulation in the USA.

10.7 Conclusion

An important aspect of designing vehicle BIW with lightweight materials is ensuring crashworthiness and occupant safety. The present chapter has provided insight into how structural components can be designed with aluminum by ensuring similar levels of stiffness and plastic strength as in a comparable steel component, while simultaneously achieving significant weight reduction. The dependence of occupant injury parameters such as Chest g and HIC on structural crash performance has been discussed. The importance of CAE comprising finite element modeling and analysis in vehicle safety design has been highlighted and an example of CAE-based design of an aluminum spaceframe vehicle for front impact safety has been presented. The behavior of an important class of lightweight materials, namely fiber reinforced composites, has been described in the context of failure modes, energy absorption under impact loads and susceptibility to brittle failure. Additionally, a number of other lightweight solutions such as hybrid tubes, foam padding and plastic trim with ribs have been pointed out for their potential and current role as safety countermeasures.

10.8 References

[1] J. Maxwell, *Plastics in the Automotive Industry*, Woodhead Publishing Ltd., Cambridge, UK. ISBN 1 85573 039 1, 1994.

[2] W.S. Miller, L. Zhuang, J. Bottema, A.J. Wittebrood, P. De Smet, A. Haszler and A. Vieregge, 'Recent Developments in aluminum alloys for the automotive industry', *Materials Science and Engineering A*, 280(1), 37–49 (2000).

[3] D. Mohan, 'Traffic safety and health in Indian cities', *Journal of Transport & Infrastructure*, Vol. 9(1), 2002.

[4] A. Deb and C.C. Chou, 'Vehicle front impact safety design using a hybrid methodology', *International Journal of Vehicle Safety*, 2(1/2), 44–56 (2007).

[5] C.B. Chavan and J.S. Karve, 'Design and Development of a Robust Aluminium Based Vehicle Platform for a Mini Car', *MDes Project Report*, IISc, Bangalore, 2002.

[6] A. Deb, M.S. Mahendrakumar, C. Chavan, J. Karve, D. Blankenburg and S. Storen, 'Design of an aluminum-based vehicle platform for front impact safety', *Proceedings of 8th International Symposium on Plasticity and Impact Mechanics (IMPLAST 2003)*, Ed. N.K. Gupta, 488–498 (2003).

[7] W. Abramowicz, and N. Jones, 'Dynamic progressive buckling of circular and square tubes', *International Journal of Impact Engineering*, 4 (1986), 234–270.

[8] A. Deb, M.S. Mahendrakumar, C. Chavan, J. Karve, D. Blankenburg and S. Storen,

'Design of an aluminum-based vehicle platform for front impact safety', *International Journal of Impact Engineering*, 30(8–9), 1055–1079 (2004).

[9] P.K. Mallick, *Fiber-reinforced Composites: Materials, Manufacturing, and Design*, Third Edition, Talyor and Francis, 2007.

[10] I.M. Daniel and O. Ishai, *Engineering Mechanics of Composite Materials*, Second Edition, Oxford University Press, Oxford, UK, 2008.

[11] A. Tabiei, W. Yi and R. Goldberg, 'Non linear strain rate dependent micro-mechanical composite material model for finite element impact and crashworthiness simulation', *Int J Non-linear Mech*, 40, 957–970 (2005).

[12] R. Velmurugan, N.K. Gupta, S. Solaimurugan and A. Elayaperumal, 'Experimental and theoretical studies on effect of stitching of FRP cylindrical shells under axial compression', *Proceedings of 8th International Symposium on Plasticity and Impact Mechanics (IMPLAST 2003)*, Ed. N.K. Gupta, 432–443 (2003).

[13] H. El-Hage, P.K. Mallick and N. Zamani, 'A numerical study on the quasi-static axial crush characteristics of square aluminum–composite hybrid tubes', *Composite Structures*, 73, 505–514 (2006).

[14] J.M. Babbage and P.K. Mallick, 'Static axial crush performance of unfilled and foam-filled aluminum–composite hybrid tubes', *Composite Structures*, 70, 177–184 (2005).

[15] A. Reyes, O.S. Hopperstad and M. Langseth, 'Crash behavior of obliquely loaded foam-filled aluminum extrusions', *Proceedings of 8th International Symposium on Plasticity and Impact Mechanics (IMPLAST 2003)*, Ed. N.K. Gupta, 395–402 (2003).

[16] S.R. Guillow and G. Lu, 'Quasi-static axial compression of thin-walled circular metal tubes including effects of foam filling', *Thin Walled Structures Research and Development*, Eds. N.E. Shanmugam, J.Y. Richard Liew and V. Thevendran, Elsevier, 771–778 (1998).

[17] A. Deb, N.K. Gupta, U. Biswas and M.S. Mahendrakumar, 'Designing for head impact safety using a combination of lumped parameter and finite element modeling', *International Journal of Crashworthiness*, 10(3), 249–257 (2005).

Index

FSC
www.fsc.org
MIX
Papier | Fördert gute Waldnutzung
FSC® C083411

Kolibri 360 GmbH
Ferdinand-Jühlke-Straße 7
99095 Erfurt, Deutschland
produktsicherheit@kolibri360.de